Science Advice to the President

Second Edition, enlarged

Science
Advice
to the
President

Edited by William T. Golden

AAAS
PRESS
A publishing division of the
American Association for the
Advancement of Science

Second Edition, enlarged

PHYS

Copyright © 1993
AAAS Press, a publishing division of the
American Association for the Advancement of Science
1333 H Street, NW, Washington, DC 20005

Library of Congress Cataloging-in-Publication Data

Science advice to the president / William T. Golden, editor.
    p. cm.
    ISBN 0-87168-509-4
1. Science and state—United States. 2. Technology and state—United
States. 3. Presidents—United States. I. Golden, William T., 1909- .
    Q127.U6S29 1993
    338.97306—dc20   92-36149
                                   CIP

AAAS Publication: 93-12S
International Standard Book Number: 0-87168-509-4
First edition published by Pergamon Books, Inc. 1980
(previously ISBN 0-08-025963-4)

Printed in the United States of America on acid free paper.

Contents

Acknowledgments to the Second Edition

William T. Golden

Above all, I express my gratitude to our country which provided shelter and opportunity to my immigrant grandparents, little over 100 years ago, as it has done for countless others.

And I thank my friends who have helped me in preparing this second edition. First, of course, are the authors of the new articles: D. Allan Bromley, John H. Gibbons, and John P. McTague. Notable are my resourceful and devoted assistant Christie Van Kehrberg, and my staff members Ellen Rosenblatt and Eugene Gorman. And invaluable has been the scholarship and judgment of David Z. Beckler, associate director of the Carnegie Commission on Science, Technology, and Government, most experienced of advisers to Science Advisers; William G. Wells, Jr., associate professor of Management Science, School of Business and Public Management, George Washington University, formerly chief of staff to President Bush's Science Adviser, D. Allan Bromley, and adviser to President Clinton's transition team; and Dr. J. Thomas Ratchford, professor of International Science and Technology Policy at the George Mason University (with whom I have been happily associated in the American Association for the Advancement of Science and elsewhere), formerly associate director for Policy and International Affairs of the Office of Science and Technology Policy under Dr. Bromley, and now consultant to Dr. Gibbons.

William T. Golden

New York, NY
July 1, 1993

Science Advice to the President was first published as a special issue of Technology In Society Vol. 2 (1980)

Acknowledgment to the First Edition

In founding *Technology In Society*, the avowed aim of the editors was to illuminate in the pages of this journal timely issues and important interactions in the realm of science, technology, and society. The majority of articles in this field offer useful, albeit sometimes controversial, conclusions on the social implications of a particular scientific development or technology, or both. To identify, much less predict, what—and how—events and attitudes will ultimately affect the course of science and technology through policies and legislation is more difficult to achieve with comparable clarity and accuracy.

As nations increasingly come to regard continued technical development as essential to economic growth and strength, as well as necessary to the improvement of the quality of life, an understanding of the social management of science and technology takes on paramount importance. In the United States, the most visible agency for such decisions has been the Presidential science advisory mechanism in its various forms. Hence, we felt it would be important to present the views of former Presidential Science Advisers as a first step in bringing to the readers of *Technology In Society* an increasing number of insightful articles on the complex social processes which affect the course of science and technology.

To identify the person who could undertake this task was not difficult. After events during the Second World War had confirmed the vital stake of the US government in the growth of science and scientific institutions, President Truman in 1950 had the wisdom and foresight to ask William T. Golden to become his special consultant to review the government's activities in that area. Golden's efforts led to the creation of a Science Advisory Committee of the Office of Defense Mobilization. As adviser to the Director of the Bureau of the Budget, he played a key role in the organization of the National Science Foundation. In the years since that time, which he has devoted to public service, Golden has worked in a unique way on behalf of people and institutions dedicated to education and research in the sciences. As Guest Editor of this issue, he brings to that post a lifetime of acquaintance with many of those who, within the past 40 years, have played a vital role in the growth of science in America.

We are deeply grateful to Bill Golden for having assumed so willingly, and for having carried out so thoughtfully, the task of creating this issue of *Technology In Society*. We hope that it will be of considerable value to the large and growing group of people, worldwide, who have to deal with issues of policy in science and technology in their professional lives.

George Bugliarello
A. George Schillinger
The Editors

Introductory Remarks to the Second Edition

William T. Golden
Editor

Plus ça change, plus ç'est la même chose.

Much has changed since the first edition of this book was published in 1980. But the essentials remain the same, intensified by the growing recognition of the increasing influence of science and technology in everyday life.

The first edition consisted of articles especially written for it by outstandingly well-qualified individuals. The articles were timely in their time and are timeless in their wisdom and in their recording of the mutations and evolution of the office of Science Adviser to the President of the United States of America and of a President's Science Advisory Committee since their authorization by President Truman early in 1951.[1,2] All these articles are republished unchanged.

In addition, for this second edition, John H. Gibbons, the incumbent Science Adviser to President Clinton, D. Allan Bromley, the outstandingly effective Science Adviser to President Bush, and John P. McTague, who served for a few months as acting Science Adviser to President Reagan in 1986, have graciously written special articles of timeliness and enduring importance. These articles are published here for the first time.

The essays in this edition, along with those in the intervening 1988 volume, *Science and Technology Advice to the President, Congress, and Judiciary,* which is now also republished in a second edition,[3] include especially written articles by every Science

William T. Golden (b. 1909) designed the first Presidential Science Advisory organization for President Truman in 1950. He is Chairman of the American Museum of Natural History and Past Chairman of the New York Academy of Sciences, and is Cochairman (with Joshua Lederberg) of the Carnegie Commission on Science, Technnology, and Government. He has served in the U.S. Navy (World War II), the Atomic Energy Commission, the Department of State, and the Executive Office of the President. He received the Distinguished Public Service Award of the National Science Foundation (1982) and a Special Tribute of Appreciation (1991). Mr. Golden is an officer and trustee of the American Association for the Advancement of Science, the Mount Sinai Hospital and Medical School, and the Carnegie Institution of Washington. He is also a member of the National Academy of Public Administration, the American Philosophical Society, and the American Academy of Arts and Sciences.

Adviser to the President over the period of more than 40 years since 1951, with the exception of Oliver Buckley, President Truman's first adviser, who died not long after leaving office in 1952.

What has changed? The President's Science Advisory Committee (PSAC), which was abolished by President Nixon in 1973, remained non-existent until its restoration by President Bush in 1989 (consistent with a recommendation by the Carnegie Commission on Science, Technology, and Government[4]) with a change of name to President's Council of Advisers on Science and Technology (PCAST). In the interim, even after the re-establishment of the position of Science Adviser to the President by President Ford, there was no PSAC (or PCAST) until President Bush's action. However, a White House Science Council assisted President Reagan's Science Advisers, but they were not presidential appointees.

The activity of the Carnegie Commission on Science, Technology, and Government, created under the sponsorship of David Hamburg,[5] President of the Carnegie Corporation of New York, was a major factor in President Bush's decisions to restore the Office of Science and Technology Adviser to the President[4] with his appointment, after some delay, of D. Allan Bromley to that position, in its originally recommended high stature, and to re-establish a presidentially appointed Science Advisory Committee[1] — with change of name from PSAC to PCAST (President's Council of Advisers on Science and Technology). Reference is made to my introductory article to the first edition of this volume (pp. 5–9). Then President Clinton, with admirable promptness, appointed John H. Gibbons (who for more than 10 years had been Director of the Congressional Office of Technology Assessment) as his Science and Technology Adviser, with Cabinet rank. At this writing, it has not yet been determined what form will be assigned to a President's science and technology advisory committee function in the Clinton administration.[6]

The stature and influence of the President's Science Adviser (as he is generally called) has fluctuated over the years, responsive to the personality of the President, his recognition of the changing and growing importance of science and technology in domestic and foreign affairs, and his appetite and receptivity for outside ideas and advice. It cannot be stressed too strongly that the science adviser's function is to serve the President. He or she cannot be an advocate for science or for any other special interests. The adviser is an advocate for the best interests of the United States, expressed in information and opinion to the President and his Cabinet and staff.[7]

Much has changed in the more than 40 years since the outbreak of the Korean War in 1950 led to the creation of the science advisory apparatus by President Truman, which was strengthened, in accordance with the original recommendation, by President Eisenhower in 1957 responsive to Sputnik. The American conception of a science and technology advisory organization to the highest level of government has radiated to more than 50 countries, including all the major countries. The establishment of the Carnegie Group of science and technology advisers and ministers to the heads of governments of the G-7 countries (United States of America, Canada, United Kingdom, France, Germany, Italy, and Japan), Russia, and the European Community, responsive to a proposal in *Worldwide Science and Technology Advice to the Highest Levels of Governments*,[8] has been a major coordinating event. The Carnegie

Group, sponsored by the Carnegie Commission on Science, Technology, and Government, first came together in February 1991, with D. Allan Bromley (Science Adviser to President Bush) and Juriy Ossipyan (Science Adviser to President Gorbachev) jointly chairing the initial meeting. The Carnegie Group has been an outstanding success, the more effective because of its informality and the intimacy of its staff-free meetings. It has now held five meetings semi-annually in different host countries, each independently financed and chaired by the science adviser (or minister) of the host country. The group plans to continue this procedure.

The cold war has ended (but the nuclear weapons have not yet been eliminated). Smaller wars, including brutal civil wars, erupt in response to ancient conflicts and those aspects of human nature and enduring human rivalries first expressed when Cain slew Abel. The emphasis on military uses of science and technology that primarily motivated President Truman has diminished. Issues of global economic competition are ascendant. They influence the quality of life in our country and everywhere. Underlying our ability to compete are matters of the health of our people, their educational levels, and their motivation to work creatively and productively. Improving the motivations and education of our large underclass, to provide them with hope for and ability to obtain and hold jobs in an increasingly technological society, is a duty and a challenge for those of us who are more fortunate. This is not just for reasons of compassion or as expressions of gratitude, but indeed as a matter of enlightened self interest.[9] Thus, the function of the President's Science Adviser continues to expand, and his or her opportunity and importance continues to grow.

Notes

1. Detlev W. Bronk, "Science Advice in the White House: The Genesis of the President's Science Adviser and the National Science Foundation," *Science*, Vol. 186, pp. 116–121, 11 October 1974. Reprinted in this volume, pp. 307-318.
2. William T. Golden, "Government Military-Scientific Research: Review for the President of the United States, 1950–51," basic documents, including precept, report, and memoranda, including interviews with about 150 appropriate individuals. Unpublished, 432 pages. Copies available at the Harry S. Truman Library, Independence, Missouri; the library of the American Institute of Physics, New York, NY; the Herbert Hoover Library, West Branch, Iowa; the Dwight D. Eisenhower Library, Abilene, Kansas; and the Library of Congress, Washington, DC.
3. William T. Golden, editor, *Science and Technology Advice to the President, Congress, and Judiciary* (New York: Pergamon Press, 1988; Washington, DC: AAAS Press, 1993). xxiii+330 pages.
4. Carnegie Commission on Science, Technology, and Government, *Science & Technology and the President* (New York: October 1988), ii + 26 pp.
5. David A. Hamburg, "Foreword," Carnegie Commission on Science, Technology, and Government, *Science, Technology, and Government for a Changing World* (New York: April 1993), pp. 4–5.
6. William T. Golden, *Ibid.*, p. 14.
7. David Z. Beckler, "Science and Technology in Presidential Policy-Making: A New Dimension and Structure." Golden, William T., "Science and Technology Advice to the President, Congress, and Judiciary," pp. 31–38.
8. "Introduction," William T. Golden, editor, *Worldwide Science and Technology Advice to the Highest Levels of Governments* (New York: Pergamon Press, 1991; Washington, DC: AAAS Press, 1993), p. 4.
9. "Symposium on the Underclass," *Proceedings of the American Philosophical Society*, Vol. 136, No. 3 (September 1992). Papers presented by William T. Golden, William Julius Wilson, Leon M. Lederman, David A. Hamburg, and Walter E. Massey at the American Philosophical Society meeting in Philadelphia, April 24, 1992.

Contours of Wisdom:
Presidential Science Advice
in an Age of Flux

Introductory Remarks to the First Edition

William T. Golden

Whirl is king, having displaced Zeus
    Aristophanes, c. 400 B.C.

The only permanent condition is change
        Heraclitus, c. 500 B.C.

Omnia mutantur nihil interit
    Ovid, Metamorphoses, *c. 5 A.D.*

You can lead a horse to water
but you cannot make him drink
    *ascribed to Bucephalus,
    before the battle at the Hydaspes, 326 B.C.*

Introduction

At this time of tectonic mobility in the social, political, military, economic, and cultural organization of the teeming surface of the Earth's crust, progress in science accelerates and technology influences increasingly the life of every living creature (including the unborn and to-be-conceived) of every species. Much of the world is in flux, and our own country is struggling in uneasy adjustment to the dilution of our post World War II cultural, military, political, and economic primacy.

William T. Golden (b. 1909) designed the first Science Advisory apparatus for President Truman in 1950. He is an officer and trustee, or director, of several scientific, educational, and industrial organizations, including the American Association for the Advancement of Science, the American Museum of Natural History, and the Carnegie Institution of Washington. He was on active duty in the US Navy throughout World War II and has served in serveral agencies of the Federal government, including the Atomic Energy Commission, the Department of State, and the Executive Office of the President.

The President of the United States can exert a major leadership influence on the course of international political and cultural evolution. It is gratifying and challenging, therefore, to serve as Guest Editor for an issue of *Technology in Society* devoted entirely to the why and the how of science and technology advice for the President. No President can function effectively without considering these elements in a broad range of policy formulation. And no President after Thomas Jefferson can be his own Science Adviser.

Purpose and Objectives of This Issue

The purpose of this extraordinary collection of original and independent essays is to help the United States, and therefore, the world, by helping the President. It aims to do this by drawing attention to and stimulating discussion of the important and growing involvement of technology, and its parent, science, in the domestic and foreign policies of our government and in the future of mankind.

The focus of United States policy-making is the President. He receives advice from many sources, is influenced by many factors, and is the final integrator. Harry Truman said it economically: "The buck stops here."

Consequently this issue of *Technology In Society* is concerned with the process of advising the President on scientific and technological matters as they impinge on and are interwoven with the fabric of domestic and foreign policy formulation. It is concerned with how to be most effective in encouraging the President, every President, to seek, consider, evaluate, and utilize such wise and public-spirited advice. It aims to stimulate a thirst for such wisdom. It is not concerned with the welfare of science and technology, important as that is, except secondarily in the very real sense that, as our country flourishes, science and technology will flourish.

The essays in this volume are addressed to the President of the United States and to his successors; to the Congress; to the career civil servants of both the Executive and the Legislative Branches of our government, and to the transient appointed officials of all ranks: men and women of patriotism and dedication. They are addressed also to the media and, above all, to the concerned citizenry, including the academic community and the students in its colleges and universities from whom will emerge, as the years go by, the future managers of our country.

The Authors and Their Selection

Frank Press, Science Adviser to President Carter and Director of the Office of Science and Technology Policy, has graciously contributed a concise prefatory paper. Seven of his predecessors as Presidential Science Adviser have thoughtfully expressed their individual observations, emphasizing the future, based on their experiences and their ruminations. Of the Science Advisers, only the first—the late Dr. Oliver Buckley, appointed by President Truman in April, 1951—and Dr. George Kistiakowsky, who declined an invitation to publish his views, are not represented here.

Each of the other authors is also especially qualified by experience, ability, and motivation. Several have been members of the President's Science Advisory Committee (PSAC) during its years of existence from 1951 to 1973. Others have served on the staff of the Science Adviser, or have been, or are, career civil servants intimately involved with science and technology at higher levels of government in both the Executive and the Legislative Branches. And several are distinguished scholars and industrialists with government experience. The late Detlev W. Bronk's article, "Science Advice in the White House—the Genesis of the President's Science Advisers and the National Science Foundation," based on contemporary interviews and documents and regarded as the definitive history of the early years, is republished to provide relevant background.

President Ford writes of the restoration of the President's Science Advisory Office in 1976, by Congressional enactment proceeding from his recommendation, after its abolition by President Nixon in 1973. The latter was invited to contribute an article setting forth the reasons for his action, but declined to do so.

Each essay has its own character, viewpoint, insights, and artistry. One is reminded of the numerous renderings of the Annunciation, the Nativity, and other crucial events by the devout Italian painters of the Renaissance and the Flemish masters of the 15th Century. Each portrays the identical subject, yet each is different and each rewarding.

The authors were selected for their distinction and relevant experience, without regard to political or intellectual orientation. Their only instructions were to think deeply and write freely. Each will see the papers of the others and of the Guest Editor for the first time in this publication. The concurrence of opinions and recommendations on major issues is therefore the more noteworthy.

A Program for the Future of Science and Technology Advice to the President

As is evident from the preceding comments, the Guest Editor is confident of the timeliness and perennial importance of science and technology in governance, progress, and survival. His more specific judgments, which are consistent with the preponderant views of the other authors, may be summarized as follows:

1. Science and technology advice are essential to the President as inseparable ingredients in a broad panoply of policy-making on domestic and foreign issues. The principles are unchanged from the World War II and post-World War II periods. However, domestic concerns, including economics and what may be called social issues (such as health care, ecology, energy, education, transportation) have grown in importance over the decades; while defense considerations, including both military matters and the quest for arms control, remain vital to national and human survival.

2. Congress also needs advice on scientific and technological matters. It has various sources available to it, including (under Public Law 94-282, The National Science and Technology Policy, Organization and Priorities Act of 1976) the Director of the Office of Science and Technology Policy, who plays the dual role of Director of that Office and Science Adviser to the President. The incumbent, Dr. Press, appears to have been remarkably successful in the delicate procedures of achieving the confidence of Congress while maintaining the privity of his intimate relationship with the President. It has been a noteworthy accomplishment.

3. The value of a Science Adviser to the President has been amply demonstrated in practice. His or her effectiveness could be enhanced by restoring responsibilities for national defense issues which were included in the duties of the early occupants of the position. This is important even though Dr. Press has, informally, achieved a measure of involvement.

4. (a) The President's Science Advisory Committee should be re-established. It should be composed of generalists of stature, wisdom, discretion, independence, and patriotic dedication. There are too many specialties to have all the important ones adequately represented. Nor should reliance be placed on one expert in each special field. Pluralism and diversity of ideas would be attainable under a small PSAC structure by the use of supplementary ad hoc committees (not Presidentially-appointed) on specific issues, consisting of qualified outsiders with the participation of one or more members of PSAC itself. Perhaps standing subcommittees would evolve in major subject areas.

Keenly aware of Occam's Razor and the Principle of Parsimony, I believe the Committee should consist of a small number, perhaps nine to not over 15 men and women. The small number will encourage collegiality with independence, and enhance the individual sense of responsibility. All the members should be chosen for merit, or meritorious members will not serve. They must understand the practical processes of politics and government. They must be courageous, but not foolhardy. Indeed, were they not courageous, they could not effectively serve the President. The prestige they bring with them will be enhanced by their Presidential appointment, and their usefulness and survival will depend on their wisdom.

Broadly, the physical, biomedical, and perhaps certain social sciences should be represented; but not every member should be a practicing scientist. Loyalty to the President, but not subservience, is essential. Esprit de corps should develop, and the expression of divergent views, within the intimacy of the group, should be encouraged. Criticism is more valuable than praise if expressed privately. Resignation is the honorable condition precedent to public disagreement with the policies of the President one serves. It should be self-evident, and history demonstrates, that to be effective any such committee must have the confidence, both for judgment and for loyalty, of the President it exists to help.

(b) To assure the requisite stature and visibility, both within and outside the government, the members of PSAC should be Presidential appointees. Study of the Federal Advisory Committee Act of 1972 and the Freedom of Information Act yields little hope that exemption from the rigid requirement for public meetings, pre-published agendas, and the like can be achieved at this time. Even location within the White House Office apparently would not provide a sheltering membrane sufficiently impermeable to insulate a PSAC from the relentless restraints of those Acts. However, though effectiveness will be impaired at formal meetings, such a committee and its individual members can still function usefully. On classified matters the restraints would be suspended, and on unclassified topics its well-planned deliberations and its availability for consultation, by the President or by the Science Adviser, as a group or as individuals, could be of major service even within the limitations.

(c) It should be noted that Congress thoughtfully provided for the "periodic revision and adaptation" of PL 94-282 (e.g., Sec. 102 (a) (6)). This could provide alternative means for achieving re-establishment of a President's Science Advisory Committee as well as other improvements in the Act. Perhaps exemption for a PSAC from the Federal Advisory Committee Act and the Freedom of Information Act will be possible by Congressional action only in the event of a crisis situation. But crises do arise and it would be well to keep discussions alive and plans in readiness for enactment should the political climate change.

Reasoning supportive of these views will be found in sundry forms in the papers which follow, and need not be separately stated here. Most important are the preponderant recommendations for the re-establishment of a PSAC. The "argument of the work," in Winston Churchill's phrase, is that the office of Science Adviser to the President should be retained and that, before or promptly after the 1980 Presidential election, a President's Science Advisory Committee should be re-established, with privacy if exemption can be won but without it if it cannot.

President's Science Advisers and Members of the President's Science Advisory Committee and alternative committees 1951-1993

President's Science Advisers

President	Science Adviser	Years
Truman	Oliver E. Buckley[b]	1951-52
	Lee A. Dubridge[b]	1952-53
Eisenhower	Lee A. DuBridge[b]	1953-56
	Isidor I. Rabi[b]	1956-57
	James R. Killian, Jr.	1957-59
	George B. Kistiakowsky	1959-61
Kennedy	Jerome B. Wiesner	1961-63
Johnson	Jerome B. Wiesner	1963-64
	Donald F. Hornig	1964-69
Nixon	Lee A. DuBridge	1969-70
	Edward E. David, Jr.	1970-73
	H. Guyford Stever	1973-74
	(also Director NSF)	
Ford	H. Guyford Stever	1974-77
	(also Director NSF 1974-76)	
Carter	Frank Press	1977-81
Reagan	George A. Keyworth II	1981-86
	John P. McTague	1986
	William R. Graham	1986-89
Bush	D. Allan Bromley	1989-93
Clinton	John H. Gibbons	1993-

The Science Advisory Committee, as established by President Truman with a letter, dated April 20, 1951, reported to the Director of the Office of Defense Mobilization and also to the President. That letter also said, "I shall welcome the recommendations of the Committee and shall call upon it from time to time." But it was not until November 29, 1957, in the aftermath of Sputnik, that President Eisenhower elevated the Committee to become the President's Science Advisory Committee—or PSAC. The original recommendation in Golden's December 1950 Report to President Truman had called for the Committee to report to the President directly. In fact, the Chairman of the Committee was also referred to as the "Presidential Science Adviser."

President Eisenhower's announcement was the sole basis for the Committee's status; no Executive Order was ever issued, then or later. The Committee continued through the years, through the Kennedy and Johnson Administrations and the first Nixon Administration. Then, early in the second Nixon Administration, PSAC was swept away along with the Office of Science and Technology under Reorganization Plan No. 1, July 1973.

P.L. 94-282, "The National Science and Technology Policy, Organization, and Priorities Act of 1976" in Title III, established the President's Committee on Science and Technology. The nine members were appointed by President Ford in August 1976; they were barely into their assignment when Jimmy Carter won the Presidency in November. One of President Carter's early actions was to abolish the Committee via Reorganization Plan No. 1, July 15, 1977. He appointed Frank Press as his Science Adviser.

The White House Science Council was established in February 1982 by George Keyworth, OSTP Director under President Reagan, as an advisory group for his office, and it was retained by his successor, William R. Graham. It had no statutory basis, and its members were not appointed by the President. It came to an end at the conclusion of the Reagan Presidency.

The President's Council of Advisers on Science and Technology (PCAST) was established by Executive Order 12700, January 19, 1990. It was to be composed of not more than 15 members, 14 of whom should be distinguised individuals from the private sector to be appointed by the President. The Director of the Office of Science and Technology Policy (D. Allan Bromley, who was also Assistant to the President for Science and Technology—Science Adviser to the President) was designated as Chairman of the Council. The Council was "to advise the President on matters involving all areas of science and technology." On June 28, 1991, the Executive Order was modified to extend the duration of the Council until June 30, 1993.

President's Science Advisory Committee and alternative committees 1951–93

Name	Affiliation[a]	Years	Field
Agnew, Harold M.[c]	Los Alamos National Lab.	1982-89	Physics
Alvarez, Luis W.	Lawrence-Berkeley Laboratory University of California at Berkeley	1973	Physics
Anlyan, William G.[c]	Duke University	1988-89	Surgery
Bacher, Robert F.	California Institute of Technology	1953-56 1957-59	Physics
Baker, William O.	Bell Telephone Laboratories	1957-59, 1976-77	Physical chemistry
Baldeschwieler, John D.	Stanford University	1969-73	Chemistry
Bardeen, John[c]	University of Illinois	1959-62, 1982-85	Physics
Beadle, George W.	California Institute of Technology	1960	Biology
Bennett, Ivan L., Jr.	Johns Hopkins University	1966-70	Pathology
Berkner, Lloyd W.	Associated Universities, Inc.	1957-58	Physics
Bethe, Hans A.	Cornell University	1956-59	Theoretical physics
Borlaug, Norman	Texas A&M University	1990-93	Microbiology

President's Science Advisory Committee and alternative committees 1951–93 (cont.)

Name	Affiliation[a]	Years	Field
Bowen, Otis R.	Governor of Indiana	1976-77	Politics
Bradbury, Norris E.	Los Alamos Scientific Laboratory	1955-57	Physics
Branscomb, Lewis M.	Joint Institute for Lab. Astrophysics	1965-67	Physics
Bromley, D. Allan[c]	Yale University (Science Adviser to President 1989-93)	1982-93	Nuclear physics
Bronk, Detlev W.	The Rockefeller Institute	1951-62	Physiology, biophysics
Brooks, Harvey	Harvard University	1960-64	Physics
Buchsbaum, Solomon J.[c]	Sandia Laboratories (1971)	1971-73,	Physics
	Bell Telephone Laboratories (1972-73)		Physics
	AT&T Bell Labs (1982-89)	1982-89	
Buckley, Oliver E.[b]	Bell Telephone Laboratories (Science Adviser to President 1951-52)	1951-55	Electrical engineering
Cairns, Theodore L.	E.I. DuPont de Nemours & Co.	1971-73	Chemistry
Calvin, Melvin	University of California	1963-66	Organic chemistry
Campbell, Glenn W.	Hoover Inst. on War, Revolution, and Peace, Stanford Univ.	1976-77	Economics, education
Chance, Britton	University of Pennsylvania	1959	Biophysics, biochemistry
Coleman, James S.	Johns Hopkins University (1971-72) University of Chicago (1973)	1971-73	Sociology
Conant, James B.	Harvard University	1951-53	Chemistry
Cowan, George A.[c]	Los Alamos National Lab.	1982-85	Nuclear chemistry
David, Edward E., Jr.[c]	Bell Telephone Labs (1970-73)	1970-73,	Physics, electrical
	Gould, Inc. (1976-77)	1976-77	engineering
	EED, Inc. (1982-89) (Science Adviser to President 1970-73)	1982-89	
Deutch, John M.[c]	Massachusetts Inst. of Technology	1985-89	Physical chemistry
Doolittle, James H.	Shell Oil, Inc.	1957-58	Aeronautical engineering
Doty, Paul M.	Harvard University	1961-64	Biochemistry
Drake, Charles L.	Dartmouth College	1990-93	Geophysics
Drell, Sidney D.	Stanford Linear Accelerator Center	1966-70	Theoretical physics
Dryden, Hugh L.	NASA	1951-56	Physics
DuBridge, Lee A.[b]	California Institute of Technology (Science Adviser to President 1969-70)	1951-56, 1969-73	Physics
Ference, Michael, Jr.	Ford Motor Company	1967-70	Physics
Fisher, William L.[c]	University of Texas	1988-89	Geology, paleontology
Fisk, James B.	Bell Telephone Laboratories	1951-60	Physics

President's Science Advisory Committee and alternative committees 1951–93 (cont.)

Name	Affiliation[a]	Years	Field
Fitch, Val L.	Princeton University	1970-73	Physics
Fletcher, James C.	University of Utah	1967-70	Physics
Frederickson, Donald S.[c]	DSR Associates, Inc.	1982-89	Cardiology
Friedman, Herbert	E. O. Hulburt Center for Space Research, U.S. Naval Research Lab.	1970-73	Physics
Frieman, Edward A.[c]	Scripps Inst. of Oceanography	1982-89	Plasma physics
Garwin, Richard L.	IBM Corporation	1962-65, 1969-72	Physics
Gell-Mann, Murray	California Institute of Technology	1969-73	Theoretical physics
Gibbons, John H.	Office of Technology Assessment (Science Adviser to President 1993-)	1993-	Physics
Gilliland, Edwin R.	Massachusetts Inst. of Technology	1961-64	Chemical engineering
Goldberger, Marvin L.	Princeton University	1965-69	Physics
Gomory, Ralph E.[c]	IBM Corporation (1986-89) Alfred Sloan Foundation (1990-93)	1986-93	Applied mathematics, computers
Good, Mary	Allied Signal, Inc.	1991-93	Inorganic chemistry, radiochemistry
Graham, William R.	R&D Associates (Science Adviser to President 1986-89)	1986-89	Electrical engineering
Gray, Paul E.[c]	Massachusetts Inst. of Technology	1982-86	Electrical engineering
Haggerty, Patrick E.	Texas Instruments, Inc.	1969-71	Electrical engineering
Handler, Philip	Duke University Medical Center	1964-67	Biochemistry
Haskins, Caryl P.	Carnegie Institution of Washington	1955-58	Genetics, physiology
Healy, Bernadine P.[c]	Cleveland Clinic Foundation and AHA	1988-91	Cardiology
Hewlett, William R.	Hewlett-Packard Company	1966-69	Electrical engineering
Hornig, Donald F.	Princeton University (1960-64) (Science Adviser to President 1964-69)	1960-69	Chemistry
Hunter, Robert O.[c]	Western Research Company	1982-89	Laser Physics
Kerman, Arthur K.[c]	Massachusetts Inst. of Technology	1982-84	Theoretical physics
Keyworth, George A. II	Los Alamos National Laboratory (Science Adviser to President 1981-86)	1981-86	Physics
Killian, James R., Jr.	Massachusetts Inst. of Technology (1951-57, 1960-61)	1951-61	Administration

President's Science Advisory Committee and alternative committees 1951–93 (cont.)

Name	Affiliation[a]	Years	Field
	(Science Adviser to President 1957-59)		
Kistiakowsky, George B.	Harvard University (1957-59, 1961-63)	1957-63	Physical chemistry
	(Science Adviser to President 1959-61)		
Land, Edwin H.	Polaroid Corp.	1956-59	Physics
Lauritsen, Charles C.	California Inst. of Technology	1952-57	Physics
Leduc, Elizabeth H.	Brown University	1976-77	Cell biology
Likens, Peter W.	Lehigh University	1990-93	Dynamics, control systems
Loeb, Robert F.	Columbia University (1951-55)	1960-62	Internal medicine
Long, Franklin A.	Cornell University	1961-62, 1963-66	Physical chemistry
Lovejoy, Thomas E.[c]	Smithsonian Institution	1988-93	Ecology
MacDonald, Gordon J.F.	Inst. for Defense Analyses (1966-68) University of Calif. (1968-69)	1965-69	Geophysics
MacLeod, Colin M.	New York University	1961-64	Microbiology
Massey, Walter E.	University of Chicago	1990-91	Theoretical solid state physics
McElroy, William D.	Johns Hopkins University	1963-66	Biology, biochemistry
McTague, John P.	Brookhaven National Lab (1986) Ford Motor Company (1990-93) (Science Adviser to President 1986)	1986, 90-93	Physical chemistry
Moynihan, Daniel P.	Harvard University	1971-73	Economics
Murrin, Thomas J.	Duquesne University	1991-93	Defense and aerospace systems
Nathans, Daniel	Johns Hopkins University	1990-93	Microbiology, molecular biology
Old, Bruce S.	Arthur D. Little, Inc.	1951-56	Metallurgy
Olsen, Kenneth H.	Digital Equipment Corporation	1971-73	Electrical engineering
Oppenheimer, J. Robert	Institute for Advanced Study	1951-54	Physics
Packard, David [c]	Hewlett-Packard Company	1982-93	Electrical engineering
Pake, George E.	Washington University	1965-69	Physics
Panofsky, Wolfgang K.H.	Stanford University	1963-64	Physics
Pierce, John R.	Bell Telephone Laboratories	1963-66	Electrical engineering
Piore, Emanuel R.	IBM Corporation	1959-62	Physics
Pitzer, Kenneth S.	Rice University	1965-68	Physical chemistry
Press, Frank	California Inst. of Technology (Science Adviser to President 1971-81)	1961-64, 1977-81	Geophysics

President's Science Advisory Committee and alternative committees 1951–93 (cont.)

Name	Affiliation[a]	Years	Field
Purcell, Edward M.	Harvard University	1957-60, 1962-65	Physics
Rabi, Isidor I.[b]	Columbia University (Science Adviser to President 1956-57)	1952-60	Physics
Ramo, Simon	TRW, Inc.	1976-77	Physics, electrical engineering
Robertson, H.P.	California Institute of Technology	1957-59	Mathematical physics
Russ, Fritz J.	Systems Research Laboratories	1976-77	Electrical engineering
Seaborg, Glenn T.	University of California	1959-61	Chemistry
Seitz, Frederick	National Academy of Sciences	1962-70	Physics
Shapiro, Harold	Princeton University	1990-93	Economics
Simon, Herbert A.	Carnegie-Mellon University	1968-71	Psychology, computer science
Singer, Isadore M.[c]	Massachusetts Inst. of Technology	1982-89	Mathematics
Slichter, Charles P.	University of Illinois	1964-69	Physics
Smith, Cyril	University of Chicago	1959	Physical metallurgy
Smith, Lloyd H., Jr.	Univ. of Calif. at San Francisco	1970-73	Physician
Stever, H. Guyford	Carnegie-Mellon University (Science Adviser to President 1975-76)	1975-76	Aeronautical engineering and astronautics
Tape, Gerald F.	Associated Universities, Inc.	1969-73	Physics
Teller, Edward[c]	Livermore National Laboratory	1982-89	Physics
Thomas, Charles A.	Monsanto Chemical Corp.	1951-55	Chemistry
Thomas, Lewis	NYU Medical School (1967-68) Yale Univ. Med. School (1969-70)	1967-70	Physician
Townes, Charles H.	Massachusetts Inst. of Tech. (1966-67) Univ. of Calif., Berkeley (1967-70)	1966-70	Physics
Truxal, John G.	State University of New York, Stony Brook	1970-73	Electrical engineering
Tukey, John W.	Princeton University	1960-63	Mathematics
Turner, Howard S.	Turner Construction Company	1972	Chemistry
Waterman, Alan T.	National Science Foundation	1951-56	Physics
Watkins, Dean A.[c]	Watkins-Johnson Company	1988-89	Electrical engineering
Webster, William	New England Electric System	1951	Physics
Weinberg, Alvin M.	Oak Ridge National Laboratory	1960-62	Nuclear physics
Weiss, Paul A.	The Rockefeller Institute	1958-59	Biology
Westheimer, F.H.	Harvard University	1967-70	Chemistry

President's Science Advisory Committee and alternative committees 1951–93 (cont.)

Name	Affiliation[a]	Years	Field
Whitman, Walter G.	Massachusetts Inst. of Technology	1951-55	Chemical engineering
Wiesner, Jerome B.	Massachusetts Inst. of Technology (Science Adviser to President 1961-64)	1956-64	Electrical engineering
Wiley, W. Bradford	John Wiley and Sons, Inc.	1976-77	Publisher
Wood, Harland G.	Case Western Reserve University	1968-72	Biochemistry
Wyngaarden, James B.	Duke University Medical School	1972-73	Medicine, biochemistry
York, Herbert F., Jr.	Livermore Laboratory	1957-58 1964-67	Physics
Zacharias, Jerrold R.	Massachusetts Inst. of Technology	1952-58, 1961-64	Physics
Zinn, Walter H.	Combustion Engineering, Inc.	1960-62	Physics

[a] Affiliation is at time of service.

[b] Chairmen of the Science Advisory Committee, ODM, were also Presidential Science Advisers, including Oliver E. Buckley (1951-52), Lee A. DuBridge (1952-56), and Isidor I. Rabi (1956-57).

[c] Member, White House Science Council

The editorial assistance of Georgia Phelps for the first edition is gratefully acknowledged, as is her compilation, from fragmentary sources, of the first comprehensive tabulation of PSAC members, 1951–1973. Subsequent additions for the second edition have been provided by William G. Wells, Jr., to whom I am grateful also for other historical information and advice.—*W.T.G.*

Science and Technology in President Clinton's First 100 Days

John H. Gibbons

The early actions of the Clinton Presidency reveal the Administration's grasp of the power of science and technology to enable the future. President Clinton and Vice President Gore believe investing in science and technology is investing in America's future: a growing economy with more high-skill, high-wage jobs for American workers; a cleaner environment where, for example, energy efficiency increases profits and reduces pollution; a stronger, more competitive private sector able to maintain U.S. leadership in critical world markets; an educational system where every student is challenged to reach his or her full potential; and an inspired scientific and technological research community focused on ensuring not just our national security but our very quality of life. Both the President and the Vice President will support science and direct technology to improve the quality of life and the economic health of our nation.

The President's proposals for deficit reduction and economic stimulus continue to reflect a recurrent theme of his campaign—science and technology are the engines of economic growth. One month into his presidency, Mr. Clinton made a serious proposal to put science and technology to work for all sectors of the American economy. In *Technology for America's Economic Growth: A New Direction to Build Economic Strength*, he has offered a comprehensive blueprint to focus American science and technology on three central goals:

- long-term economic growth that creates jobs and protects the environment

- making government more efficient and more responsive

- world leadership in basic science, mathematics, and engineering

John H. Gibbons (b. 1929) is President Clinton's Science Adviser and Director, Office of Science and Technology Policy. From 1979 through 1992, he directed the Congressional Office of Technology Assessment. Earlier, he devoted a decade to experimental studies of the origin of the solar system at Oak Ridge National Laboratory (ORNL), after which he led important work on the environmental impacts of energy production and the technological potential for energy conservation at ORNL and the University of Tennessee.

With this initiative, we take a critical step toward ensuring that the federal investment in science and technology becomes a key instrument for promoting U.S. economic growth and for satisfying other national goals.

Long-Term Investment Package

Advances in technology created two-thirds of the productivity growth in the United States over the past 60 years. The knowledge-based, growth industries of the future depend on continuous generation of new technological innovations and rapid transformation of those innovations into marketable products. We can promote technology as a catalyst for economic growth by

• *directly supporting the development, commercialization, and deployment of new technology.* The President has announced his intention to increase the civilian share of government for Research and Development (R&D) funding to more than 50% by 1998 and to encourage industry consortia to improve our competitive posture.

• *creating a world-class business environment for innovation and private sector investment.* We must ensure that our tax, trade, regulatory, and procurement policies support the objectives of industrial and technological leadership and high-wage jobs.

• *investing in life-long learning.* We intend to use new technology to increase the productivity of learning in all environments, to make instruction and resources available to schools, businesses, and homes throughout America, and to turn schools into high-performance workplaces.

• *supporting information superhighways.* Accelerating the introduction of an efficient, high-speed communications system can have the same effect on U.S. economic and social development as investment in the railroads had in the nineteenth century.

• *upgrading the transportation infrastructure.* A strategic program to develop new technologies for assessing the physical condition of the nation's infrastructure, together with techniques to repair and rehabilitate those structures, could lead to more cost-effective maintenance of the infrastructure necessary for economic growth.

Reinventing Government

Present circumstances demand that we use technology to reinvent government— to make it work better, quicker, smarter, and cheaper. The federal government can become more responsive to the needs of its clients by making better use of information technology, making government more cost-effective, efficient, and user-friendly; by operating energy-efficient buildings, thereby ending the senseless waste of money and setting a good example for the nation; and by reforming procurement policy, thus increasing government's efficiency and creating new markets for innovative products.

World Leadership in Science, Mathematics, and Engineering

Our basic science program provides an ongoing sense of adventure and exploration while improving the knowledge base. It also lays the foundation for new technologies. The federal government has invested heavily in basic research since the Second World War, and this support has paid enormous dividends. None of the innovations in technology proposed in our initiative will be funded at the expense of basic science. Our budget proposal ensures that support for basic science remains strong and that stable funding is provided for projects that require continuity. But stable funding requires setting clear priorities. Improved management of basic science can ensure sustained support for high-priority programs.

Can Science Advice Flourish in a Receptive Environment?

The fates have conspired to place many of the tools my predecessors wished for at my disposal. I work for a President who committed to a premier role for science and technology in national affairs when he nominated his Science Adviser simultaneously with his Cabinet Secretaries. There is congressional support for a fully operational Office of Science and Technology Policy (OSTP) in the Executive Office of the President to help realize the President's goals for science and technology. Only a few weeks into my appointments as Assistant to the President for Science and Technology and Director, OSTP, I cannot imagine completing the marathon ahead without this running start . . . and the benefits of my experiences at the Congressional Office of Technology Assessment (OTA).

Creating and nurturing an organization reflective of the nature and personality of its several masters necessarily consumes a Science Adviser's early days—perhaps an entire term. OSTP exists primarily to advise and assist the President, the Vice President, and immediate White House staff, particularly the Office of Management and Budget, the National Economic Council, and the National Security Council. My early arrival on the scene ensures OSTP's cognizance of the various agendas and the opportunity to present the options offered by science and technology at a point in the policy process when they can still make a difference.

The White House also depends on OSTP to lead the entire Executive Branch toward a coherent science and technology policy. We provide the fora—through the Federal Coordinating Council for Science, Technology, and Engineering and through other mechanisms, such as the newly established S&T (for "Science and Technology") Deputies Group established to implement the technology initiative—for developing and implementing sound policies and programs to link science and technology to national goals. OSTP, a small office, catalyzes activities within and among the much larger agencies, and, importantly, in the economy at large. We coach a potentially world class team—Executive agencies, laboratories, research universities—that still needs to learn each player's strengths.

Congress relies on OSTP's coordinating function as well. Congressional members expect OSTP to help them exercise oversight of the R&D budget. They also expect OSTP to bring the import of science and technology to bear on each major governmental initiative. Maintaining the appropriate balance between responsiveness to the President's need for flexibility and the Congress's demands for OSTP's attention taxes my equanimity along with that of my clients. It is a healthy tension, one deserving of respectful monitoring by all parties.

OSTP will also be a service for the private sector, for the S&T users—from researchers, to industrial leaders, to teachers, to consumers—who elected this Administration. We are building a feedback loop into every element of science and technology policy, from development, to implementation, through evaluation. The venues for private sector involvement in OSTP's work, including foreign and domestic stakeholders in our science and technology policies, are plentiful and diverse—ranging from formal advisory committees to increased interaction with industry and professional associations.

The majority of my attention falls to the urgent task of reinvigorating OSTP to attend to the immediate needs of the Clinton Administration. However, C.P. Snow wisely cautioned, and I think President Clinton would agree, "A sense of the future is behind all good politics. Unless we have it, we can give nothing—either wise or decent—to the world." Thus I continue to wrestle with the problem I had the privilege to raise in an earlier essay for this series[1], namely, how to build continuity, quality control, and long-range analytical capability into a highly politicized environment. I have yet to discern how, or whether, analysts and activists can peacefully and productively cohabit, but it is on my agenda.

I draw few conclusions from my experiences to date. The President's and Vice President's appreciation of and attention to science and technology are gratifying. The enthusiasm for technology development and commercialization programs, for tax, fiscal and regulatory policies that encourage innovation, and for investing in research and development with an eye toward the social rate of return makes this White House an exciting place to ply my trade. For this Administration, if mechanisms were not already in place for science and technology policy analysis and advice, they would have quickly been established. But the test remains to devise a modus vivendi that successfully integrates S&T consciousness into the rough-and-tumble, fast-moving, and complex world of White House political decisions. Early signs are encouraging, but the jury is still out.

Reference

1. Gibbons, John H., "Science, Technology, and Law in the Third Century of the Constitution," in Golden, William T., ed., *Science and Technology Advice to the President, Congress, and the Judiciary* (New York: Pergamon Press, 1988; Washington, DC, AAAS Press, 1993), pp. 415-417.

Science and Technology
in the Bush Administration

D. Allan Bromley

There is little in this world more useless than unwanted advice. Nowhere is this more the case than in the White House. The success of a Presidential Science Adviser thus rests crucially upon the personal trust and chemistry that exists between the Adviser and the President and the extent to which that trust and personal support is manifested in signals—both small and large—as read by the other members of the President's inner circle of Advisers and by the senior members of the federal government. I was singularly fortunate to work for a President who was entirely convinced that he needed continuing, reliable input on science and technology— even when he himself did not necessarily recognize that need initially; who was remarkably sensitive to the importance of science and technology in almost every aspect of our modern society; and who more than any President in memory was prepared to support science and technology as an investment in our national future.

The scientific and technological enterprise in the United States is still by far the largest in the world. The United States spends more on research and development activities (both civilian and military) than do France, Germany, Japan, and the United Kingdom combined. Federal investment in the support of basic research has grown in real terms by over 325% since 1960, by more than 40% since 1980, and by more than 30% since 1989. This latter growth reflected the Bush Administration's strong and continuing commitment to this nation's long-established policy of investing simultaneously, through the federal government's support of basic research, in the creation of new knowledge and in the training of young minds to use that knowledge creatively.

D. Allan Bromley (b. 1926) was the Assistant to the President for Science and Technology and Director of the Office of Science and Technology Policy throughout the Bush Administration. He is currently the Sterling Professor of the Sciences at Yale University and was a member of the White House Science Council and the National Science Board during the Reagan Administration. He has served as President of the American Association for the Advancement of Science and of the International Union of Pure and Applied Physics. A member of the National Academy of Sciences and the American Academy of Arts and Sciences, he has served on a number of governmental and private sector boards and committees.

More than ever before, science and technology play important roles in almost every aspect of American society; this is reflected in an ever increasing need for timely and effective advice concerning both to the highest levels of decision-making in our federal government.

Historical Background

Formal mechanisms for providing science advice to Presidents of the United States has a lengthy history, and the relationship between the Presidents and their respective Advisers and Advisory councils has been anything but a smooth one. In the modern era, the Adviser tradition effectively dates back to Vannevar Bush and President Roosevelt in the late 1930s and, subsequently, throughout the years of World War II. The first person to actually hold the Science Adviser title, however, was Oliver Buckley. Appointed to that office at the suggestion of William Golden in 1951, although Buckley was Science Adviser to President Truman, he also chaired the Science Advisory Committee and reported to Truman primarily through the Office of Defense Mobilization, rather than directly. He was succeeded in that position by Lee DuBridge and subsequently by I.I. Rabi. As far as the records indicate, Truman made relatively little use either of the Science Advisory Committee or of his Science Adviser.

All this was to change dramatically in 1957 with the almost hysterical U.S. response to the launching of the Soviet Sputnik. Because the American public had enjoyed an overwhelming sense of scientific and technological superiority dating from the war years, the fact that the Soviets were able to put a satellite in orbit before any American success on that frontier led to a gross overreaction. The general sense was that somehow the Soviets had gained an overwhelming lead on us in almost every aspect of science, technology, and education. When Edward Teller was asked, during a television interview, what we might expect to find on the Moon he answered, succinctly, "Russians!" When Rabi, as Chairman of the Science Advisory Committee, suggested to President Eisenhower that he needed a full-time Special Assistant for science and technology—reporting directly to him within the White House as William Golden had also recommended to President Truman—Eisenhower, perhaps not surprisingly, agreed immediately; he also accepted immediately James Killian's suggestion that the Special Assistant should be backed up by a President's Science Advisory Committee (PSAC), as Golden had also recommended to President Truman, with members drawn from among the most distinguished scientists and engineers in the United States—also reporting directly to the President. It is a measure of what has happened to our society that it took President Eisenhower and Killian nine days to bring the PSAC into being, and to a meeting in the White House, whereas it took George Bush and me almost precisely nine months to wade through the various legal thickets to accomplish this same thing in 1989.

Eisenhower met frequently with his PSAC and referred to its members as "my scientists." Kennedy also used PSAC and his Science Adviser, Jerome Wiesner, extensively and, in the Reorganization Plan Number 2 of 1962, Kennedy moved Weisner (who, up to that time, as Special Assistant for Science and Technology, had been a Confidential Assistant within the White House structure) into the Executive Office of the President as Director of the Office of Science and Technology (OST). President Johnson had neither the time nor the interest to interact significantly with PSAC and President Nixon fundamentally distrusted scientists generally and some of his PSAC members in particular because they had chosen to go public in opposition to his administration's positions on the supersonic transport and the antiballistic missile system. In 1973, Nixon abolished both the position of Special Assistant for Science and Technology and the President's Science Advisory Committee within the White House—requesting that the advising responsibility be taken over, to the extent necessary, by Guy Stever, then the Director of the National Science Foundation.

Although President Ford, from the time he was sworn in as President in mid-1974, was very much interested in reestablishing a Science and Technology presence in the White House, because of his Congressional background, he was convinced that it was important to involve the Congress in the establishment of any new Advisory mechanism. Also, because he was very much interested in having his Vice President, Nelson Rockefeller, substantially involved in that mechanism, whatever it might be, it took some two years before a science Advisory apparatus was reestablished in 1976 through Public Law 94-248. This legislation established the Office of Science and Technology Policy (OSTP), the Federal Coordinating Council for Science, Engineering, and Technology (FCCSET), and an Intergovernmental Council for Science, Engineering, and Technology (ICSET). It was assumed that the Director of the Office of Science and Technology Policy would serve as Science Adviser to the President, as well as Chairman of the Federal Coordinating Council and the Intergovernmental Council, thus building in a fundamental tension in that, as Science Adviser to the President, the Adviser operated under executive privilege while, as Chairman of OSTP and of these Councils, he was answerable to the Congress. Although there has been much discussion of the potential for problems in this arrangement, these have been much more theoretical than real and the advantages inherent in the double-hatted position, at least in my opinion, have far outweighed any potential disadvantages.

Although the 1976 legislation provided for a President's Science Advisory Committee, neither Presidents Ford nor Carter saw fit to establish such committees. In part this was undoubtedly a reflection of the fact that the Freedom of Information Act (FOIA) was now in place and many felt that it would be simply impossible to have a PSAC function with any part of its deliberation open to the public. As we shall see, this is a question that continues to cause some level of problem but is nowhere nearly as serious an impediment to effective action as was originally feared. Beyond that, it may well be that the Ford and Carter senior staff were uncomfortable about the idea of bringing private sector individuals directly into the highest levels of decision-making without having any control over what issues they might raise or what positions they might support.

To the distress of many in the scientific community, when George Keyworth was appointed as Reagan's initial Science Adviser, he chose to establish the White House Science Council (WHSC), a group modeled somewhat along the PSAC lines, but reporting to the Science Adviser rather than to the President. This WHSC mechanism was continued during the latter years of the Reagan Administration when William Graham served as Science Adviser. Many in the scientific community felt that this very substantially downgraded the visibility and importance of science and technology in the White House. As someone who was privileged to be a member of the WHSC during the eight Reagan years, I am not at all convinced that President Reagan would have made greater use of a committee reporting directly to him, as opposed to the Adviser, but I can also vouch for the fact that the ambiance was quite different. Although the WHSC membership was a very distinguished one and although it carried out a number of very high-quality studies, there was never the same sense of direct involvement at the highest level of government that comes from personal participation with the President of the United States in meetings of the Advisory group.

Early Science and Technology Activities in the Bush Administration

Commitments by George Bush prior to his election. During the election campaign in 1988, in a speech to the Ohio Society of Broadcasters, George Bush made the commitment that, if elected, he would upgrade the position of his Science Adviser to that of The Assistant to the President for Science and Technology, placing that individual formally at the same level as The Assistant to the President for National Security and making him one of the inner circle of Presidential Advisers. He also promised to put in place a Presidential Science Advisory Council that would report directly to the President, something that had been missing from the White House Science and Technology apparatus for some 20 years.

In April of 1989, George Bush asked me to take on the joint responsibilities of The Assistant to the President for Science and Technology and the Directorship of the Office of Science and Technology Policy. In my discussions with the President, I asked for three specific promises. First, that I would have access to him when I needed it—on the condition that I not abuse this privilege; second, that once we had agreed on a particular course of action relating to science and technology, that he would provide his personal support for its implementation; and third, that for the first time in the history of OSTP he would nominate all four of the Presidentially-appointed, Senate-confirmed Associate Directors called for in the 1976 legislation. He agreed to these three requests and subsequently followed through completely on all of them.

Responsibilities of the Science Adviser. As we discussed my responsibilities, it became clear that they divided naturally into two subsets. First, was that relating to *science and technology for policy* where my responsibility was that of providing the President and his senior associates with the best possible scientific and technological input to decisions where frequently science and technology would be relatively minor components of the overall decision but, in many cases, still very important ones. Second was that concerning *policy for science and technology* where fundamentally my responsibility was that of insuring that the United States science and technology enterprise remained in a healthy and productive state, contributing effectively to such national goals as economic competitiveness, national security, environmental stewardship, and an improved quality of life for all our citizens.

Access was obviously critical to both of these categories, not only to the meetings and discussions where scientific input was required, but equally important to those where, in the absence of someone with a scientific and technological background, decisions could well be made that might have an unfortunate impact on the U.S. science and technology enterprise. Such access had always been a problem for my predecessors as Science Advisers but, as The Assistant to the President for Science and Technology, I served on the Domestic Policy Council, the Economic Policy Council, and the Space Council, attended all Cabinet meetings automatically and worked closely with both the National Security Council and the Competitiveness Council when their discussions had scientific or technical implications. As one of the thirteen Assistants to the President, I was a member of his inner circle of Advisers; as the only one of them holding Senate confirmation I found myself frequently testifying before Congressional Committees—42 separate times in all—whenever members of the Congress wanted to focus on a particular area of science or technology or indeed when they simply wanted to beat up a little on either the Administration or the President. Maintenance of Executive Privilege was never really a problem because members of the House and Senate were aware of it and respected it generally.

A problem that I shared with all my predecessors was the fact that some members of the scientific community felt that the Presidential Science Adviser was their lobbyist in the White House. It bears emphasis that the moment the Adviser is perceived to be functioning as such a lobbyist or engaging in special pleading on behalf of the scientific community rather than functioning as a member of the President's inner circle—committed to implementing the President's goals—his effectiveness and the cooperation that he receives from the other senior staff members disappears almost instantaneously.

Although I accepted the position as Science Adviser in April of 1989—and functioned essentially full time, on a consulting basis, thereafter—it was not until August 4 that I was formally confirmed by the Senate. This delay represented the sum of a whole series of small but singularly frustrating delays at both ends of Pennsylvania Avenue. Such delays are of critical importance because if the Science Adviser is not in place early in the Administration it is impossible for him to participate in the selection of the large number of persons throughout the upper level of the federal government who require scientific and technical credentials and it is impossible for him to participate with the other senior Advisers during the crucial initial period of the Administration when the interactional chemistry and communication channels are forming. Having missed this critical period, it was necessary for me to devote substantial effort to earning my way into the inner White House loop, since that loop was already well established. I am delighted that my strongest recommendation to the Clinton transition team, namely that my successor as Science Adviser be appointed as early as possible, was taken seriously and John H. Gibbons was appointed on Christmas Eve of 1992 so that he could be involved from the outset of the Clinton Administration.

Rebuilding the OSTP staff

At the request of President Bush, I spent a large fraction of my time between April and August of 1989 in pulling together nominations for senior positions at OSTP, on the President's Council of Advisers for Science and Technology, and for a large number of the still unfilled science and technology positions throughout the federal government. I was highly gratified to find that scientific credentials were the important criterion in the Bush Administration and that only in the most extreme cases did questions of political affiliation even arise. Although I do not know for sure, since I never asked, I suspect that a substantial fraction of my staff in OSTP were Democrats as far as their politics were concerned.

Although the 1976 legislation founding OSTP specifies that there be up to four Presidential-appointed, Senate-confirmed, Associate Directors, it is silent on the functions of these individuals. After considerable thought, I selected *physical sciences and engineering, life sciences, policy and international affairs,* and *industrial technology* as the four areas that would be placed under the direct responsibility of Associate Directors. The industrial technology assignment was a completely new one and was selected both because of the intrinsic importance that both the President and I attached to it and because we wished to send a message to the community that the Bush Administration was very serious with respect to its interest in the industrial technology field. Having selected four Associate Directors, an Assistant Director was selected to work closely with each of the Associate Directors, and an additional Assistant Director was appointed with primary responsibility for each of national security affairs, environmental affairs, social science, and for management—both of the OSTP office generally, as well as the three Presidential Councils—PCAST, FCCSET, and InterSET. These individuals are listed in Appendix 1.

The five Presidential-appointed, Senate-confirmed senior members of the OSTP group brought a remarkably broad range of professional expertise, experience, and background to the Office and it was our goal from the outset to function in a collegial fashion rather than as a group of isolated, disciplinary, or interest divisions. To that end, following the senior staff meeting that I attended every morning at 7:30 a.m. in the White House, the senior OSTP staff met every morning for at least half an hour to discuss the major items under consideration throughout the Office and, because it was difficult to retain perspective in the brush-fire atmosphere of the White House, we held regular retreats off-site where we would disappear from the Office for an entire day and devote ourselves to discussion of the Administration's longer-range goals, the stages of our major projects and undertakings, and the state of U.S. science and technology generally. These collegial discussions were among our most important activities throughout the entire Administration in that they allowed us to bring the best of our combined knowledge and experience to bear on the issues before us.

When I arrived in 1989, the total staff of OSTP numbered 11, and, when I left in January of 1993, it numbered 65. Of this total, 43 were approved full-time equivalents and the remainder were detailees on short-term assignments from other Federal agencies, liaison representatives from other agencies, and Fellows under a variety of sponsorships. Over this same period, the OSTP budget increased from about $1.5 million to $6.25 million. Congress throughout was generous in its support of OSTP's activities and the effective limitation on our budget was that imposed by objections from other offices within the Executive Office of the President, where members were unhappy about OSTP's relative rate of growth; this unhappiness reflected itself in limitations imposed by the Office of Management and Budget in its final preparation of each year's Presidential budget.

Relocating the OSTP staff

For decades, the Science Adviser to the President has traditionally occupied the northeast corner suite on the third floor of the Old Executive Office Building. I followed in this pattern, but when I arrived in Washington, the remaining OSTP staff and the space available for expansion was all in the New Executive Office Building on 17th Street. For all practical purposes—in terms of ease of communication—Pennsylvania Avenue could have been five miles wide and so one of my very early initiatives was that of bringing as much of the OSTP activity as possible into the Old Executive Office Building. Since it is well known that space in the West Wing of the White House and in the Old Executive Office Building is carefully guarded and that ownership of such space changes only as a consequence of direct Presidential intervention—if not, indeed, mortal combat—it took several months to complete arrangements wherein I traded all of the OSTP space in the New Executive Office Building for just about one-half of the equivalent floor space on the fourth and fifth floors of the Old Executive Office Building together with one of the townhouses on Jackson Place. In retrospect, despite the relative crowding that was involved, this move was probably one of the most effective actions taken during my service in Washington, because it greatly increased the amount of real time interaction, communication, and collegiality within the Office. Reluctantly, it was necessary to keep the full staffing for the Presidential Councils—PCAST, FCCSET, and InterSET—in the Jackson Place townhouse, and I have no question whatever but what this reduced our efficiency and made it much more difficult, than would otherwise have been the case, to insure a continuing integration of OSTP's scientific and technical expertise in the activities of the Councils. I have made a very strong recommendation to Jack Gibbons that he attempt to integrate the OSTP and Council activities further.

An Auto-Immune Disease

It may be worthwhile adding a parenthetical comment about one of the major personnel difficulties experienced throughout my four years in Washington and a difficulty that is becoming more and more serious throughout the entire federal government. Because of ever more stringent interpretation of conflict-of-interest regulations, it has become effectively impossible to bring someone from a private sector organization into government in mid-career because, in order to do so, the individual is typically required to give up his or her pension rights, seniority, and the option of returning to the same organization. I took the point of view that any one crazy enough to agree to these regulations was not someone that I wanted in my Office. But, in fact, the federal government pays a very high price for this grossly overdrawn emphasis on potential conflict-of-interest and in this arena, in a very real sense, we, as a nation, have fallen victim to an auto-immune disease. Since it is not the regulations themselves, or the legislation, that is primarily at fault, but rather their interpretation by generations of ever more nervous lawyers, it may be that one partial solution to this problem could be obtained by having the Congress add language to the pertinent legislation spelling out, once and for all, what it intended to achieve by the legislation in the first place.

It is also worth noting that, while American industry currently spends about $76 billion each year on research and development, the National Association of Manufacturers estimates that this same industrial sector spends $118 billion on *outside* legal services. This does *not* include payments to in-house counsel, damages paid to litigants, or any indirect litigation costs of diverting this enormous resource away from value-adding activities. Legal costs have doubled in each of the last two five-year periods.

In September 1992, after more than a decade of effort on the part of many groups, the U.S. senate failed to pass meaningful Tax Reform legislation.

If we continue to pay roughly 1.5 times as much on litigation as we do on the creation of new wealth in American industry i.e. research and development, we are on a trajectory to economic disaster.

Rebuilding bridges

For a great many reasons, many of them irrelevant, communication linkages between OSTP and many other components of the federal government had tended to wither and decay during the latter years of the Reagan Administration so that one of our major objectives in the early days of the Bush Administration was the rebuilding of these relationships and communication bridges. I shall touch on them very briefly below.

With the Office of Management and Budget (OMB). It has always been true that those who control the finances have substantial voices in policy. I therefore devoted a substantial amount of time to meetings with Richard Darman, the Director of the Office of Management and Budget, and Robert Grady, his Associate Director for the Sciences and Technology. Whereas in the past OMB had worked with the agencies in developing the President's annual budget and then asked OSTP for its input toward the end of the process, we agreed that, during the Bush Administration, staff members in OSTP would work closely with those of OMB in the ongoing interactions with the agencies throughout the entire year. This insured that there would be no surprises and that OSTP would be able to be as helpful as possible in providing input concerning priorities and balances as the budget evolution process went forward. Without an unprecedented level of cooperation between OSTP and OMB, it would also have been impossible for us to develop the FCCSET mechanisms for the crosscut analyses leading to Presidential Initiatives in areas of national importance that I shall return to below.

With other components of the Executive Office of the President. I was fortunate in having senior colleagues who were comfortable with science and technology. Darman was a Harvard PhD who had substantial high-tech experience and interests, Sununu was a MIT PhD in mechanical engineering, Boskin was a Stanford PhD who had spent much of his professional life studying the economic consequences of technology, and Porter was a Harvard graduate and Rhodes Scholar with strong ties to education and to the business sector. Although we differed from time to time on matters of emphasis, we were fortunate in sharing an unusually broad fraction of our backgrounds. Despite much press speculation to the contrary, Sununu was the strongest ally and supporter that I had in the entire White House—other than the President.

A large fraction of the President's senior staff met regularly around the staff table in the Executive Mess at lunch each day and a significant fraction of all the business of the Bush Administration was carried on, discussed, and communicated over those luncheons. The abolition of the Executive Mess will make this kind of easy, friendly, and regular communication much more difficult in the Clinton Administration.

With Agency Heads. The United States is unique in that more than 20 agencies support substantial science and technology programs, and, thus, the cooperation of the heads of these agencies is essential if any effective coordination is to be achieved. It is also understandable that agency heads tend to be extremely wary about what they view as efforts to control their agencies by anyone in the White House and personal relationships and trust are enormously important. I therefore made a point of visiting each of the agency heads, early in the Administration, to discuss with them what they viewed as their major problems and to talk to them about the revitalization of the Federal Coordinating Council (FCCSET) that I had been discussing with the President. Only because they had a relatively clear idea of what we were attempting to achieve, and an equally clear understanding that the President fully supported the activity, were we able to attain anything like the unprecedented level of cooperation that eventually developed under the FCCSET.

With the Congress. Over the later years of the Reagan Administration, considerable antagonism developed between OSTP and the Congress as a result of an unfortunate downward spiral. Because OSTP did not have adequate staff to produce the reports demanded by the Congress, and thus did not respond to these demands, the Congress administered punitive cuts to OSTP's budget. This in turn further reduced the staff and made OSTP appear even less responsive. During my first few months in Washington, I met on a one-on-one basis with more than 25 senators and representatives from key committees in both Houses, in addition to holding a number of meetings with groups of staffers from Authorization, Appropriations, and Science and Technology committees and subcommittees. These were primarily of a get-acquainted nature and provided fora where unfulfilled expectations, festering misunderstandings, and general unhappiness could be placed on the table and discussed openly. These meetings also uncovered a tremendous reservoir of goodwill and support, which was of enormous value to us throughout the Administration. Additionally, a number of senior members of both the House and the Senate, on a somewhat irregular basis, organized breakfast meetings to which I was invited and at which any member present felt free to raise any issue that he or she wished to discuss. It is to everyone's credit that no one ever attempted to override the appropriate bounds of executive privilege in any of these remarkable open discussion, although the fact that I was engaged in them did give significant pause and concern to some of my senior colleagues in the White House.

This led to a rather amusing shuffle early in my Washington tenure. Since I was the first Science Adviser to have the rank of Assistant to the President, someone decided that I really should have an office in the White House and, in short order, I was presented with such an office on the second floor of the West Wing. It was a very small office; there was space for only a single secretarial support staff member; and the file cabinets were overflowing with the files belonging to other Assistants who, desperate for space, had filled every available cubic inch.

For about six weeks, I spent a few hours each day in this office and spent an unreasonable fraction of that time clearing the files and trying to maintain the new collegiality and communication with my OSTP staff that had been achieved by moving them near to my other office in the Old Executive Office Building.

During this period, I appeared a number of times before both House and Senate committees who were interested both in meeting the new Science Adviser on their turf and finding out as much as they could about Bush Administration policy on science and technology. Interested in the results of these hearings, a few Senators raised the question of whether it might not also be interesting to hear from some of the other Assistants to the President—forgetting for the moment the question of executive privilege. This sent a *frisson* of real concern throughout the West Wing and in short order the Counsel to the President and the Chief of Staff were discussing with me the possibility that perhaps after all I should respect decades of tradition and use my Old Executive Office exclusively. I was happy to agree, although the media had a few days of active speculation as to whether my resignation was imminent.

With the Professional Societies. I had long felt that the professional societies that I had come to know very well during my Presidency of the American Association for the Advancement of Science (AAAS) represented a resource that had been much too little used by the federal government. To that end, I organized a series of regular meetings between the Presidents and Executive Directors of the professional societies in the science and technology area and the senior members of OSTP. For convenience, we divided the societal representatives into four groups representing physical sciences, life sciences, social sciences, and engineering; each of the meetings typically occupied some two hours, and at each I invited selected Presidents to make short opening presentations to highlight major issues that they wished to discuss. We found these meetings to be extremely effective communication vehicles and, flowing from them, a number of the societies carried out extensive activities in support of particular programs or activities underway in OSTP and FCCSET. As a particular example, the Materials Research Society (MRS) organized regional meetings across the nation and collected vitally important grassroots information for use in our Material Science and Processing FCCSET crosscut activity.

With the Private Sector. Unfortunately, there has been an erosion of mutual trust between the federal government and the private sector in the United States and repeatedly, in discussions with senior industrialists, we found that, at best, they considered their relationship with the federal government to be neutral and more frequently adversarial. On many occasions, I organized meetings between representatives of major industries—including the semiconductor industry, the computer industry, the food industry, the aerospace industry, and the automotive industry—with senior members of the White House staff. These meetings would normally involve dinner in the White House Mess and several hours of intensive discussion, usually followed by homework assignments from us that provided us with very important input for use in the development of legislation, regulations, and the annual Presidential Budgets.

At the same time, the various FCCSET committees and crosscut groups made a major effort to involve the private sector in their planning activities to obtain the private sector validation and calibration that would otherwise be missing from an all governmental activity. Here again, in general, the private sector responded enthusiastically and effectively to our requests for such constructive criticism and their input played an important part in the development of the Presidential Initiatives that appeared in each of the Bush Budgets. One of my major regrets is that we were unable to evolve even greater interaction between the private sector and our Administration.

With the Media. As a result of some unfortunate experiences, my predecessors had tended to minimize their contact with media representatives and, in some quarters, a substantial ill will had developed. I decided from the outset that it was important, in so far as possible, to be as open as possible with media representatives and that in so doing I could help them to get our messages across to the public. I therefore, as a matter of course, set aside time for several lengthy interviews each week. With few exceptions, I found the reporters and other media representatives to be interested, responsible, and willing to take substantial extra time to really understand the issues involved. In general, in return for our being prepared to take the time to work with them, we were well treated by the media. Indeed this was a source of some resentment on the part of some of my senior colleagues but I have long felt that members of the scientific community are on shaky footings when they protest loudly about the lack of scientific literacy among the public and, yet, at the same time, are unwilling to devote time themselves to working with scientific journalists who are our best linkage to that public in the short run.

Revitalizing FCCSET

The United States is unique in that its scientific and technological activities are, in general, built from the bottom up rather than the top down as is characteristic of the rest of the developed world. Instead of a single agency responsible for science and technology, we have more than 20 federal agencies each of which supports a significant part of our overall science and technology enterprise. This plurality of support channels is one of the major sources of our science and technology strength because, as a result of it, we have been able to say, that with very rare exceptions, no really good idea has gone without support for any significant period. At the same time, with many agencies working on different components of a given field, in the absence of effective coordination and communication, there is a great potential for overlap, inefficient use of resources, and lack of complete coverage. The need for such coordination had been recognized for a long time and one of the early actions of President Eisenhower's PSAC was to recommend the formation of a Federal Council that would be charged with such coordination and integration of federal programs into truly national programs rather than a heterogeneous collection of agency programs in a given area.

Although the Federal Council was formed in the late 1950s—and even remained functional to some degree during the period when there was no scientific organization in the White House—during the Nixon Administration—it never fulfilled the hopes of its originators. The 1976 legislation that reestablished a White House science and technology presence, in addition to creating the Office of Science and Technology Policy, also restructured the Federal Council into the Federal Coordinating Council for Science, Engineering, and Technology with the immediate, and somewhat unhappy, acronym FCCSET. Despite the change in title, FCCSET, too, did not live up to expectations for it and, in my early discussions with President Bush, we decided that it deserved a careful reexamination. The problem soon became clear; the FCCSET did not function as anticipated primarily because its membership was at too low a political level so that no matter what decisions were made, in the final budget crunch each year more senior political figures in the agencies could—and indeed did—disavow the FCCSET decisions. With this diagnosis in hand and with strong support from the President, FCCSET was restructured so that its membership, during the Bush Administration, consisted of the Secretary or Deputy Secretary of 12 of the 14 Cabinet agencies together with the Heads of all the major independent agencies including NASA, EPA, NSF, CIA, as well as, OMB. One of our very important early rulings was that there could be no substitution in FCCSET meetings; agencies were either represented by their senior members or they were not represented at all. This, together with strong pressure from the President, kept FCCSET attendance at a high and effective level. As an additional incentive to senior people having deep interest in matters of science and technology but whose schedules were so full that they had little occasion to be exposed to either during their Washington stay, I made a practice of inviting a very distinguished scientist or engineer to make a half-hour presentation, on a topic of high current interest, to the FCCSET at each of its meetings and then answer the member's questions. Many of the members have subsequently told me that these sessions represented some of the most exciting and stimulating half-hours that they had encountered in government.

Under this structure, we established eight standing committees spanning the entire spectrum of science and technology and each of these in turn established *ad hoc* panels and working groups to address particular issues and problems.

Under the FCCSET structure, two major categories of activities were dominant— one budget related and one not budget related. The first category was in response to a Presidential request for advice on the selection of a limited number of areas of major national importance and the development of strategic planning for those areas. These subsequently became known colloquially as *crosscuts* because they cut across the boundaries of many agencies and, in each case, they subsequently became formal Presidential Initiatives in the annual budget submissions to the Congress.

In identifying the crosscut topics, the President's request was put to the FCCSET membership and each member was asked to submit two or three candidate topics with the requirements being that they cut across the mission of more than a single agency, that they were of national importance, and that they were worthy of Presidential attention. With some 20 to 25 such topics in hand, the FCCSET Council met and, on each occasion, in a remarkably short time, whittled the list down to five or six; it always struck me as remarkable that so diverse a group would so quickly converge and agree on the most important items on such a list. Having selected five or six, we then requested that the original proposer—together with several other agencies that were prepared to volunteer, at the meeting, to work with the proposer, prepare a few-page proposal fleshing out the suggestion that the particular topic met the three selection criteria noted above. With these proposals in hand, the FCCSET Council met once again and made the actual selection of the crosscut topics. During the first three years of the Bush Administration, the Council selected global change research, high-performance computing and communication, material science and processing, biotechnology, and mathematics and science education as the crosscut topics and in 1992 added advanced manufacturing to this list.

Once a crosscut topic had been selected, as Chairman of FCCSET, I met with the appropriate standing committee chairman and together we selected a chairperson for the crosscut. Then in consultation with that individual we selected two vice-chairman from complementary agencies, and, finally, in consultation with the FCCSET members, we appointed the crosscut members, one drawn from each of the participating member agencies. As a matter of instant tradition, the staff and other support for the crosscut activity was provided by the agency supplying the chairperson and the central FCCSET staff structure, reporting to my OSTP Assistant Director for Management and the Science Councils, was provided on a rotating basis by all the member agencies. Although the central staff was critical to the successful functioning of the activity, it was relatively small in size. Each of the Presidential Councils had an Executive Secretary reporting to the OSTP Assistant Director and each had two or three senior policy analysts and support staff who worked directly with the eight standing committees and with the crosscut groups.

In each case, the crosscut group began by developing an inventory of federal activity in the area in question. Initially, everyone involved was extremely wary, protective of their home agency, and of information that might be unique to that agency. It was almost always the case that, at the initial meeting of the crosscut groups, none of the members had ever met previously despite the fact that many of them had been working on the topic in question for years—if not decades. We were repeatedly reminded of the woeful lack of communication among the scientists and technologists working in the federal government, even at senior levels. In every case, too, the inventory was the first of its kind and always came up with surprises, usually including the fact that there was vastly more activity within the federal government in the particular area in question that anyone had anticipated. With this inventory in hand, the crosscut group then undertook to plan a research program for the area for the next five years. Recognizing that only OMB has the authority to request budgetary information and projections from the agencies, we worked closely with OMB in developing so-called Terms of Reference (TOR) for the crosscuts. Normally the groups were asked to develop a number of contingency alternative budgets so that whatever funding level the Congress might end up appropriating we would have the benefit of the expert planning for how to most effectively use that level of funding. Each of the crosscuts required a major amount of work on the part of the agencies because, particularly in the first year of each crosscut, the individual agency programs had to be restructured in major ways to make them come together as parts of a coherent national program. In a few cases, it was found necessary to completely scrap agency programs and start over fresh and it is a tribute to the Cabinet secretaries involved, and to their recognition of the degree of Presidential interest involved, that they were willing to permit and support such major revision in their programs.

It soon became clear that five or six crosscuts was the maximum number that could be reasonably handled in any one year without overloading the individual participating agencies. That being the case it was clear that, if we were to be able to continue adding new crosscut areas, a formal mechanism had to be found into which to move programs from the crosscut and Presidential Initiative status to make room. To that end, the FCCSET developed a new entity called a National Research Program (NRP), where the integration and coordination continued as before but where the agencies were no longer asked to prepare contingency alternatives and where the emphasis was on evaluation and assessment of the extent to which the goals set forth during the Presidential Initiative phase were, in fact, being accomplished, rather than on expansion of the program as was typical during the Presidential Initiative phase. I return to the NRP below.

By early December of each year, the crosscut groups had completed their inventories and planning, and the participating agencies in each crosscut came together and, under the FCCSET aegis, made a formal joint presentation of the crosscut to the senior staff of OMB. This normally took a full half day and involved a great number of detailed questions, clarifications, and fine tuning. At the end of this activity, we had an agreed set of plans for the coming five years with specified activities for each of the participating agencies, as well as each agency's projected budget level. At no time was pressure applied to any agency to induce them to participate in a given crosscut, and individual Cabinet members could, and did, on occasion, choose not to participate. But once an agency had agreed to be part of a crosscut at a specified budget level, then the OMB froze (colloquially, "fenced") that part of the agency's budget so that it was not possible for the agency to request funding as part of a Presidential Initiative and subsequently move it to some other part of the agency program i.e. a Washington Monument play.

Each of the crosscut groups prepared booklets giving the results of the inventory and the five-year plan to accompany the President's annual budget submission to the Congress. These booklets have now become recognized worldwide. President Bush took the first of them—on global change research—to the 1989 Economic Summit in Paris. They are now recognized as models for national research programs in major areas and have been translated into Russian, Chinese, Japanese, German, and French, and possibly into other languages. Beyond that, one of the nation's major business schools has chosen the global change research program as a case study for its students!

Reflecting their status as Presidential Initiatives in areas of national importance, the Budget each year during the Bush Administration requested annual increases in funding for them in the 10 to 40% range. Inasmuch as this was a period when the total domestic discretionary budget was frozen by agreement with the Congress, this meant that we necessarily had to identify other components of the domestic discretionary budget that could be phased back or terminated. This provides a very concrete measure of the Administration and Presidential support behind the Initiatives.

Because of the Balkanization of the Congress, however, once these highly integrated, coordinated Initiatives reached the Hill, they were taken apart and the various agency components considered—again in isolation—by the particular Authorization and Appropriation Subcommittees having jurisdiction. Frequently, these acted quite differently so that the crosscut group had to devote major effort to reassembling as coherent as possible an activity based on the results of independent subcommittee actions. The Congress recognizes this as a problem, but the subcommittees are singularly loathe to give up any turf or influence. Although we in OSTP tried, repeatedly, to convince the appropriate Subcommittees to allow us to make a joint presentation to them similar to that made earlier to OMB on behalf of each Initiative, we succeeded in only a single instance when Chairmen Ford and Brown allowed us to make a joint presentation of the Mathematics and Science Education Initiative.

Redistributing authorization and appropriation jurisdiction in the Congress to allow comparison of apples with apples or, even better, elimination of many of the subcommittees all together, is one of the most urgent and important challenges that the nation faces in working toward more effective science and technology policies and more effective and efficient utilization of our national investment in science and technology. Nor is this problem in any sense restricted to science and technology.

Over the years of the Bush Administration, however, truly unprecedented cooperation and communication has developed among the agencies as the participants in the FCCSET activities learned to trust one another. This is evident in the widespread flowering of bilateral and multilateral interagency agreements and memoranda of understanding quite outside of the FCCSET structure. It also bears noting, that peer pressure among the agencies has developed strongly so that no agency is now prepared to table less than its best effort before its peers around a FCCSET table, and thus the FCCSET forum has had a very positive impact on the overall quality of the agency participation in areas subjected to crosscut review.

As noted above, budgetary activities are only one of the major categories of FCCSET activities. We chose to build on the trust developed through the budget activities by asking FCCSET to undertake a wide range of nonbudget related studies in areas of common interest to a large number of agencies. Among these were such topics as risk assessment and management, scientific misconduct, the structure of international agreements in science and technology, environmental technology, and the relationship between the federal government and the research-intensive universities. Typically 15 to 20 agencies were represented in each of these studies.

Again, as noted above, only a limited number of crosscut activities can be effectively handled in any given year, and so it became necessary to develop a trajectory along which maturing Presidential Initiatives could move—after the few years of rapid expansion and development that characterized their status as Initiatives. For this purpose, again as noted above, FCCSET developed the National Research Program (NRP) status, and last year moved the Global Climate Change research program to that status. This, we emphasized, clearly indicated no lessening in interest in the area, no reduction in the level of interagency coordination and integration involved, but the amount of work required was reduced substantially by dropping the requirement that several alternative contingency budgets be developed. Furthermore, we protected the freedom of action of the agency heads by removing the fencing requirement on the agency funding commitment. Again, in the spirit of protecting the freedom of action of agency heads, although only very rarely exercised, we had insisted from the outset that if and when an agency head found himself, or herself, in complete disagreement with what the FCCSET crosscut group had planned for his, or her, agency, the agencies head's wishes automatically were followed. This fact, in itself, was of very real importance as a safety valve and, in no small measure, insured the remarkable level of cooperation actually achieved.

Early in the FCCSET process I learned a very important lesson reflecting a fundamental difference between academic and governmental activities. In the academic world, in a meeting of senior faculty, even though many may have had no prior exposure to the topic on the table, all are prepared—and indeed determined—to engage in active discussion, hopefully learning as they go. Senior federal officials, for what should have been obvious reasons, feel much less free to discuss a topic in a rather public meeting until their staff has examined it carefully, briefed them in detail, and, with them, thrashed out an agency position as at least a starting point for discussion. Finally recognizing this difference, I arranged for each FCCSET member agency to identify a so-called *"point-of-contact"* who was given the formal responsibility within the agency for seeing that this staff work and briefing was actually carried out prior to each FCCSET meeting. As Chairman of FCCSET I also arranged for each point-of-contact to receive detailed briefing books—prepared by the central FCCSET staff—some two weeks prior to each FCCSET meeting, and I then met with all these points-of-contact to clear up problems, misunderstandings, or concerns at least a week prior to the FCCSET meeting itself. I also found it extremely useful to meet with the FCCSET committee chairman and vice chairman, as a group, on a regular basis and to participate in at least one meeting of each committee and of each major working group at least once each year.

By law, all participants in FCCSET activities must be federal employees so that it was essential from the outset that we get input from the private sector. Not for nothing has Washington been referred to as 69.2 square miles surrounded by reality, and the validation and calibration by representatives of the private sector of plans developed entirely within the government I have found to be essential as a reality check. In this activity we were much helped by the professional societies.; in each of the crosscut activities we received essential input as well as the reality check from the relevant societies. I have already mentioned the outstanding work done by the Materials Research Society, and as a second example, in the high-performance computing and communication crosscut, the group of computer manufacturing CEOs was of major assistance and their input heavily conditioned some parts of the planning in this particular crosscut.

Earmarking

I noted earlier the growing concern about the impact of the "Balkanization" of the Congress on the orderly consideration of the Administration's budget proposals—particularly those that have been integrated and coordinated by the FCCSET mechanism and are then dismantled for consideration by the many committees and subcommittees having jurisdiction.

There is growing concern too, both in the Administration and in the Congress about the process of earmarking. This is the process whereby the Congress specifically adds items to the final appropriations that were neither requested by the President nor submitted to the detailed and multi-level quality and priority review that is characteristic of all the items that become part of the President's budget submission.

It should be emphasized that there are three types of earmarking, one entirely appropriate and the two destructive. In the first, the Congress recognizes that some programs of international importance has either been omitted or underfunded in the President's budget and corrects this in the appropriations process; in the second, a constituent approaches a Representative or a Senator and requests direct funding of a specific program or facility—effectively end-running all the normal service procedures; and in the third, a lobbyist, hired by a constituent, makes the contact to the same end. The latter two strike at the heart of the quality control on all federally supported activities and increasingly are acted on during House-Senate conferences where the action is protected from any public scrutiny.

OSTP undertook a review of the earmarking process in fiscal years 1990, 1991, and 1992. Funding that would otherwise have been devoted to peer-reviewed items in the Presidential budget but that was diverted to earmarked items in these three fiscal years, was found to be roughly $800 million, $1 billion, and $1.7 billion, respectively, more than a doubling from 1990 to 1992. If this trend continues, it will ultimately destroy the quality upon which our leadership in world science and technology depends.

It should be noted that the purely academic infrastructure component of this earmarking, which many claim reflects the lack of an adequate federal competition infrastructure program, has remained roughly constant between $400 and 450 million over these three years. Contrary to this claim, and to frequent arguments that earmarking benefits those institutions who cannot compete with prestige institutions, the OSTP analyses show that it is the latter who receive the lion's share of the earmarked funds. It should also be noted that our OSTP numbers refer only to research and development and thus are not directly comparable with these published, for example, by *The Chronicle of Higher Education*, which focused especially on education.

The President's Council of Advisers in Science and Technology (PCAST)

It was the recognition of a major need for high-level private sector input to White House decision-making that was, in part, responsible for President Bush's decision to reestablish a Presidential Advisory Group—reporting directly to him—as had not been the case for some 20 years. It bears emphasis, however, that the environment within which PCAST operated is vastly different from that which Eisenhower's original PSAC found in Washington. In the late 1950s, there was still very little scientific or technical talent distributed either throughout the agencies or, in fact, throughout Congressional or Committee staffs. I well recall Edward Purcell telling me how many times he, as a member of the original PSAC, had to explain to people throughout Washington why it was that a satellite did not simply fall like a rock straight toward the center of earth. The situation today is, of course, very different in that highly talented scientists and engineers are distributed throughout the federal government and the Congress. It is also true that whereas Eisenhower's PSAC devoted itself almost entirely to matters of national security and space, the spectrum of major areas of scientific and technical interest has expanded greatly, and there are very few areas of national interest today that do not have major scientific or technological components. While physicists totally dominated the membership of the early PSAC's, a much broader representation was appropriate for the Bush Administration. As it turned out, PCAST had among its members three engineers, two physicists, two chemists, two biologists, one ecologist, one physician, one mathematician, one geologist, and, very important, one economist. All members held Presidential appointment under Executive Order; all held general waivers from the President—something original to PCAST—that relieved them of all the normal conflict-of-interest restrictions on the grounds that their advice to the President was of such value that he was prepared to waive conflict-of-interest restrictions, recognizing that each member had provided full disclosure prior to appointment. Appendix 2 provides a listing of the members and their affiliations.

After PCAST members had been appointed and sworn in by the Vice President, I again emphasized to the President that it was essential—from the outset—that it be recognized that PCAST reported to him directly and not to the Science Adviser as had been the case during the Reagan years with the White House Science Council. I suggested that perhaps the President would wish to emphasize this changed status by hosting a dinner for the PCAST at its first meeting. He countered with a better idea, namely that he and Mrs. Bush would host the first PCAST meeting at Camp David, and this meeting was held on February 3, 1990. It occupied more than three hours on a Saturday morning and, in addition to the President, Dick Darman, Mike Boskin, Roger Porter, and John Sununu were in attendance. Discussion continued over a leisurely luncheon hosted by the President and Mrs. Bush and all-in-all the President spent more than four hours in active, enthusiastic conversation with the group. It was already evident in this first meeting that the chemistry among the PCAST members was excellent as was that between the members and the President. Subsequently, PCAST met on a monthly basis for a day and a half, and the President participated in between a third and a half of all of these meetings, usually whenever he was in town.

PCAST responded directly to Presidential requests for information, it reviewed FCCSET reports and commented on them prior to their being designated Presidential Initiatives, and it also kept a watching brief so that it could identify issues in science and technology to the President before they became problems. Much of its activity was carried out by panels established to address particular topics; each of these panels was chaired by a PCAST member with one or more fellow members participating as well as experts drawn from the private sector, depending upon the specific topic under discussion. In special cases, as, for example, in PCAST's study of research-intensive universities, the Council functioned as a committee of the whole. It reported to the President in many ways; directly, during its meetings with him; indirectly, through me as the PCAST Chairman, and formally, through specific reports. Normally these reports were made available only to the President, but, at his last meeting with PCAST in December of 1992, the President requested that the PCAST reports that had been forwarded to him in 1992 be made public so that they could form the basis for broader discussion of the topics involved. These reports are entitled, *Achieving the Promise of the Bioscience Revolution: The Role of the Federal Government; High-Performance Computing and Communications Panel Report; LEARNING to meet the Science and Technology Challenge; MegaProjects in the Sciences; Science, Technology, and National Security; Technology and the American Standard of Living;* and *Renewing the Promise: Research-Intensive Universities and the Nation.* All are available on request from OSTP.

Many had feared that the Freedom of Information Act (FOIA) and the Federal Advisory Committee Act (FACA) legislation and their requirement for public meetings would effectively preclude useful discussion in a PCAST-like group. This turned out not to be the case. In each of the monthly PCAST meetings, a substantial part of the first day was devoted to open discussion with the media and interested members of the public in attendance. Subsequent closed sessions focused on classified material, discussion of specific individuals as candidates for appointments, and other topics that were legally precluded from discussion in open sessions. Unclassified summary minutes of all meetings, including those with the President, were available to those interested.

In 1992, three print organizations, the *Bureau of National Affairs Inc, Science and Government Report*, and the Washington Bureau of *Nature*, joined together in an attempt to force all PCAST meetings, including those with the President, to be opened to the public. The Department of Justice successfully defended PCAST in this action before Judge Thomas R. Hogan. Those bringing the action have subsequently claimed that by doing so they forced PCAST to release information that would not otherwise have been publicly available. This is not the case; all that was made available had either been intended for public availability in response to FACA requirements, i.e. minutes, or were released at the request of the President, i.e. reports. I am convinced that it is possible and proper for a substantial part of the PCAST activities to be conducted in open meetings, as has been done throughout the Bush Administration, but at the same time, I, and many members of PCAST, believe that a requirement that *all* PCAST meetings be open would very seriously damage and reduce its effectiveness.

Although I have been delighted with PCAST and its activities—as was the President, as he emphasized on many occasions—in retrospect, I believe that it could have been used considerably more effectively. When I originally contacted the members, at the President's request, I told them that we would expect one week per month from each of them devoted to PCAST activity and all agreed that this was not only feasible but acceptable. As things turned out, we did not initially challenge them with large enough or demanding enough tasks to fill that large a time commitment and, since all the members were enormously busy individuals, other pressures on their schedules rapidly took care of any available time. Had I to do it over again I would have given them broader and more demanding tasks more appropriate to the wealth of experience, expertise, and wisdom represented by the PCAST membership.

The combination of PCAST and FCCSET made it possible for me to provide the President and the senior members of his Administration with much better and more complete *science and technology for policy* as well as *policy for science and technology* than would otherwise have been possible.

The Intergovernmental Council on Science, Engineering, and Technology (InterSET)

The 1976 legislation establishing OSTP and FCCSET also made provisions for yet another Presidential Council—that on intergovernmental activities in science, engineering, and technology. This Council had indeed been established in the 1970s and continued to function until the early 1980s when it was abolished by executive action. Its performance had been disappointing because it had not succeeded in focusing on specific actions where the federal government could cooperate effectively with the state, regional, and local authorities to make available federal information, know-how, and programs that would otherwise simply not be utilized. During the Bush Administration it became increasingly clear that on occasion after occasion, taxpayers were calling on consultants—often at enormous cost—to provide them with information and recommendations that in fact were already available in one or other of the federal agencies and that had already been paid for by the U.S. taxpayer. Unfortunately, no effective communication linkages were available to bridge the gap.

Two specific examples in which I was involved early during the Administration involved Geographic Information Systems and Personal Identification techniques. In the former, when a region requires, let us say, a new reservoir for its water supply, it is customary to call in a consultant who makes extensive measurements on population density, land use patterns, hydrology, transportation, water utilization, and a variety of other parameters appropriate to the area. What is generally not recognized is that the U.S. G.e.ological Survey and the U.S. Geodetic Survey have much, if not all, of that information already on hand and indeed in convenient, up-to-date, computer-readable form. By overlaying these various data sets, augmenting them, if and when necessary by specific spot measurements, it very quickly becomes apparent where the possible reservoir sites remain in the area, and the cost is greatly reduced. In the second example, one of our major western cities has a welfare budget that exceeds the total budget of well more than half of the world's nations. Unfortunately, it is known that a significant number of its clients draw benefits under a variety of identities, and the city managers approached OSTP and the federal government for assistance. As it turned out, the Sandia National Laboratory, as part of its security program for nuclear weapons, has devoted substantial resources and technical effort to precisely this question of improving techniques for unambiguously identifying humans. By bringing representatives of the city and the Laboratory together, we have been able already to make substantial progress.

With this as typical background, in 1992, the President authorized the creation of a new Presidential level Council, the Intergovernmental Council on Science, Engineering, and Technology, with the acronym, InterSET, with membership drawn from past and present state governors, large city mayors, state legislators, county officials, and a limited number of federal employees. This group was tasked with selecting a number—less than five—of issues relating to science and technology that were of major importance to the states, municipalities, and regions and were such that the federal government might be able to provide substantial assistance. With those recommendations in hand, I established a permanent FCCSET Committee, the Intergovernmental Science, Engineering, and Technology Committee (ISET), with membership drawn from across the whole spectrum of the federal agencies. This latter committee was chaired by the Assistant to the President for Intergovernmental Affairs and was tasked with taking the recommendations of InterSET and pulling together a federal interagency response.

It is too early yet to report on the performance of these two new entities other than to note that there is very real enthusiasm on the part of both groups; those outside the federal government enthusiastic about obtaining access to the resources of the federal government and those within the federal government enthusiastic about being able to find convenient communication channels so that their resources and expertise can be brought to bear on real problems across the nation.

In all of these activities, it has been critically important that OSTP be viewed as the honest broker. For that reason, I believe that it is important that OSTP handle no funding, that it provide no real or apparent competition with the agencies, and that it not attempt to force any specification, or any OSTP chairman, on FCCSET, PCAST, or InterSET activities. It is vitally important also for agency heads to be in no doubt whatever about the President's personal interest in the participation of their agencies in these Presidential Council activities.

Emphasis on Technology

At the same time that he appointed James Killian as his Special Assistant for Science and Technology, President Eisenhower changed the name of the White House Science Office to that of Science and Technology Policy. In fact, however, the Office remained largely one of *science* until the beginning of the Bush Administration. Even before arriving in Washington, I took it as a substantial part of my mission to emphasize the T in OSTP, and I learned from my discussions, both with the President and with the Chief of Staff, John Sununu, that they were in general agreement with this thrust.

I was extremely fortunate in being able to attract as my Associate Director for Industrial Technology, Dr. William D. Philips. He had previously been a Vice President for R&D at DuPont, an Executive Vice President for R&D at Mallinkrodt, and between those positions, had served as Chairman of the Chemistry Department at Washington University, and had spent a sabbatical year at MIT working with Alex Rich to become familiar with the burgeoning biotechnologies. At the time he joined OSTP, Dr. Philips was the Adviser to the Governor of the state of Missouri on matters of science and technology and was very actively involved in establishing science and technology centers throughout the that State.

We agreed that, as our first major task under the technology rubric, we should attempt to articulate U.S. technology policy, something that had never been done formally before. We decided on this primarily because of the confusion that was rampant not only within the Administration, but in the Congress, as to what we meant by technology policy and by industrial policy. The extreme points of view were typified by the right wing Republican view that the federal government should have no contact whatever with the private sector and, therefore, no real role in the development of technology, and that of the left wing Democrat who believed that there was a major role for the federal government, both in cooperation with, and in competition with, the private sector in areas that were judged to be of national importance. Because of the wide spectrum of opinion, even within the White House, on these matters, it took us almost a year to develop a formal policy statement that met the approval of everyone—from the President down—within the White House complex, and, in September 1990, we published the formal document, *U.S. Technology Policy*.

The rationale for federal involvement in the development of generic or enabling technologies defined as those that have broad application across all sectors of our society is precisely the same as that for federal involvement in the support of basic research. In both cases, because of the very nature of the work involved, it is impossible to predict where, when, or to whom the benefits will flow and, for that reason, no single institution or organization can guarantee that it will be able to claim a sufficient return to justify the necessary investment. This leaves the federal government as the only source of investment of sufficient magnitude to insure that the U.S. remains in an economically competitive posture.

The lack of clearly articulated definitions has posed continuing problems in all discussions of federal support of research and development. It has been traditional, for example, to discuss basic and applied research, but I, for example, find it totally impossible in many cases to distinguish between basic and applied research as quite separate activities, although I believe that I can distinguish good from bad research at first sight. On the basis of our work in OSTP, we have proposed that ,for discussion of funding purposes, the following four definitions be used.

• *Thematic Research:* Funding intended by the funding agency to achieve a coordinated research program with many investigators working on different aspects on a single research theme, goal, or application.

• *Human Resources:* Funding intended primarily to develop increased research capacity, broadened participation in research, or increase the number and skills of qualified researchers.

• *Instrumentation:* Funding intended by the funding agency specifically for the acquisition replacement, development, or maintenance of research equipment.

- *Disciplinary Support:* Funding intended primarily to advance the state of knowledge in a particular field of science and engineering.

We have further suggested that the following three definitions be used to describe different types of research.

- *Fundamental Research:* Research designed to build the core of knowledge a field or subfield with the results intended at the time that the research is funded for dissemination to other researchers and educators.

- *Strategic Research:* Research designed to build the base of knowledge and skills in an area of evident interest to a broad class of users both internal and external to the research community that can identified at the time the research is funded.

- *Directed Research:* Research directed toward gaining knowledge for the particular missions or products, processes, or services that are related to the specific needs of the sponsoring organization.

Obviously no set of definitions can be applied without ambiguity in some cases, but we believe that these would clarify much of current discussion.

In the Bush Administration we defined technology policy as that covering the entire process leading from initial discovery of the underlying principles or phenomena through the development of technologies to the point where individual private sector entities can ask, and answer, the question, what can this technology contribute to our activities? In other words, it has been our belief that the federal government has a substantial role to play through the entire precompetitive phase of the technology development process. Once the individual private sector entity is able to identify a specific use for the technology, however, we believed that market forces should govern, and we have not believed that we, in the federal government, had either more information or greater wisdom than those in the private sector to make the competitive judgments required.

We have defined industrial policy as covering situations where the federal government attempts to select winners and losers in the marketplace by providing federal support where it is judged necessary to prevent a selected company from becoming a loser. In the Bush Administration we have not felt this to be an appropriate use of taxpayers dollars again because we have not been convinced that federal bureaucrats had better information or greater wisdom that would enable them to out-predict the market.

On March 7, 1991 in a speech to the American Electronics Association—and on many subsequent occasions—President Bush pointed specifically to the importance of cooperation between the federal government and the private sector in the development of generic, precompetitive technologies. As he put it, "This Administration is committed to working with you in the critical precompetitive development stage where the basic discoveries are converted into generic technologies that support both our economic competitiveness and our national security. Here, again, we can help to level the international playing field on which you operate."

Development of a formal statement of our U.S. technology policy was a particularly interesting case because it involved moving OSTP into a new and polictically sensitive area, that of technology policy, and, in so doing, it raised the full spectrum of difficulties that might be expected within the Executive Office of the President. There were those who objected on ideological grounds, as noted above; others objected because they extrapolated this new initiative to additional funding and staffing for OSTP which would increase its apparent stature with respect to their own offices in future; and yet still others were concerned about matters of turf and what they viewed as undue publicity and visibility for OSTP in the media. Throughout some singularly unpleasant meetings and discussions, the President and John Sununu stood firm in their support. But in some other quarters, the level of childishness, jealousy, and pettiness would have done justice to an elementary school class.

The technology initiative in OSTP had, from the very outset, strong support within the Congress not only in the science and technology committees but also in the authorization and appropriation committees and subcommittees. Congressman Robert Roe, then Chairman of the House Science and Technology Committee, indeed proposed legislation that would have changed the position of the Director of OSTP from Executive Level II to Executive Level I as a specific and high visible indication of his support and that of his colleagues; this would have been passed had it not been for a determined campaign in opposition mounted from within the Executive Office of the President.

One of the strongest supporters in the Congress for technology activity throughout the Bush Administration was Jeff Bingaman, Chairman of the Armed Services Appropriations Subcommittee and a Democratic Senator from the State of New Mexico. In 1991, he arranged for OSTP to be legislatively instructed to establish a Critical Technologies Panel—drawn equally from the private sector and from within the federal government—that was charged with examining the various lists of critical technologies that had been prepared by the Department of Commerce, by the Department of Defense, and by various private organizations and then, incorporating its own input, with coming up with a definitive list of technologies that had very broad potential applicability across both civilian and military sectors. This activity was chaired by Bill Philips and reported out its master list of 22 critical technologies in March 1991.

Because this release was widely covered in the *Wall Street Journal, The Washington Post,* and the various trade publications, a considerable brouhaha again developed around the suggestion that OSTP was attempting to move the Bush Administration into industrial policy.

Unfortunately, all this occurred while I was in Novosibirsk in central Siberia negotiating various agreements with the leaders of this, then-Soviet, Academic City. In my absence, the amount of misinformation, misunderstanding, and, again, childishness, escalated until finally, after many failures, I managed to complete a telephone circuit from the office of an old friend in Novosibirsk to Gorbachev's Science Adviser in Moscow, also an old friend, to the U.S. Embassy in Paris and then to OSTP. By the time I returned to the U.S. a week later, all had returned to normal In 1992, Senator Bingaman extended the legislation relating to OSTP to include the establishment of the so-called *Critical Technologies Institute* (CTI) with the intent of providing the resources and staffing that would make possible long range strategic planning necessary to facilitate implementation of the critical technologies in the industrial sector. Although we were in full agreement with the intent, there were fundamental difficulties with the Institute as legislated. Among them was the requirement that funding be channeled through OSTP, which, as noted previously, in my opinion, would have seriously damaged our honest broker role. The Governing Board of the Institute was legislated to be a mixed one, drawn jointly from the private sector and from the federal government, and our experience in attempting to bring private sector representatives into PCAST and FCCSET activities had convinced us that this would cause enormous difficulties and delays. After almost a year of negotiation involving Senator Bingaman and his staff, OMB, and OSTP, we were able to resolve the difficulties. The financial aspects are now handled by the National Science Foundation which contracted with the Rand Corporation to establish the Critical Technologies Institute as an FFRDC within the Rand Corporation. While Rand thus supplies the support structure, the activities of the Institute are under the control of a Governing Board consisting of Cabinet Officers and senior members of the Executive Office of the President. This Board, at the request of President Bush, was chaired by me as Director of OSTP, and the OSTP Assistant Director for Industrial Technology was appointed as its Executive Secretary. Decisions of the Board are transmitted by the Executive Secretary to the contracting agent in NSF and thus to the Rand Corporation for action.

The Critical Technologies Institute was designed to fulfill three very important functions, not only for OSTP, but also for a number of the other EOP Offices and the White House generally. First of all it will provide the opportunity for longer range strategic studies which are essentially impossible within the brush-fire environment of the White House. Secondly, as structured, it makes possible to bring into the Critical Technologies Institute senior academics and senior mid-career industrial scientists and engineers for extended periods where they can become involved in the strategic studies as well as in shorter term activities where their expertise and experience can be invaluable and where such input is impossible to obtain directly to OSTP for the conflict-of-interest reasons noted above. Finally, CTI can provide the support and infrastructure required for many of the activities in OSTP that are congressionally mandated or that originate within the Office itself, but where in the past it has been difficult to obtain adequate infrastructure support to carry out the work in timely fashion.

The second Critical Technologies Panel report in the biennial series called for in the legislation was issued in January 1993 and was prepared under the chairmanship of Dr. Frederick Bernthal, Deputy Director of NSF, with full support from the CTI. Rather than attempt to develop a new list, since very little has changed since 1991 regarding that list, this second Panel has focused on evaluating where we in the United States stand relative to the rest of the world in each of the 22 critical technologies and on strategies for improving our competitive position.

It bears noting that this need for strategic support for OSTP, was recognized very early in the Bush Administration. Immediately after President Bush asked me to take on the responsibilities as Director of OSTP and as the Assistant to the President for Science and Technology, Tom Ratchford (whom I had already identified as one of my potential Associate Directors) and I met, on an informal basis, with the Co-Chairmen of the Carnegie Commission to explore the possibility of a private sector supported organization that might provide the kind of strategic studies that we believed would be essential. The Carnegie Commission, in turn, commissioned Professor William Wells of the George Washington University to carry out a detailed study on the feasibility of establishing such a support activity reporting to OSTP; Wells carried out a detailed series of interviews with a broad spectrum of persons experienced in White House activities in past Administrations. His conclusion was that, although there was no question as to the need for the kind of strategic support in question, a realistic evaluation of the White House environment led him the conclude that a strategic support structure reporting only to OSTP would not be accepted by the other components of the White House, and he therefore advised the Carnegie Commission against its establishment. Subsequent events fully supported his analysis.

Obviously much of the thinking that had gone into these early discussions resurfaced during the negotiations surrounding the Critical Technologies Institute and, as now constituted, the CTI has the potential to provide most of the desirable aspects while minimizing potential difficulties. Indeed, it is the hope of all concerned that, once established, the CTI will be recognized as a valuable adjunct to many components of the White House.

Although it is not generally recognized that there are some 726 federal laboratories currently active in the United States at a total annual expense of roughly $22 billion, they represent an absolutely unique resource in terms of know-how, manpower, technology, and facilities — one that is not matched anywhere in else in the world. It is, however, increasingly recognized that many of these laboratories have far outlived their original missions, and, in consequence, are not contributing to the nation at a level commensurate with the current investment in them. In particular, as part of our emphasis on technology, we undertook a determined effort in the Bush Administration to more effectively utilize the resource represented by these laboratories and obtain a better return on our investment in them. The National Technology Initiative (NTI), which represented the spearhead of this activity, had its origin, in 1991, in a series of national meetings organized by the Secretary of Commerce to highlight the potential for American industry in greatly enhanced exports. Because of the success of this series of meetings in 1991, in 1992 the Commerce Department decided to hold a similar series of meetings across the nation bringing researchers from the federal laboratories together with local industrialists in the hope of opening communication channels and of facilitating technology transfer from the laboratories to these local industries. This Commerce initiative very quickly caught the attention of a number of other Cabinet Departments, including the Department of Energy, the Department of Defense, the Department of Transportation, and, of course, NASA and OSTP. As a result of these activities across the U.S., something over 1500 Cooperative Research and Development Agreements (CRADAs) were signed between representatives of the federal laboratories and industries, ranging from the smallest to the largest. President Bush gave a major address on the technology policy of his Administration at the meeting of the National Technology

Initiative held in Chicago in October 1992, and the Initiative by all measures has been judged extremely successful in terms not only of the CRADAs signed but also the communication channels opened between individual federal laboratories and the industries in their surrounding regions. Perhaps most important, it developed an entirely new mind set on the part of many laboratory scientists and engineers.

In late 1990, building on our other technology activities, we in OSTP had, after several discussions with the President and with John Sununu, begun organizing a White House Conference, wherein the President was to have met with a cross section of industrial CEOs, labor leaders, and federal laboratory personnel to highlight the importance that his Administration attached to technology and its effective utilization both for economic competitiveness and toward a more productive America. This concept had the enthusiastic support of the President, the Secretaries of Commerce and of the Treasury, as well as the Chief of Staff and, indeed, would have looked very much like the Economic Summit held by President Clinton shortly after his election. Unfortunately, with John Sununu's departure from the White House, it was never again possible to pull together the coherent support needed to move forward with this Conference. Had we been able to do so, I am convinced that there would have been little if any defection of the high-technology industrialists from the Bush Camp. It is unfortunately true that this is but a single illustration of cases where we failed to convey to the private sector or to the public generally any appreciation of what the Bush Administration had already accomplished in the way of bringing technology to bear on our economic competitiveness and the improvement in the quality of life of our citizens, or any sense of the President's deep commitment to this area and his concern about the economic well-being of our industrial sector.

Education

It was clear from the outset of the Bush Administration that the nation faced a major crisis in the area of education. Our focus in OSTP, obviously, was on education in mathematics and science, but the crisis extends much more broadly across the entire spectrum of contemporary education. We find ourselves in a paradoxical situation. We still lead the world in higher education, and something like one-half of all the young people who leave their own borders for higher education come to American institutions. At the college level, because we are the only developed nation that has no centralized standards for what constitutes a college education, we have peaks of excellence that far exceed world standards as well as vast swamps of mediocrity that defy description. On the average, however, we remain competitive. It is at the pre-college level where our educational system, by all measures, including all recent international comparisons, has essentially collapsed and where for the first time in the history of this nation our children are receiving an education that is inferior to that which we received. When something approaching 50% of the high school graduates from our major urban high schools are functionally illiterate and innumerate, it is obvious that we are far from developing a work force appropriate to the American 21st Century and far from establishing the base of American young people equipped to go on to higher education and to play leading roles in the science and technology of our future.

The National Commission on Excellence in Education appointed by Secretary Bell of the Department of Education in August 1981 produced a stirring call to arms in its April 1983 report, *A Nation at Risk,* that for the first time alerted a significant fraction of the American media and public to the deterioration of our precollege educational system. Unfortunately, however, little concrete action or improvement followed the release of this excellent report.

Recognizing the growing seriousness of this situation, in October 1989, President Bush called together the Governors of all the States for an Education Summit in Charlottesville. This was only the third time in the history of the nation that such a convocation had occurred, the last time being in 1933 when President Roosevelt called the Governors together to confront the banking crisis. I had the privilege of attending the Summit with the President and for two days participated in the intense discussions that resulted in the now famous six education goals articulated by the President and the Governors. Of these goals, number four was the one of greatest import to OSTP, although numbers three and five also refer to science and mathematics. Goal number four states simply that "U.S. students will be first in the world in science and mathematics achievement by the year 2000."

Returning from Charlottesville, I moved quickly to organize a meeting with the senior members of the National Science Foundation and of the Department of Education because, for years, the National Science Foundation had been producing very high-quality educational material in mathematics and science but had been unable to convince the Department of Education to distribute this material through the extensive distribution system that it already had in place reaching all of the nation's high schools. It was a fascinating meeting, not least because I found that apart from myself, the Director of the National Science Foundation, and the Secretary of Energy, none of the participants had ever met one another before. Within this half-hour meeting, it was, in fact, possible to arrive at an agreement whereby the Department of Education would in future distribute NSF materials to the nation's high schools, and this marked the beginning of a totally new level of cooperation between these two agencies.

Since we were at the same time actively engaged in restructuring the Federal Coordinating Council, we devoted special attention to the FCCSET Committee on Education and Human Resources (FCCSET/CEHR), and I was successful in convincing Admiral Watkins, the Secretary of Energy, to accept the Chairmanship of this Committee with Vice Chairmen, Cross, the Assistant Secretary of Education, and Williams, the Assistant Director of the National Science Foundation responsible for educational programs. Because Admiral Watkins had a long history of interest and effective participation in educational matters, and because he, as a Cabinet Member, represented a very major agency, he was able to avoid any of the conflict and jealousy that might well have otherwise arisen primarily between the Department of Education and the NSF, and at the same time, was able to bring in very senior members of some 16 other agencies to participate actively in furtherance of the President's Education Initiative.

FCCSET rapidly accepted mathematics and science education as a suitable topic for a crosscut analysis, and the inventory that was a part of all such analyses rapidly showed that there was much more educational activity underway within the federal government than had been anticipated, but that out of a total national expenditure of some 690 billion dollars annually on education, the federal government spent only approximately 6%. What this made clear was that the federal role necessarily had to be a catalytic one that recognized the statutory responsibility of the states and localities. But it is also true that we in the U.S. spend more per student on elementary and secondary education than does any other nation or region, with the exception of several Swiss cantons. It is clear that the amount spent in the U.S. is not the problem but rather how it is spent. There has been a continuing shift in the past from spending money on actual classroom teachers to spending it on growing bureaucratic overlays and on highly specialized teachers serving only a relatively small fraction of the total student population. There is an urgent need for change here.

Over a period of three years, Admiral Watkin's Committee did an outstanding study of our educational system, focusing immediately on the pre-college area as that in greatest difficulty and—within that sector—on elementary education. It further concluded that the fundamental problem was the lack of objective standards and accepted methods for measuring performance against those standards of what children would be expected to know in mathematics and science at the conclusion of grades 4, 8 and 12, respectively. It was also recognized that less than 50% of those teaching mathematics and science in the nation's elementary schools had any formal education in either field, and further that we simply could not afford to wait for another generation of teachers but had to work with those now in place. Lastly, it was recognized that the fundamental requirement for effective elementary education remains parental involvement and interest, and the need to substitute for that in those unfortunate situations where, for whatever reason, parental involvement was either absent or impossible.

With stimulus from the FCCSET/CEHR activity and exemplary cooperation from the professional societies in mathematics and the sciences, the objective standards mentioned above have now been almost completely developed and will very soon be available for public comment; so also are the first drafts of the performance measurement procedures. During the summer of 1992, some 45 thousand elementary school teachers of mathematices and science were taken through a refresher course in their subjects, with the goal being that of taking all 1.5 million such teachers through similar refresher courses before the end of the decade.

The FCCSET Committee also produced a detailed catalogue, *Guidebook to Excellence,* giving the names, addresses and telephone numbers of contact persons in each of the federal facilities and in many private sector industries, so that educators across the nation could make contact with the resources available in such institutions. It was fully recognized from the outset that neither federal laboratory employees nor industrial employees can be expected to show teachers how to teach, but by bringing students and teachers into their laboratories on week-ends, during vacations and in summers they can accomplish two very important goals. The first is that of motivating the students. In a great many cases, inner-city students have never had an opportunity to see a working scientists or engineer or any of the facilities and equipment with which such persons work, and it is wonderful to see how frequently, after just a few weeks of exposure, these students all at once catch fire, and exclaim, "I could do that!" That in itself—giving them a sense of the opportunities and challenges open to them if they continue their education—may be the most important thing that we can do for such students. In the case of teachers, such contact gives them a linkage with the field in which they teach and apart from providing them with specific assistance can provide an enormously important morale improvement.

Incredible as it may seem, even in recent years such activity was considered, by some, to be outside the mission, and therefore not permissible activity for federal laboratory employees. Only after extensive effort on the part of many individuals was it possible, on November 16, 1992, to have President Bush sign Executive Order 12821 which, for the first time, makes education and the training of an adequate work force a part of the mission statement for *every* federal agency.

Just prior to the end of the Bush Administration, FCCSET/CEHR released a report entitled, *PathWays to Excellence* which lays out, again for the first time, a federal strategy for science, mathematics, engineering and technology education design to achieve the goals established at the 1989 Education Summit. This represents a major achievement and one that would have been totally impossible without unprecedented cooperation among the sixteen involved agencies of the federal government.

Although the initial FCCSET/CEHR focus was on the K-12 sector of our educational enterprise, it has shifted to include K-14. It is in the 12-14 segment—in the two-year colleges—that the technicians essential to the operation and maintenance of our increasingly technological society typically receive their training. We in the U.S. have fallen far behind all of our major industrial competitors in terms of recognizing and rewarding such individuals. We spend only 2.8% of our federal education investment on the two-year colleges that we will need to fill a growing shortage of technicians in the years ahead. FCCSET/CEHR has focused national attention on this challenge.

Again, although the focus throughout the Bush Administration has been on precollege education, it has been recognized that we simply cannot take our world leadership in higher education for granted, and it is also becoming increasingly clear that the relationship between the federal government and the Research-Intensive Universities is under substantial stress. For that reason, in early 1992 I requested that both PCAST and FCCSET undertake detailed examinations of this relationship—viewing it, respectively, from outside and inside the federal government. Their reports, *Renewing the Promise: Research-Intensive Universities In the Nation* and *In the National Interest: the Federal Government and the Research-Intensive Universities*, respectively, provide detailed analyses of the present problems, challenges, and opportunities in this critical educational area and make vitally important recommendations to all the communities concerned. As I shall note below, this set of reports was intended as the first of a triad of reports designed to address the overall question of the appropriate character of, and demands on, the research and development enterprise in the United States as we enter the 21st century, and they deserve the serious consideration of all concerned citizens.

Environmental Matters

In October 1989, President Bush requested that I take on the chairmanship of the Domestic Policy Council's Working Group on Global Climate Change. This group had been in existence for several months but had not coalescenced or reached any agreement as to its function and direction. Its membership comprised Cabinet members, and Heads of independent agencies, and the charge given to it by the President was that of formulating a national Global Climate Change policy. I immediately established three study panels. The first, led by the Council of Economic Advisers, was charged with attempting to establish an economic base for discussion of Global Climate Change. The second, headed by the Department of Justice, was charged with carrying out a detailed study of the structure of international agreements of the kind that we might be asked to consider in the Global Climate Change area, and the third, headed by the Department of the Interior, was charged with compiling an inventory of the work on Global Climate Change already carried out by companies in the U.S. private sector—none of which had previously been collected or analyzed in any coherent fashion. At the same time, we brought in expert witnesses—from the major agencies such as NASA, NOAH, EPA and NCAR and from a number of major universities having study programs in the economics and science related to Global Climate Change—to brief the members of the Working Group.

In November 1989, William Reilly, the Administrator of EPA, and I co-headed the American delegation to the Nordwijk Conference on Global Climate Change. More than 100 nations were represented, and almost all were prepared to commit to stabilizing their emissions of greenhouse gases at the 1990 level by the year 2000. Reilly and I were unwilling to make this commitment on behalf of the United States because neither we, nor anyone known to us, had any detailed economic or technical understanding of what would be involved in achieving this projected level of emissions. The lack of economic analysis was astonishing, as was the equivalent lack of technical understanding in a great many instances. I asked the head of one of the major European delegations how exactly they intended to achieve the projected emission goals and was told, "Who knows, after all its only a piece of paper, and they don't put you in jail if you don't actually do it." This attitude, time and again, makes it extraordinarily difficult for the U.S. to participate in international meetings because as a matter of principle we believe that when we commit to a course of action we must be prepared to follow through on it.

Reilly and I reported to the President on our return from Nordwijk, and his response to our description of the lack of adequate economic and scientific understanding was to direct that we do something about it. Out of this came the April 1990 White House Conference on Science and Economic Research Related to Global Climate Change which I co-chaired with Michael Boskin, the Chairman of the Council of Economic Advisers, and Michael Deland, the Chairman of the

Council on Economic Quality. The President both opened and closed the Conference and again more than 100 countries were represented. In retrospect, the Conference was successful in achieving its primary goals. But at the time, the Bush Administration received major negative press treatment concerning the Conference; in large measure this reflected the fact that we had been forced to take on some public relations consultants whose track record previously had been with political conferences and conventions. The problem was that in a political convention the losers disappear after the convention, while casting an international scientific conference in a winners and losers framework, inevitably results in a group of very unhappy "losers" who do not disappear after the conference but who instead go home to foment all manner of national and international difficulties.

The Conference did, however, inject economics solidly into all subsequent national and international discussions of Global Climate Change. It had a major impact on the scientific studies as well, because the economic study carried out under the DPC Panel demonstrated very clearly that some scientific information was much more pertinent to reducing economic uncertainties than was other, making it possible for us to fine-tune the global change scientific research program, being developed under FCCSET, to focus more clearly on information that would be of greatest policy value in the short term. As an indication of the enthusiasm with which some countries adopted the economic component of the global change program, the Japanese, immediately following the White House Conference, established their own economic group on Global Climate Change with 70-100 senior staff members.

In his closing remarks at the Conference, President Bush suggested the formation of a global network that would foster environmental stewardship on a worldwide basis. What he had in mind was an Institute in the Americas, one in the Western Pacific and one in Europe and Africa which could subsequently be tied together with latitudinal information channels to form the global network. In May of 1992, after a series of preparatory meetings involving twenty-two countries of North and South America, the Inter-American Institute for the Environment was formally established during a meeting in Montevideo, Uruguay, with eleven nations signing on the spot and the other eleven expected to sign within the subsequent year. Recently, the Japanese have hosted an organizational meeting for the Western Pacific Institute with enthusiastic participation from Australia, New Zealand, Indonesia, Korea and China, while in Europe, Italy has taken the lead in establishing an Institute that would be based at Ispra and would focus on environmental problems of Europe and of Africa. Although the net envisaged by President Bush is not yet complete, very substantial progress has been made towards its establishment, and I am optimistic that the momentum is present to carry it through to completion. Such a worldwide network will be essential if we are to hope to build—and effectively use—the information base on which responsible environmental policies can be developed internationally.

I have just mentioned the activities of the Economics Panel of the DPC Working Group. The Panel under the leadership of the Justice Department not only carried out a detailed examination of the various kinds of possible international agreements with which we might be confronted but also did an outstanding piece of work in developing the so-called *comprehensive approach* to the greenhouse problem wherein all greenhouse gases are treated on the common basis of their *greenhouse potential index* (GPI), which is based on a variety of scientific parameters including molecular efficiency, residence time in the atmosphere, distribution in the atmosphere and so on. The comprehensive approach has the tremendous advantage that it makes possible the negotiation of an overall framework agreement which then does not require modification for each separate greenhouse gas, nor change as new research results suggest the necessity for change in the indices assigned to the various greenhouse gases. This comprehensive approach has been generally accepted in the international community as appropriate for future negotiations.

With the approach of the Rio Summit in June of 1992 and with memory of the preceding conference in this series—held in 1972 in Stockholm—we recommended the appointment of a senior ambassador who would be responsible for all planning for the Rio Conference as Senator Howard Baker had been responsible for that for Stockholm. After the fact, Baker reported that the eighteen months that he had devoted to planning for Stockholm had been too short a time and that he recommended a similar arrangement but with longer lead time for any subsequent conferences of the same kind. For a variety of reasons, no such appointment was made prior to Rio and so much of the coordination responsibility was shared between a State Department Group headed by Ambassador Ryan and a new Strategy Panel formed under the DPC Working Group on Global Climate Change. Of course, as we came closer to the actual Rio Conference more and more of the total White House staff became involved, and Clayton Yeutter, then Chairman of the Policy Coordinating Group that had replaced both the Domestic Policy Council and the Economic Policy Council, took over the primary responsibility for coordinating U.S. policy.

Here again, George Bush was not well served by his Advisers, in that, despite substantial accomplishments in the area of Global Climate Change—including unprecedented clean air legislation, world leadership in the phasing out of the chloro-fluoro-carbons and a major forestry initiative, George Bush received very little credit in the media, and time and time again when he made a major step forward in environmental action or policy, the environmental activists—fed by a remarkable series of leaks from within the White House—would take a position well ahead of that reached by the President in his most recent advance, and then beat up on him for not going as far forward as the activists would have liked.

At Rio, Bill Reilly did an outstanding job of representing the United States, despite hostile national and international media, despite unfortunate leaks of highly confidential memoranda between Rio and the White House, and despite internecine warfare within the U.S. government. We did sign the framework convention on Global Climate Change after having successfully negotiated the removal of unacceptable clauses; we refused to sign the bio-diversity initiative presented at Rio because it had fatal flaws which would have seriously injured U.S. pharmaceutical companies and would have required U.S. taxpayers to contribute to an international fund over whose disbursements we had no control whatever.

In December 1992, we were the only country that was able to table, in Geneva, the national action plan that had been promised by all those nations that signed the

Framework Convention in Rio. This national action plan represented an enormous amount of work on the part of many U.S. agencies and outstanding work on the part of Robert Reinstein, of the U.S. State Department—who had been our primary negotiator throughout the entire Framework Convention negotiations—as well as on the part of the members of the Strategy Panel.

I am convinced that history will be kind to the Bush Administration with respect to its environmental activities. There are times, both nationally and internationally, when exerting leadership is not synonymous with being popular. I believe, that the Rio Conference was a case in point.

One area where, despite the fact that the United States currently provides about 45% of the funding worldwide, we were unable to ever engage in meaningful public discussion was that of world population. This is a problem rapidly approaching crisis dimensions but, because we have been unable to gain broad public acceptance of the fact that abortion is a signature for the failure of family planning, rather than a method of family planning, we have been unable to give this problem the attention that it deserves.

Many in the developed world have decided that population growth is no longer a problem because they assume—correctly—that biotechnology holds the promise of a substantially expanded world food supply and because they have noted—again correctly—that population growth rates in the developed world have fallen significantly since 1970.

Such a view, however, is very much in error, given the much higher growth rates in many of the developing countries, and, unless action is taken to reduce the rates of growth, worldwide, to about 2.5 children per woman in 2025, the current world population of about 5.5 billion will not saturate below something like 12 billion. If the rate of growth is just 2.8 children per woman in 2025, world population will not saturate below 20 billion. At present, every four days world population grows by one million people. In a high fertility country like Pakistan, the population grew from 40 million in 1950 to 123 million in 1990 and is projected to reach 267 million in 2025. More important, some parts of the world will witness population growth that is totally inconsistent with either political stability or acceptable social and economic conditions. The projections are particularly grim in the case of Africa and of India.

Cities provide the most striking evidence of population pressures and by 2000 about half of the world's population—about 3 billion persons—are expected to live in cities. By the year 2000, cities like Sao Paulo in Brazil and Mexico City will have expanded to populations of 31 and 26 million respectively, having more than doubled in size in the last quarter century, and will become two of the largest and most environmentally unacceptable cities on the entire planet. In Mexico City, smog and ozone levels violate Mexico's own national standards during more than 300 days each year. In Karachi, Pakistan the municipal service systems were designed for a population of 400,000 but now the city houses 9 million; nearly half are without access to any basic city service.

Without effective international population control, all efforts toward improving quality of life are doomed. We already know at least some of the actions that must be taken, including, for example, the universal education of women and the much more ready availability of long-term contraceptives. Here, again, we have the technology, but, unfortunately, for a combination of moral, ethical, religious, and political reasons, we have as yet seen no global response to this population crisis in any way commensurate with its importance or its seriousness.

Unless effective control actions are initiated, while the poor and hungry of this world will be those, both as nations and as individuals, who will experience unprecedented suffering, the wealthier developed nations and their citizens cannot hope to escape the political, economic, and moral ripple effects. This is a problem worthy of the best efforts of all of us.

National Security Activities

In the early days of Science Advising to the President of the United States, the Advisory apparatus devoted a very large fraction of its time to matters of national security and the space program, as indeed is required under Public Law 94-282, which states, "the Director (OSTP) shall advise the President of scientific and technological considerations involved in areas of national concern including national security"..."For the purpose of assuring the optimum contribution of science and technology to the national security, the Director, at the request of the National Security Council (NSC), shall advise the National Security Council in such matters concerning science and technology as relate to national security." Recognizing both that my two predecessors as Science Adviser had spent an increasing fraction of their time in office on matters relating to national security and that my good friend, Brent Scowcroft, would hold the position of the Assistant to the President for National Security in the Bush Administration, I made the decision from the very beginning that I would de-emphasize OSTP's focus on national security matters and instead be prepared to assist Scowcroft and the NSC whenever and wherever necessary.

This, however, does not imply that OSTP did not have remaining responsibilities in the national security area. In particular, for example, the Director of OSTP is charged with the responsibility of directing the exercise of the President's war power functions in telecommunications under Section 706 (a, (c-e)) of the Communications Act of 1934 as amended and delegated by Executive Order 12472 of April 3, 1984, "assignment of national security in an emergency preparedness functions." In brief, in the event of a national emergency—either of a wartime or non-wartime character—the Director of OSTP, by statute and Executive Order, assumes responsibility for the nation's telecommunications system and carries out these responsibilities through his Chairmanship of the Joint Telecommunications Resources Board (JTRB). Maintaining this capability requires a significant staff, and continuing exercise of the entire national communications system. Much of this work of necessity is very highly classified and cannot be delegated. A second example of a somewhat different nature is that the Director (OSTP) is formally charged with exercising the President's final sign off authority in the launch of any U.S. space vehicle carrying any significant radioactive material—as for example, the plutonium-based deep space power units. This too requires a substantial staff activity.

With the end of the Cold War, the question of assistance to scientists and engineers in the former Soviet Union and of course the related question of attempting to minimize proliferation, either through dispersal of trained personnel or of actual weaponry from the former Soviet Union, was a matter of increasing concern. OSTP devoted substantial effort to evaluating the level and scope of unemployment among scientists and engineers in the CIS with expertise relating to weapons of mass destruction and their delivery systems. We also focused on the prospects for their seeking employment with nations of proliferation concern (either by immigration or through *in situ* contractual arrangements) and on helping to create incentives for them to turn their talents to peaceful purposes.

I remain convinced that only when we get aggressive deployment of U.S. private interests in the former Soviet Union will we really have a permanent impact on the economic well-being of that set of Republics. At the present time there are three major impediments to such involvement—impediments that we have discussed in detail with senior members of the Soviet and then the Russian governments but which still remain. There is no legal structure as yet in the former Soviet Union that defines ownership, making it unacceptable for American enterprises to invest substantially in the development of technologies, for example, in cases where they have no firm assurance of ultimate ownership of what they produce. It is also true that there is no adequate intellectual property legislation, as yet, nor is there a legal infrastructure which would suffice for the resolution of conflicts if and when these arise in commercial negotiations. We have been assured repeatedly that such legislation is about to be passed but, as yet, it does not exist.

The breakup of the Soviet Union and the uncertainty that still surrounds the future of the Russian and other Republics—together with the emergence of regional threats to peace—have profound implications for the intelligence resources that the United States will require for our future security and national objectives. OSTP has devoted substantial effort to assisting the NSC in compiling long-term national intelligence requirements to help insure that future Presidential and national science and technology intelligence requirements are appropriately tasked. OSTP, acting through its Assistant Director for National Security, worked very closely with the intelligence community to seek ways in which intelligence can support economic as well as science and technology policy making. In particular, foreign targeting of the U.S. research and development base and other private sector commercial and business data is an area of growing concern. Understanding the objectives and methods of foreign governments and the vulnerabilities of American business and industry is essential to safeguarding the nation's security and to devising policies and programs to deal with unfair or illegal threats to America's ability to compete.

Over the years, the intelligence community has devoted an enormous amount of effort, creativity, and resources to developing frontier technologies for intelligence collection—and with impressive results. However, relatively little has been invested in technology to support the analytic process subsequent to collection or to dissemination of the results of the analyses to the ultimate American customer. As the intelligence community redirects its resources and responses to changing requirements, it is critical that we use the advances in technology now available to support intelligence analysis. OSTP's work with the intelligence community has highlighted areas of major success in the use of frontier technologies but has also emphasized areas where major progress is still possible. We have worked closely with the intelligence community to explore new avenues for effectively and efficiently upgrading the technological infrastructure for data storage, retrieval, and processing to help improve the analysis of all intelligence data. We have also worked aggressively to make available those parts of these technological structures that can be made available to the broad civilian sector without damaging any aspect of our national security.

One of the major defense related areas of concern involving first-line science and technology is that involving potential proliferation of weapons of mass destruction. These concerns—reinforced by our experience in Desert Storm and complicated as never before by the breakup by the former Soviet Union—have occupied an impressive amount of OSTP activity in areas of physical science, life science, and national security. Accounting for Soviet-made weapons and nuclear materials, as well as chemical and biological agents, is an extremely difficult task; guarding against proliferation of weapons expertise as scientist and engineers with critical skills seek to emigrate is in many ways even more difficult, and, as long as economic hardships prevail in the new republics, as seems probable for the foreseeable future, there will always be the temptation to exploit Soviet investments in the world's most extensive weapons infrastructure to generate hard currency returns. A central concern that will continue with the new Administration is that of broadening international cooperation in controlling the spread of ballistic missile technology, of chemical and biological warfare capability, and as well as nuclear capabilities and other hightechology weaponry to hostile or terrorist states.

We find ourselves in an age where national security allies have become overnight economic competitors with their own special interests in America's technology and wealth, and their own varying methods for pursuing those interests. OSTP has served as an Advisory member of the Committee on Foreign Investment in the United States—CFIUS—attempting to understand and assess those special interests.

While the efforts of industrialized countries or foreign commercial firms to access U.S. research and technology have largely been legal, open, and direct, they have not been exclusively so. In the judgment of the U.S. intelligence community, we presently lack the evidentiary basis for establishing any overall trend toward increased economic espionage among advanced industrial countries. Nonetheless, economic intelligence collection by such countries is potentially damaging to our economy because the nations involved are strong economic competitors as compared to the former Communist states which are not. OSTP has worked closely with the intelligence community in responding to this growing threat to American economic competitiveness. A particular vulnerability in all of these areas concerns our telecommunications and computer systems, both government and private sector that have become increasingly vulnerable to espionage from both foreign governments and foreign business competitors.

During the Bush Administration, we benefited in a major way from close collaboration between OSTP and the intelligence community, and one of my early arrangements was that of obtaining a senior liaison officer from the intelligence community who was stationed for a large fraction of his time in the OSTP offices to act as a real-time communication link. We benefited from the real time information that we were able to obtain which was pertinent to our international activities and to many other frontier technology activities within our own government. In return, the public also benefited very substantially, both from the new public availability of major breakthrough technological systems, such as those mentioned above, and from the release of decades worth of highly pertinent satellite data invaluable for obtaining long-time environmental data sets spanning the entire globe. This recently released data represents an absolute treasure trove for environmental scientists and, indirectly, environmental policy makers. Indeed, it is an excellent example of the benefits that can accrue to the entire society through careful and realistic reexamination of the extent to which formerly classified information and technologies can now be made much more openly available.

International Science and Technology

Science and technology have always been among the most international of human activities, with scientists frequently having closer connections with colleagues on the other side of the planet than with those on the other side of the hall. It is also unquestionably true that with burgeoning communication capabilities the planet is indeed being converted into a global village, and nowhere is this contraction more obvious than in science and technology.

Traditionally, OSTP has had the responsibility for the technical aspects of international agreements in science and technology. These responsibilities expanded very substantially during the Bush Administration. I had been heavily involved in the Reagan Administration in our Bilateral, Head-of-State agreements with India and with Brazil and carried over these responsibilities on my arrival in Washington. In 1987 and 1988—under the leadership of my predecessor, William Graham, a very successful agreement had been signed with Japan creating a high-level Science and Technology Commission whose chairmanship I took over as well. During 1990, we established a U.S./Soviet Joint Commission on basic research, as well as Joint/Consultative Groups with the European Community and with Canada. Despite the fact that our political relations were at a low ebb, I signed the renewal of the Sino/U.S. Science and Technology Bilateral with the Chinese Ambassador in 1992, and in 1992 also traveled to Israel, Jordan, and Egypt in furtherance of the peace process and to gain a personal view of the very substantial investments being made by USAID in those three countries. During my tour of duty, I traveled throughout South America with President Bush and on a number of subsequent follow-up trips. I also traveled to Japan, the Soviet Union, to almost all the European countries and to Canada to represent the U.S. in bilateral and multi-lateral science and technology agreements.

Because it had become very evident that personal contact among the senior scientific representatives of the developed countries could pay handsome dividends, I responded enthusiastically to a suggestion by William Golden, Co-Chairman of the Carnegie Commission, that we consider the possibility of organizing a very informal meeting of these individuals on a trial basis, without any staff, without any publicity before or after the event, with the intent being simply that of getting to know one another and being able to discuss items of mutual interest in a completely open fashion. The first meeting of this group was convened in Mt. Kisco, NY in February of 1991, with a representative from each of the G7 countries plus Russia, i.e., the United Kingdom, Germany, France, Italy, Japan, Canada, the United States, Russia and the European Community. This meeting was judged a resounding success, and so six months later a second meeting was held at Mt. Kisco to be followed at six-month intervals by one at Leeds Castle in the United Kingdom and then at one Rambouillet in France. The next such meetings are scheduled for Canada, Germany and Japan, in that order. Because of the personal trust and communication channels that have been established through this group, now known as the Carnegie Group, it has been possible, on many occasions, through simple telephone conversations, to resolve issues that could well have become serious problems had they not been nipped in the bud, and it has been agreed by all concerned that this staff-free meeting on a regular basis is a very important complement to the more formal, very-highly-staffed, meetings characteristic of the formal bilateral and multi-lateral agreements.

As noted above, one of the major challenges in international science and technology during the Bush Administration was that of providing assistance to the scientists and engineers of the former Soviet Union, as their society collapsed with the end of the Cold War. At my request, the National Academy of Sciences convened a broadly representative cross section of members of the American scientific and engineering communities and on a fast time scale provided me with extremely useful recommendations for effective action in this area. The American professional societies have played vitally important roles with the American Physical Society and the American Chemical Society providing major leadership in getting funding to the key scientists in the former Soviet Union without customs, tax or other bureaucratic difficulties. Private channels both in the United States and in Europe have demonstrated that their intrinsic viscosity is much lower than is that in governmental circles.

It has become clearer, with each passing year, that some of the programs and facilities needed to define and to reach the frontiers of modern science simply transcend the funding abilities of any single country, and the time has come when such megaprojects must be treated from the very outset as international programs. This is in contrast to current megaprojects such as the Superconducting Super Collider or Space Station Freedom where one nation—in these cases, the U.S.—makes the decision to move forward with the facility, carries out the detailed design, and only then approaches other nations for substantial contributions toward the construction and operating costs of the facility. This will no longer suffice, and for that reason, I am extremely pleased that in early 1992 we had an opportunity to present a U.S. proposal on internationalizing big science to the first scientific Ministerial of the OECD held in the past five years.

Our proposal was that the OECD should establish a *Mega-Science Forum* where leading scientists from around the world could come together to decide upon the outstanding unanswered questions in their field. With these in hand, groups of engineers and other instrumentation experts would be convened to decide on the kinds of facility that would be required to provide answers to these questions and finally, the senior scientific bureaucrats and politicians could come together to discuss appropriate funding, operational and other aspects of a truly international program. The assumption is that all of these things would be accomplished before any decision was taken as to the facility, its location or its funding and operating costs. The proposal was received enthusiastically at the ministerial level and the forum has already been established. In meetings thus far, astronomy and deep sea drilling have been considered as areas appropriate for internationalization and nuclear fusion is a topic soon to be tabled. I am convinced that the day of the single-nation megaprojects is past and that the Superconducting Super Collider represents the last of its species.

One of the persistent problems in the international area throughout the Bush Administration has been the concern expressed by some that technology paid for by U.S. taxpayers was flowing out of the country without any reciprocal benefits accruing to us. This protectionism was most evident in the CO-COM listing of technologies whose export to the Eastern Block and other unfriendly nations was prohibited during the Reagan years.

At the beginning of the Bush Administration, we found ourselves with the rather ridiculous situation, for example, that we were attempting to protect 486 class microprocessors, while at the same time, it was quite feasible for someone to back a truck up to a Radio Shack in Frankfurt, Germany, load up with just that technology and drive east to Vladivostock unloading the technology along the way. Recognizing that it simply was not feasible to protect much of the technology on the CO-COM list, after detailed discussions we reduced the list by 80%. Those items left were predominately of a systems character, where we had very real reasons to wish to protect them—from a national security point-of-view—and we agreed that we would protect this remaining 20% of the list more aggressively than in the past while leaving the remaining 80% completely open. We, as a nation, have benefited throughout our history from the maximum degree of openness in both science and technology, and I remain convinced that such openness—with very specific exceptions such as those just noted—is in our best interest. At the same time, however, we must learn to be more effective negotiators in our technological interactions. In the past, Americans have tended not to do their homework, have tended not to understand what their negotiating partner has that would be of value to the U.S., have not been adequately aware of the cultural differences involved, and have not been prepared to negotiate vigorously to get reciprocal benefits for American technology moving to other countries. I have become convinced that by becoming much more knowledgeable and effective negotiators we can, in almost all cases, insure a win-win situation where both we and our negotiating partner benefit from the exchange of technologies.

Even more is it important for us to keep our basic science open to the world. Unhappily, during the entire Bush Administration, I was frequently in conflict with the U.S. Trade Representative (USTR) because of her insistence that many of our bilateral science and technology agreements be held hostage pending acceptance, by our negotiating partner, of our intellectual property rights language. While agreeing in principle with her goals, I found her approach to be much too inflexible for our dealings with friendly sovereign powers—including Canada as an outstanding example. We also had difficulties with USTR in that at one point they were prepared to agree to the so-called Dunkel Text in the GATT negotiations which would have defined most federal support of research and development as a subsidy and, therefore, open to countervailing duties downstream. This situation is not yet resolved.

Finally, I must note that as was the case with all my predecessors as Presidential Science Advisers, I failed, miserably, to convince the U.S. State Department that science and technology is, and must be, an integral part of our foreign relations. Alone among the developed nations, we typically do not have highly trained scientists and engineers present in our embassies who can report back to Washington in real time on developments in science and technology of importance to us, or who can act as effective spokespersons for U.S. science and technology in the country where they happen to be stationed. In striking contrast, for example, the Swedish Government has something like 150 technology attaches—selected and paid for by Swedish industry—installed in its embassies worldwide, with 17 in the United States alone, all reporting back on a real time basis to Stockholm.

After discussing this situation with State Department representatives, without success, I attempted to arrange for the Commerce Department to task at least one of its commercial attaches in each embassy with a technology reporting responsibility and had made significant progress toward that end before the change in leadership in the Commerce Department required a totally new start. Based on my experience throughout the past four years and my continued conviction that we desperately need a much more effective science and technology component in our diplomacy, I am very much tempted to recommend that we give up on the idea of having qualified representatives in our embassies and instead, clone the ONR London office in major capitals abroad. That London office has a long history of recruiting distinguished American scientists who, on the basis of their own worldwide reputations, find laboratory doors open to them throughout any country that they may visit. Such reception is essential if real time reporting of value is to be expected.

Conclusions

Let me then conclude by noting something that I would have undertaken had I been invited to participate in a second Bush Administration.

For the last half of the 20th century science and technology in the United States has been following a vision and a blueprint largely laid down in 1945 by Vannevar Bush in his remarkable essay, *Science: The Endless Frontier*. This was written at a breakpoint in human history—the end of World War II—an appropriate time to consider the broad scope of research and development in the United States. The promise spelled out in Bush's Report—that support of fundamental research in the nation's universities would pay handsome dividends—has been more than fulfilled. But it bears noting that nowhere in his Report does Bush mention industry as a player in the Research and Development enterprise. He assumes that federal funding to universities will produce new knowledge and young minds trained to use that knowledge creatively, and that both will find their way into industry without any overt action on the part of federal government.

We now know that this last assumption is not valid and we recognize, too, that we have again come to a breakpoint in human history with the end of the Cold War, with the emergence of the European Community and the Western Pacific Rim as economic superpowers, and with enormous changes in our own society. It is no longer obvious that the vision articulated in 1945 and so successfully followed in the latter part of the 20th century is appropriate for the 21st century. To that end, it had been my plan to carry out a detailed study of the relationship between the federal government and the research-intensive universities, which we, indeed, completed in December 1992, and to follow this with a similar study of the federal laboratories, again carried out in parallel by FCCSET and by PCAST from inside and outside of the federal government respectively, and finally, in close collaboration with the private sector Competitiveness Council to carry out a detailed study in PCAST of the industrial research enterprise.

I have already commented on some of the problems associated with the federal laboratories, and our reports address those associated with the research-intensive universities in some detail. The industrial research enterprise also shows some alarming trends. For some 15 years prior to 1985, the support for research and development within the industrial community grew at an almost constant annual rate of 7.5%. Since 1986, this situation has changed dramatically and the growth rate currently is, if anything, slightly negative. More important perhaps, however, has been a simultaneous change in the characteristic time horizon of the industrial research with more and more of the resources being devoted to relatively short-term research, immediately related to production, and less and less to the long term more fundamental research formerly typical of the central corporate laboratory.

It had been my intent—with the above three studies in hand—to examine the rationale for, the structure of, the support of, and the effective utilization of research and development in the United States in the decades ahead, and to attempt to articulate a vision for the R&D enterprise appropriate to the 21st century. I have recommended that the Clinton Administration carry forward the remaining components of this program.

I would be remiss indeed if I concluded this chapter without stating explicitly what a rare privilege and honor it was for me to have the opportunity to work closely with George Bush during his Presidency, and I was also singularly fortunate in the quality of the people with whom I had the privilege and pleasure of working. Someone asked me recently what the two greatest surprises were that I encountered in coming to Washington; the answer was obvious. First, the people with whom I came into contact both in the Administration and in the Congress were of much higher personal quality than I had anticipated, and, secondly, it took infinitely longer than I had considered possible to make anything happen.

Despite this, I believe that we were successful in making some important things happen; that there is an unprecedented level of cooperation and communication in the science and technology enterprise within the federal government; that we have established a new basis for our participation in the international science and technology community; that we have begun to address a scandalous problem that exists in our pre-college educational system; and that we have exerted very real leadership in the international arena of global stewardship.

And finally, I take both pleasure and comfort in my belief that we have turned over to the Clinton Administration a science and technology enterprise that, despite its problems, is more than capable of responding to the challenges that we, as a demanding society, will continue to pose to it.

Appendix 1

Senior Staff of the Office of Science and Technology Policy
Executive Office of the President 1989–1993
(Note: The immediate prior position of each individual is shown in parentheses.)

DIRECTOR
Dr. D. Allan Bromley — *1993–*
(Henry Ford II Professor of Physics, Yale University)

ASSOCIATE DIRECTOR FOR THE LIFE SCIENCES
Dr. James W. Wyngaarden — 1989–1991
(Director, National Institutes of Health)

Dr. Donald A. Henderson — 1991–1993
(Dean, School of Public Health, Johns Hopkins University)

ASSOCIATE DIRECTOR FOR PHYSICAL SCIENCES AND ENGINEERING
Dr. Eugene Wong — 1989–1992
(Chairman, Combined Departments of Electrical Engineering and Computer Science, University of California, Berkeley)

Dr. Karl A. Erb — 1992–1993
(Assistant Director, OSTP, for Physical Science and Engineering)

ASSOCIATE DIRECTOR FOR INDUSTRIAL TECHNOLOGY
Dr. William Philips — 1989–1991
(Executive Vice President for R&D, Mallinkrodt)

Dr. Eugene Wong — 1991–1993
(Associate Director, OSTP, for Physical Science and Engineering)

ASSOCIATE DIRECTOR FOR POLICY AND INTERNATIONAL AFFAIRS
Dr. J. Thomas Ratchford — 1989–1993
(Deputy Executive Officer, American Association for the Advancement of Science)

ASSISTANT DIRECTOR FOR MANAGEMENT AND
THE SCIENCE COUNCILS
Ms. Maryanne Bach — 1990–1991
(Assistant to the Secretary, Department of the Interior)

Dr. Vickie V. Sutton — 1991–1993
(Office of the Administrator, Environmental Protection Agency)

ASSISTANT DIRECTOR FOR NATIONAL SECURITY
Ms. Michelle K. Van Cleave — 1989–1993
(Counsel, House Committee on Science and Technology)

ASSISTANT DIRECTOR FOR THE ENVIRONMENT
Dr. Nancy G. Maynard — 1989–1993
(Director, Ocean Science Division, NASA)

ASSISTANT DIRECTOR FOR THE SOCIAL SCIENCES
Dr. Pierre Perrolle —
(Division of Behavioral and Social Science, NSF)

ASSISTANT DIRECTOR FOR INDUSTRIAL TECHNOLOGY
Dr. James Ling — 1989–1991
(White House Science Council)

Mr. F. Stanley Settles —
(Allied Signal Corporation)

ASSISTANT DIRECTOR FOR PHYSICAL SCIENCE AND ENGINEERING
Dr. Karl A. Erb — 1989–1992
(Head, Nuclear Physics Division, NSF)

Dr. Lee S. Schroeder — 1992–1993
(Lawrence Berkeley National Laboratory)

ASSISTANT DIRECTOR FOR THE LIFE SCIENCES
Ms. Rachel Levinson — 1989–1992
(Office of the Director, National Institutes of Health)

SPECIAL ASSISTANT TO THE DIRECTOR FOR HEALTH AFFAIRS
Dr. William F. Raub — 1991–1993
(Acting-Director, National Institutes of Health)

SPECIAL ASSISTANT TO THE DIRECTOR FOR NATIONAL SECURITY
Dr. Charles Herzfeld — 1991–1992
(Director, Defense Research and Engineering)

Appendix 2

The President's Council of Advisers on Science and Technology 1990–1993

CHAIRMAN

D. Allan Bromley
The Assistant to the President for Science and Technology and
Director, Office of Science and Technology Policy

MEMBERS

Norman Borlaug
Distinguished Professor, Department of Soils and Crop Sciences
Texas A&M University

Solomon Buchsbaum
Senior Vice President, Technology Systems
AT&T Bell Laboratories

Charles Drake
Albert Bradley Professor of Earth Sciences and Professor of Geology
Dartmouth College

Mary Good
Senior Vice President for Technology
Allied-Signal Incorporated

Ralph Gomory
President
The Sloan Foundation

Peter Likins
President
Lehigh University

Thomas Lovejoy
Assistant Secretary for External Affairs
The Smithsonian Institution

John McTague
Vice President, Technical Affairs
Ford Motor Company

Thomas Murrin, Dean
School of Business and Administration
Duquesne University

Daniel Nathans
Professor of Molecular Biology and Genetics
John Hopkins University School of Medicine

David Packard
Chairman of the Board
Hewlett-Packard Company

Harold Shapiro
President
Princeton University

FORMER MEMBERS

Bernadine Healy
Director
National Institutes of Health

Walter Massey
Director
National Science Foundation

Science Advice to the President: Important and Difficult

Lee A. DuBridge

The United States of America and the modern Technological-Industrial Age were born at about the same time. The United States recognizes its birthday as July 4, 1776. James Watt filed his first patent for the modern steam engine in 1769. While America was going through its struggling years to become a nation, the Industrial Age was growing up in Great Britain. However, the steam engine and its various applications were soon being transferred to the United States, and the Industrial Age began to flourish here, also.

Thus, from its very beginnings the government of the United States faced scientific and technological problems, including problems of defense, of surveying the nation, of building highways, and, later, railroads. Soon it was necessary to develop standards of measurement which could be used throughout the country and to adopt laws and regulations governing the various effects of the new technological developments. Thus, the United States Government has always needed some scientific and technical advice.

George Washington was himself a skilled civil engineer, and Thomas Jefferson was surely his own Science Adviser. Though not many formal efforts to provide science advice to the Government were undertaken at first, we must recall that Abraham Lincoln in 1863 signed the charter of the National Academy of Sciences, an agency designed to advise the government on scientific and technical matters. Later Woodrow Wilson, at the onset of World War I, authorized the creation of the National Research Council as an operating research arm of the National Academy of Sciences. Thus, for the first time in United States history, a civilian scientific community was recruited to engage in the problems of design and

Lee A. DuBridge (b. 1901) was Science Adviser to President Nixon from January 1969 to September 1970 and had been a member (1951-56) and then Chairman of the original Science Advisory Committee. He was President of the California Institute of Technology from 1946 to 1969 and is President Emeritus. He was Director of the Radiation Laboratory at M.I.T. during World War II. Dr. DuBridge has been a member of the General Advisory Committee of the AEC and of the National Science Board and has been Trustee of the RAND Corporation and of the Rockefeller Foundation. He is a member of the National Academy of Sciences and of the American Philosophical Society, and a Fellow (and Past President) of the American Physical Society. He has served on numerous government advisory committees.

development of new instruments of war and advising the War and Navy Departments on their use. The NRC can be credited with many achievements contributing to our success in World War I.

Between the two World Wars there was little organized activity on the part of the scientific community in working on military technology. The National Research Council continued to advise the Government on such problems as highway design and construction, standards of measurement, and other technical matters.

Yet during this time the American scientific community was growing to maturity. Many American scientists had been studying in Europe and brought back new ideas to the American scientific establishment, greatly enhancing the US scientific productivity. The US scientific strength was further fortified by the many distinguished scientists who immigrated here to escape Nazi or Fascist dictatorships.

The Beginnings of Science Advice

Hence, at the outbreak of World War II a very powerful scientific community was ready to volunteer its services to the government to assist in the war effort. President Roosevelt authorized the organization of the National Defense Research Committee and, later, the Office of Scientific Research and Development, to mobilize this effort. Their enormous, extensive, and effective activities are well known to everyone. The successes were so great that the end of the war saw many branches of government (conspicuously, of course, the Department of Defense) seeking to establish a far more active program of scientific assistance to the Government, both in the carrying out of applied research in universities, in industry and in nonprofit laboratories, as well as in creating a multiplicity of scientific advisory groups to the various agencies within the Department of Defense, in the newly created Atomic Energy Commission, and, as time went on, in nearly every agency of government which had to deal with scientific and technical problems.

An important step forward was taken by President Truman, who, in 1950, signed the bill creating the National Science Foundation, which took over substantial responsibility for the forwarding of basic research in this counry. At the same time both the Atomic Energy Commission and the Defense Department were employing more civilian scientists to do basic research within their own establishments and through contractors. A "golden age" of science support marked the years 1946 to 1968. But the exponential growth then leveled off.

President Truman also appointed (on April 14, 1951) a committee, first headed by Dr. Oliver Buckley, known as the Science Advisory Committee of the Office of Defense Mobilization. This was the forerunner of what became the President's Science Advisory Committee. Though at first inactive, this Science Advisory Committee, after 1952, became on its own initiative more and more engaged in the growing problems of national defense, including air defense, military rockets, and guided missiles.

The Russian Sputnik created a further incentive in the country toward the support of science, since Sputnik seemed to reveal to many Americans that the United States had lagged far behind Russia in its scientific development. A new agency, the National Aeronautics and Space Administration, was created to forward the US space program, and other agencies greatly accelerated their scientific programs. President Eisenhower converted the Science Advisory Committee of ODM into the President's Science Advisory Committee and appointed, for the first time, a full-time Presidential Science Adviser, Dr. James R. Killian, Jr.* This action came at a strategic time and was widely acclaimed throughout the scientific community and the country at large. PSAC and the Science Adviser were regarded as the capstone of the Government's scientific advisory structure. PSAC could look broadly at national problems and not be confined by the special interests of individual agencies. President Eisenhower himself told some of us that he found it very helpful to have an unbiased, broadly-based and distinguished scientific group helping him to unravel the many technical problems which he faced in defense, space, and various civilian enterprises. PSAC developed some extraordinarily penetrating and far-reaching recommendations resulting from extensive studies by individual panels set up under the President's Science Adviser.

Sound Scientific Advice

PSAC flourished under the Eisenhower and Kennedy administrations. Though much of its work and many of its recommendations were highly classified because of military security, there was nevertheless a strong feeling, especially in the scientific community, that the President of the United States was now securing on an intimate basis sound scientific advice on national problems.

Difficulties appeared, however, during the Lyndon Johnson administration. The Vietnam War created divisive feelings throughout the country and within the scientific community. Members of PSAC itself were divided on the issues arising in this conflict. President Johnson was unhappy with this division of opinion within his own White House structure and thus was less encouraging and cooperative with PSAC than Eisenhower and Kennedy had been. Nevertheless, the Science Adviser and PSAC continued to pursue energetically many studies of scientific, educational, and international concern.

The Nixon Administration also found it difficult to understand the work of PSAC and the way in which it or some of its members publicly criticized Presidential policies on such matters as the antiballistic missile, the supersonic transport, and the continued war in Vietnam. Nixon also felt that the Defense Department itself had an adequate array of science advisory committees and that PSAC could well devote most of its attention to non-defense matters. This indeed had been occurring, for PSAC had already examined problems related to natural resources, to energy, and to environmental and other matters.

Nixon's staff, however, became unhappy that PSAC and its members did not always support Presidential policies and were also unhappy that PSAC did not seem to be an adequate political asset to the President. The final result was that the

---

* Actually such an appointment had been recommended to President Truman (and approved by him) in 1950—but the Director of ODM (General Clay) did not go along and Dr. Buckley acquiesced.

position of Science Adviser was abolished and with it PSAC and the Office of Science and Technology.

Shortly thereafter, however, in the Ford Administration, Congress took action to re-establish a Science Office and a Science Adviser, and the President signed the bill re-creating this advisory structure.

An Active and Effective Agency

President Carter's interest in science and technology were quickly evident after his inauguration, and his appointment of Dr. Frank Press as Science Adviser was widely acclaimed. Under his leadership the Office of Science and Technology Policy, as it was now called, became again an active and effective agency within the White House. (PSAC, however, was not re-established.)

There has been for many years, however, considerable controversy, even in the scientific community, as to the effectiveness of the White House Science Office. This is understandable, in part, because the Science Adviser and his staff and their committees normally make confidential reports to the President for his use in reaching decisions on important national issues. Thus the public at large and even most of the scientific community remain unaware of the nature and extent of the activities of the White House Science Office. It is not only that many of the recommendations to the President are classified for military reasons, but it is also because the President needs to have private advice from many sources as he reaches his important decisions. This advice should be rendered to him in a confidential way, without publicity, so that all aspects of these issues can be studied by him in private before a decision has been reached.

Thus we have a paradox: a committee placed in a very conspicuous position at the highest level of the Government is nevertheless forced by this very situation to carry out many of its activities under wraps, submitting reports and recommendations in private to the White House. These activities are publicized only when the President deems this appropriate.

During my own term of service as President's Science Adviser, I saw many articles appear in the press and in scientific publications criticizing PSAC for failing to consider or take actions on certain specific issues when, as a matter of fact, those very issues had been under intensive study and had resulted in confidential recommendations to the President.

As we approach the decade of the 1980s, it appears that the work of the Science Adviser and the Office of Science and Technology Policy is becoming ever more important and in many ways even more difficult. The problems that face the country are severe and complicated, involving a difficult mix of scientific, technological, economic, political, social, and international issues.

Our energy situation is only one example of the different problems we face. The technical problems of developing new sources of energy, especially those not using fossil fuels, are not easily resolved. There are many advocates for each of the various alternative sources: nuclear power, fusion, solar power, wind, geothermal, tidal, and other energy sources. Each of these has, of course, possibilities for producing some energy to relieve the pressure on petroleum supplies. Each, however, faces limitations, difficulties, and many unsolved technical, economic, and environmental problems.

Nuclear energy would seem to offer the greatest hope in the near future for large-scale energy production, and yet the problems it faces, both real and imagined, are all but intractable at the present time.

There is, in fact, no alternative energy source which can provide the increasing amounts of energy our society will require that does not meet formidable economic or environmental problems or both. The task of the government, therefore, in using its limited resources to encourage many investigations of different energy sources is a technical, economic, and political problem which surely the President's Science Advisers must face and carefully analyze.

Anyone can draw up a long list of other problems which our country faces, both domestic and international, both civilian and military, all of which involve important scientific and technical considerations, bound up, however, with economic and political problems.

I trust that the present and future Presidents of the United States will continue to give the White House scientific advisory structure strong support, depending on that structure to mobilize the best thinking of the scientific community and interpreting it to Government agencies in understandable ways.

The President and His
Scientific Advisers

I. I. Rabi

The problem of advising the President on scientific matters is not basically different from advising him on any other matters, except for the fact that people who become President rarely have strong scientific foundations. So it's separate from most other problem areas which confront the President. And of course it has not been conventional to have the President address these problems. Nevertheless, they have grown in importance. By science advice, of course, one does not mean, as far as the President is concerned, technical advice in the sense of detailed explanations of the operation of the laws of the universe or detailed descriptions of various devices. The advice one gives to the President must be broadly conceived and it must speak to the President in the sense of a translation into political terms of basic scientific, technical developments in all fields in which his decisions will be important, both for the national security and the national welfare.

This is then a very personal thing and a certain relation must develop between the President and his adviser. Since it's personal, the technical competence of the adviser is only part of his qualifications. He must be an individual who will interest the President, who will be able to explain things in such a way that they're usable by the President and in such a way that the President does not resent the attitude which may be a top-lofty one, as is so often the manner of some scientists in talking about science. Therefore there must be an impedance match between the President and his Science Adviser. That's of basic importance. Otherwise what the Science Adviser tells the President may not be understood; he may need an intermediary to explain it to him. That may function but it is not a satisfactory substitute for a direct relationship.

Isidor I. Rabi (b. 1898) was chairman of the original Science Advisory Committee from 1953 to 1957 and a member of PSAC until 1968. He is University Professor Emeritus at Columbia University, where he has been a member of the physics faculty since 1929. Throughout World War II he served at the Radiation Laboratory at MIT (O.S.R.D.) and at Los Alamos. He is a consultant to the Department of State and a member of the Naval Research Advisory Committee, is vice president of the UN Conference on Peaceful Uses of Atomic Energy, and a member of the United Nations Science Committee. He was a member of the General Advisory Committee of the US Atomic Energy Commission from 1946 to 1956 and its chairman from 1952 to 1956. He is a member of the National Academy of Sciences, serves on numerous boards, and has received many awards including the Nobel Prize in Physics (1944).

For an optimal relationship, the Science Adviser must know and understand the problems of the President as he sees them, as they exist in government. Therefore he has to be close to the President, see him frequently, see his aides frequently, so that he fully comprehends what the situation is and does not offer irrelevant advice or, worse yet, advice based on misunderstanding the situation. He must be a person who will like the President in the sense of really wishing him well and wishing to help him and wishing to subordinate his own preferences, political or societal, to the needs of the President for the policy which comes naturally to him. All this means that the Science Adviser must try to be a part of the President's mind-set. It's very difficult, because he must understand the President, understand his personality, and understand his own limitations since he is working not in a technical but in a political sphere, an expanded political sphere that includes those fields which he represents, which he's trained to represent, fields which may not in the past have been of importance for political consideration. This, in essence, I see as necessary for the relationship to flourish between the Science Adviser and the President.

Not An Easy Selection

The selection of such a person is not easy. Not many scientists or engineers can meet the description I've tried to give, but in a country of over 200,000,000 people there certainly exists a person who should be chosen by each President after an interview, perhaps in depth. And the President must somehow impress on his Adviser the nature of his needs. With President Eisenhower this was simple because he was an outgoing personality and he understood his needs and had a feeling for his own inadequacies. He felt he needed help and desired help, and was not diminished in his own self-esteem by receiving such help.

Now, the position of the Science Adviser is a very responsible one, because there come times when the Science Adviser will have to advise the President to make quite crucial decisions, and the President has to rely on the Science Adviser. Therefore he must have a great deal of confidence in his Science Adviser, because he may run into severe criticism and need defense. This means that the Science Adviser should preferably be a man of national prominence. This helps defend the President.

The Adviser should be backed up by a group of scientists, a President's Science Advisory Committee (PSAC), which he may or may not chair but which consists of informed people and people of judgment who, although not directly connected with the President in a day-to-day fashion, understand his problems, perhaps not to the same degree as the Science Adviser, and understand that they're there as advisers and not as experts on particular narrow fields. If those people are of that caliber and also have the reputations I've described, then the President has something to fall back on and is protected in following a course of action which he would not have the time, or sometimes maybe even the capacity, to comprehend fully, to comprehend the ramifications of this course of action in interrelationship with others. This of course he works out with his Science Adviser, who must understand that the President has many concerns which are not of a scientific nature but of a political nature, which involve information to which he may not have access, or hunches which the

President may have, a certain mind-set which will cause him to go against his Adviser's opinions and use his overall judgment. This is something which the Adviser must accept and respect.

What does the Science Adviser need to perform his functions? He is sitting next to the President, who is also the Commander-in-Chief of the Armed Forces and who executes the laws as they are passed by Congress. In fact, the President is the leader of the nation. He makes the budget; he lays down the policies with the aid of his Budget office and his staff and his Cabinet Departments, which then go to the Congress.

The Science Adviser, therefore, since he is a part of the Presidency, must have full information to do his job. He must have access to every part of the Executive Branch of the Government, including the Department of Defense, which is the user, as far as the Government is concerned, of most of the available scientific and technical knowledge that may come to exist. And then there's the Department of State, which has responsibilities for our very important relations with other countries. And the National Security Council.

The Preservation of Prestige

A great deal of the international prestige of the United States has come about because of the high level of scientific achievement in the United States in the World War II and postwar years. This prestige is something that is very important to preserve in foreign relations. It is a very important asset to trade on—it's very attractive, one of the most attractive portions of the whole American image abroad. We have on the one side this tremendous military power, and this is not the sort of thing we would like to use too much; but implicit in it is the prestige of the scientific achievement which is universally respected and desired on all sides. This is one branch of American power which is not readily understood by the political mind. The Department of State, for example, considering the personnel which goes in there, service officers, and even the political appointees, have not adequately understood the scientific community, the domestic scientific community—have not understood how to use it. This is a very important element in American policy which has been used very successfully sometimes and has been very much neglected at other times. But, as a source of influence and power, it is equivalent to a great deal of military hardware; and also it's something to trade.

Thus there is this very important other element of presenting the United States not as just a great and powerful giant but as this very highly cultivated country, in fields of culture which are universally respected and which all nations desire. So the Science Adviser has to have close connection with the President and the State Department to aid them in utilizing this fundamental strength, not just in horse-trading or haggling but in an actual important element of respect for our country, so that where the United States sits is the head of the table, not only because of power, military or commercial, but also because of this highly prized intellectual capacity and achievement.

Of course, the Science Adviser must also work well with the Office of Management and Budget. His advice is important there, because although the Office of

Management and Budget has traditionally had some very able people, they are not qualified fully to understand the effect of some of their decisions on the whole posture of the United States, or on the vigor of science in the United States which has important domestic, military, and foreign implications.

Help for the Adviser

My remarks place a very heavy responsibility on the Science Adviser, but he alone can't be everything; he can't be the sum of all wisdom. He himself needs help, needs to have reassurance and confidence in his own attitudes and ideas. He must present himself and be perceived not as an individual but as a spokesman *of* (not *for*) the American scientific community. He needs, and the President needs, to be assured that the Science Adviser is not off the track in what he is suggesting, but has the backing of a group of highly competent and prestigious people with some of the same qualities that the Science Adviser has. So he needs a committee, that is, a PSAC, a number of people who will advise him and at times as a group advise the President directly. The latter access is necessary to fortify their sense of high responsibility.

Through this President's Science Advisory Committee, this group—consisting of people who are respected in the community, both the scientific community and the public in general—and through special working subgroups, the President can tap the scientific resources of the whole country. Of course, he has access to all the scientific resources within the government itself, but these are often groups who may find it very hard, in the competition within the government, to be as objective as can be some outsiders who are independent of the competition.

And, of course, great as the government is, the country itself is greater. It is very important to keep in touch with the full resources of the United States in so far as this is possible. Furthermore, creative ideas, new insights, are very important because the world changes, the scientific view changes, and no single individual can keep sensitive to them all. Therefore, it is very important to tap broadly the resources of the United States in science. This would be through universities, through industry, through the various associations which exist in industry and scientific organizations. And of course what exists abroad should not be neglected. The United States, although of great strength in the scientific field, is not the dominant figure it was some decades ago, because other nations are coming forth very strongly. It is necessary to have access to people who will not be limited to our own situation, who realize that we must utilize the knowledge, attitudes, and understanding of the rest of the world.

Legislative Branch Relations

I have given the broad outlines of the functions and qualities of the Science Adviser, of advisers to the Science Adviser, of their national function of advising the President. Now what relationship should they have with the Legislative Branch? This is very much a matter of opinion. My own opinion is that the President is best served by having his own advisers work directly for him without any division of loyalties and with the least injection of political considerations. The politics is up to him. The relations between the Executive and the Legislative are not always smooth; they haven't been so in the last few administrations. I think, therefore, that the Science Adviser and his staff and committee should work directly for the President. If they are to have relationships with the Congress, these should be with the knowledge of and perhaps permission of the President; or by request by the Congress through the President to have the Science Adviser or his Committee members testify before them, but not to have dual responsibilities. I think it is best to keep the lines clear, so that the President will have full confidence that there are no leaks, that they are working for him and that nobody is running around and/or speaking carelessly.

This whole operation should be directly, in my opinion, in the White House Office. How this would appear to the Congress I don't know. They might resent it. In this field, which frequently requires very sensitive information, highly classified information—highly classified technically and, perhaps, politically—I think they should work for the President. How Congress should get its scientific advice is another matter. Many schemes have been set forth. I would prefer a separate structure, and in case of differences of opinion the experts might argue with one another, but privately.

Now how should the Science Adviser's office be organized? How big should it be? It could be almost any size, but to operate in this fashion, the fashion I've described, close to the President and backed up by a President's Science Advisory Committee and subcommittees, with people drawn from the whole nation, depending on the problems and their expertise, it will need a staff. If the staff becomes too big, then the operation of the staff itself will require a lot of administration, and when the administrative machinery becomes big, then of course the Adviser himself becomes occupied with it and loses freshness and contact. I believe that the staff should be quite small, I think about 25, 30, something of that sort. Such a number of carefully, selected people should be sufficient also to organize and to service and manage the advisory groups. I do not believe that the Science Adviser should be set up to answer a broad variety of job shop questions. After all, we do have a Federal government rich in scientists with a lot of information.

The Science Adviser should address himself to basic problems of concern to the President. If he is not qualified to do this in some areas, the whole country is accessible to him. He can very quickly set up ad hoc groups. He must not make a mirror image of the whole Federal structure, but restrict himself to the most important problems unless the President directs him to other fields. Such a policy will help him and PSAC maintain prestige and respect outside. He won't be competing with the whole Federal government and his advice will be more cogent. There is always a tendency to go into all fields, and I think a great weakness in the past has been to cover not only all the fields of the military but also Health, Education and Welfare, State Department, and almost everything else. Now, of course, we have the energy field, which is indeed very important. So the Science Adviser should be selective in his fields of involvement, if the President will permit him to be.

Further comment is in order about the President's Science Advisory Committee and its relationship to the Science Adviser and his staff. The Science Adviser, of course, should be full-time resident in Washington. His advisers, the PSAC, should meet frequently, certainly at least once a month. The committee should usually be chaired by the Science Adviser, but not necessarily. The Science Adviser himself is appointed by the President. He may be just right for the President, but there may come a time when the Science Adviser and the committee disagree. The committee should have access to the President as a committee. It is even conceivable that the Science Adviser should be chairman of the committee by election—that, in principle, the committee could choose one of its own members as chairman. The committee members, although part-time, should regard themselves as a part of the President's office and should therefore refrain from any semblance of using their inside knowledge in public discussions, in talking to newspapers or otherwise. If a member finds himself in disagreement with the policy of the President and the Science Adviser and, having given full voice to this disagreement within the group, finds he cannot go along with that policy, he should resign. Then after a suitable interval, he will be free to offer his own opinion. The fact that he is only part-time does not mean that he is free the rest of the time to express himself publicly, since he could not divorce himself from the prestige of the President's Science Advisory Committee and certainly would be thought to have used inside information which is not readily accessible outside. This is a very important limitation, self-limitation, but it should be understood and accepted in such office.

Each member of the PSAC has an important national responsibility and it generally takes a good deal of effort outside of committee meetings to prepare himself, read reports, and so on. He should be compensated accordingly. Selection of members for the committee will require great care. It is very difficult to get people from industry—from private industry—to disengage themselves sufficiently from their own concerns, their own competitive concerns, to insure that their advice is objective. So by and large I think the personnel has to be drawn from fields outside of industry. There may be special industrial organizations like the Bell Telephone Laboratories, enterprises of that kind, which could be objective scientifically. On the other hand, they would be less valuable on the political side. The members should be carefully selected, not only for their scientific competence, but also to be as free of bias as possible. Industry people could appear on the various sub-panels which might be set up, study groups for special subjects and so on, for their expertise is important and can't be done without.

Breadth and Wisdom

Another point with regard to the composition of the PSAC and the selection of the Science Adviser himself. It is my experience that, if you get a group in which the members of the group represent different fields of expertise, it becomes rather ineffective because each expert tends to dominate the subject in his particular field. But you need to have opportunity for broad critical discussion of all questions. So one needs people of breadth and wisdom who are not brought in just as specialized experts. Such people are hard to find, but they exist. Of course, the fields of science are intensely subdivided. I think the National Academy lists about 50 different kinds of physicists, for example. And in medicine and biology you can't really have experts; you have to have people of good scientific background who also have solid judgment in appraising information outside of their fields and know how to bring out the relevant parts. The people have to be selected for their scientific eminence but also for broad judgment. There are scientists of the highest eminence who lack scope. Such an individual's value would be restricted, useful only for special purposes, that is, to be asked in for testimony.

Suitable generalists, of course, are hard to find; and it is not easy to define the requisite specifications. My own experience, being a physicist, is that you find such people more readily among physicists, simply because physics is so fundamental and profoundly permeates all the other fields. But that is only a matter of experience and there must be many people who could thoughtfully discuss topics outside of their special fields—military education, let us say. Just from personal predilection I would not bring in sociologists for example, nor historians, nor people with a strong political background. They know a great deal and are usually very well acquainted in Washington. They know a great deal but they also know a lot which is not so. And they can be very intimidating to the scientist who hasn't moved around in those corridors.

Who Should Select Members for a PSAC?

An important question: Who should select the members of PSAC? Each President would have confidence in his own appointees. On the other hand, in changes of administration if you have a new set of appointees, then it may become political. So I believe that the Science Adviser, because of his close personal relation, has to be appointed by and serve at the pleasure of the President; but the members of PSAC, although appointed by the President, should have tenure at least to the degree that they can overlap administrations. Because, after all, the President doesn't have to take their advice. But they should not be considered by the scientific community or the world at large as creatures of a particular President or to have a particular political orientation. It is very important that they be viewed as scientists: objective, independent, and not of visible political complexion, expressing their advice in a way that is understandable to the President or the public, untinged by political considerations.

Some reference to the early years of formal Presidential science-advising may be in order. The first Adviser was Oliver Buckley, appointed by President Truman, on a full-time basis, early in 1951 along with a Science Advisory Committee of which he was chairman. Dr. Buckley had been president of Bell Telephone Laboratories and during World War II played an important role in the Office of Scientific Research and Development (OSRD). He was succeeded by Lee DuBridge, and then by me, on a part-time basis. I stayed on during the Eisenhower period until shortly after Sputnik, when Jim Killian came as full-time Science Adviser and the whole office was revitalized and reported directly and exclusively to the President. Previously, it was called the Science Advisory Committee of the Office of Defense Mobilization (ODM), and also had direct access to the President, who used it very little. During my time it reported to the President through the Director of the ODM, Dr. Arthur Flemming, who followed General Lucius Clay. The operation at that time was that various people from various departments of government, chiefly Defense and State, would meet with the committee and its chairman to testify and answer questions. The secretary of the committee, David Beckler, sat with the National Security Council and would inform us of what went on there. We were very well informed and we would meet, generally monthly, and report to the President through Dr. Flemming.

But we also had our own initiative. The committee, which at that time consisted mostly of people who knew one another well, engaged in discussions which ranged over the whole field of national policy in so far as science was involved. One very important consequence was the setting up of a study under Dr. Killian, of MIT, called the Technical Capabilities Panel, whose report went to the President and had a very important influence on policy.

We were on the whole very well informed but didn't go to the President on many topics. Naturally, as a part-time adviser, I was relatively remote from the President compared with Bobby Cutler, the Director of the National Security Council. Thus, on one highly important matter, for example, on which the latter opposed me, I recognized that I was outgunned and that it would have been futile to appeal to the President. General Cutler was there all the time, I only from time to time.

The Concern Over Sputnik

When Sputnik came in 1957, President Eisenhower became very concerned. He called me in as chairman of the Science Advisory Committee, and later the whole committee met with him. At that point the question was what to do. Sputnik was a tremendous event; not that we'd not been concerned with satellite development, but we had not realized that the Russians were achieving a workable one before we did and that this event would be of such tremendous international political importance. It was a complete surprise. It was as if we had set ourselves an ambush, because the President had stated that we were doing such a thing but adequate funds had not been provided. The Department of Defense, in effect, took away some of the funds of Project Vanguard by not letting us use a launch-pad free of charge. It was actually in the Navy Department, the Naval Research Lab was running it. The President said that we in the United States were making a satellite. It was really a challenge. And at that point this project was undercut by the military who had reasons of their own which I don't understand. So we were left with inadequate funding, and there we were. The Russians had the Sputnik and our name was mud. So the President called in his Science Advisory Committee and asked, ''What should I do about this?'' It was a real case of departmental jealousy.

''A Part of the President's Brain''

I advised him in the presence of the committee that what he needed was a man whom he liked, who would be available full time to work with him right in his office, to help by clarifying the scientific and technological aspects of decisions which must be made from time to time. He would be a part of his brain, so to speak. President Eisenhower readily agreed. In a short time, by Executive order, he transferred this Science Advisory Committee to directly within the White House and he appointed Dr. Killian as his Science Adviser. This worked very well and he was very pleased; and I remember his saying he didn't know how he had functioned adequately before. He said that he had all sorts of inputs from the various departments, particularly the Department of Defense, with its tremendous resources, but he actually had no direct staff support in that field—a qualified independent individual whom he could ask to look into questions which were bothering him. He didn't fully trust the advice he was getting from the departments because of their natural bias and competition. This basic arrangement of the Science Adviser full-time in the White House, reporting directly to the President, as originally approved by President Truman but not put into effect in 1951, served very well through the years until extinguished by President Nixon in 1973.

Moving from the past to the present and future, I would like to point out that, whether recognized or not, science advice for the President has always been of very great importance. Of course, its importance was more obvious when we were engaged in the Cold War. But there has been no time I know of during which the President needed more and better scientific advice than he does now.

Twenty or 30 years ago, there was no question about the position of the United States as far as science was concerned: We were Number 1, and that was it. We could make mistakes and we could fail to take advantage of our opportunities, but they were in no sense fatal mistakes. But now, with other countries coming along with developments in science and technology, we are hardly even Number 1. A friend of mine who attended a meeting of economists in Hawaii was told by a friendly Japanese, "Why don't you Americans do what you can do best—raw materials and agriculture—and let us do the high technology." It's bitter to hear this, but there is a great deal of truth in it; we have lost our high position through neglect and lack of recognition of what was happening in the whole field, internationally, in science and technology and their applications. In pure science we are still well ahead, but the others are coming along very rapidly, particularly Germany and France, which are spending large sums of money on basic research. This is a situation which is deteriorating rapidly from the American point of view, and our industrial productivity relative to other countries, our chief competitors, is declining. We are no longer premier in inventions, instrumentation, and the like. This directly affects our standard of living, as is increasingly becoming apparent. We have been driven out of high technology fields like electronics by the Japanese and the Germans. We are beginning to lose out in aviation to the French and British with the air bus and other developments of that sort. To help preserve our standard of living domestically, and our position in international competition, we will need very basic policy guidance in the scientific and technical fields.

The President has not been able to provide this policy leadership for the last several administrations. The Presidents have been occupied otherwise and have failed to understand the great importance of science and technology. And of course the lag is significant. We lost much ground because of the Vietnam War. During those years, while the President and the country were paying attention to other things, and we thought we were able to make both guns and butter, we suddenly began to find that perhaps our guns were not so good and that we could make butter but not electronics, not automobiles. Now we're in a position that is almost crisis with respect to the applications of science and technology; and our posture in the world has become too much a matter of special interests, special narrow interests, which involve changing our culture. The environmentalists, for example, are very high-minded but do not represent the comprehensive interests of the country. There are other groups which do not place their particular interest, their particular predilection, in proportion. And the President hasn't been able to do it.

There is a tremendous amount of sentiment that has grown into American policy, in which people act and feel as if we were alone in the world and can write our own ticket. But we are in a world of competition; and unless there are clear policy decisions made in full recognition of that, we are bound to go down. We have been going down; and as you look at the curve, that trend has been accelerating. Our decline is becoming precipitous. One can see that in many fields. And our national prestige, insofar as it affects real foreign policy, has declined. In the fields of nuclear energy, people don't pay any attention to the United States as far as policy is concerned except for our power to regulate materials and their export, and to regulate manufacture and usage. Our views are not respected. Thus the President needs science and technology advice badly, and he needs it in virtually all fields. It would be hard to find a field in which science and technology aren't important.

The Role of the Military

The Congress has chosen to keep the President's Science Adviser out of military matters. Yet that is the most important source that we have of new inventions, new applications, new science. And the reason is a simple one: the military ask for the impossible, but they can pay for it. It's a rather odd way of getting scientific and technological advance, but this has been the situation ever since the end of World War II. The military supplied the money because they had needs, or fancied they had needs, for very superior science and technology. The military pressed for the development of computers, the development of great computers, and the reason we are so far ahead in the world so far in computers is just that. The military were consistently impressed by their need for aircraft. And the reason we have these transports—the wide-bodied and other kinds of aircraft and engines—is that the military paid for their development, with our taxpayers' money. They wanted them and they paid for their development. It's a strange way of getting industrial advantage, but this has been our history in recent years.

Now the President needs informed and independent people who will advise him about these matters, who will look at this policy and other policies and show what each will do, what it can do and what it won't do. It appears that the President is responding to uninformed popular opinion in such matters as solar power and cutting down on our immediate needs, which may be nuclear. He seems to be responding to protests of special groups who don't understand the problem at all. Of course, there will always be divisions within the scientific community. If one eminent scientist stands out against the rest, the public is likely to judge the truth to be somewhere in between. People of competence and prominence are needed to advise the President. So I think he needs such advice and needs it very badly. The Science Adviser must be a man with a strong voice who can stand up and, when necessary, slug it out with the other people around the President. He must not only have ability and eminence but also a certain kind of inner strength.

That is what we need. But I don't see it happening. I see the slide just continuing. The cause is psychological: the lack of leadership on the part of the President and the Congress, the influence of special interests, and so on. We can see what happened to Britain. It's happening to us almost as if we were copying the British, in the same sense. Examples are the decrease of productivity and the loss of value of the currency.

Advice to the Congress

Finally, I'd like to comment about scientific advice to the Congress. The Congress certainly needs such advice; and it is getting it abundantly from the staff of the National Science Foundation. However, it is fragmented; it is not presented in such a way as to suit the national interest. There are many Congressmen and Senators, many committees and subcommittees, and each one, lacking strong central leadership, can become a law unto itself. That is the tendency. I think the Congress needs scientific advice, it needs authoritative advice, but optimally it needs it from a single highly respected source. It should not be exclusive, it cannot take away the privileges of a Congressman or a committee, but there ought to be some channel through which advice can come which by its origin and its backing would have a greater prestige than the particular hit-or-miss observations that Congressmen and Senators pick up from their favorite scientists, or favorite military men, or favorite engineers, or favorite demigods.

So I believe that somehow, in the Office of the Speaker of the House or by some organizational device which would make it more representative than just the Speaker, there should be an Office of Science and Technology Assessment. There is such an entity now; it has had tough sledding. It should, however, be clearly representative of either the House of Representatives or the Senate. It is hard to believe that they would get together with a single organization but maybe they would. The Speaker of the House or the President of the Senate (who would be the Vice President) might appoint some oversight committee selected for this purpose. That could be a rather large office, but that is a Congressional problem, rather different from the Presidential one. Nevertheless, there should be some centralization of advice so that a Congressman can get such information and use it with confidence and good conscience. The Library of Congress has been such a source of information for some Congressional needs. In scientific fields I would like to see something of that sort, a prime person, not appointed politically, who would be the chairman of a committee; or some variation. This does not take away any of the prerogatives, but if a Congressman has his favorite economist, if a senator has his favorite scientists, he would also have the prestige of the established scientific line to buck. But I'm sure people will say, "Well, they're not always right," and of course that is true. Sometimes the statement that "They're not always right" slops

over into meaning ''They're usually wrong; the establishment is usually wrong.''
That's often right.

Those are my thoughts for the future based on consideration of the past in the
light of the present.

The Origin and Uses of a Scientific Presence in the White House

James R. Killian, Jr.

Since I was the first to bear the title "Special Assistant to the President for Science and Technology," I recall in this article the origins of formal arrangements in the White House for Science Advisers to the President. President Eisenhower was the first President to arrange for such advice, largely as suggested by a report prepared by William Golden, and every President, save one, continued uninterruptedly to use science advice.

In his second term, President Nixon dismantled the White House science arrangements, but legislative action by the Congress aided by strong advocacy from the scientific community led to the restoration of the science presence in the White House by President Ford; and President Carter has in Dr. Frank Press his Science Adviser and an Office of Science and Technology.

Let me recall that Presidents before Eisenhower used science in policy-making. There was Thomas Jefferson, who had by far the best Science Adviser any American President ever had—himself. Together with other Founding Fathers, notably Washington, Franklin, Madison, and John Quincy Adams, he infused into the American system the concept that there should be a true marriage between science and politics. While they were thwarted in their hopes to establish a national university to promote "useful knowledge and discoveries" for the new Republic, they still succeeded in introducing into the American system an intellectual outlook that in Hunter Dupree's words, "made science a formative factor in making both the Federal Government and the American mind what they are today."

The other great period when science and engineering served our society and the

James R. Killian, Jr., (b. 1904), was Science Adviser to President Eisenhower and Chairman of the President's Science Advisory Committee (1957-59); he was a member of the original Science Advisory Committee from its inception in 1951 until 1957. He has been associated with the Massachusetts Institute of Technology through virtually all of his career and was active there in the administration of government military scientific work throughout World War II. He recently retired as its Honorary Chairman and was previously its Chairman and its President. He is Chairman of the Board of the MITRE Corporation and has been a director of American Telephone & Telegraph, General Motors, and other major corporations. He is a member of the National Academy of Engineering and has served on numerous governmental and educational boards and commissions. He is the author of Sputnik, Scientists, and Eisenhower *and drew upon the contents of this memoir in preparing this article.*

whole Free World with decisive brilliance was, of course, World War II. The superb accomplishments of American science through the Office of Scientific Research and Development was facilitated by the fact that Vannevar Bush, in influence if not in title, was Science Adviser to President Roosevelt. This wonderfully effective relationship, aided and abetted by Harry Hopkins, was not only a major factor in the winning of the war but in devising new ways for our government to insure the prosperity of American science after the war.

I was launched into what was for me the outer space of the White House in October, 1957, when the Soviets orbited Sputnik I. This technological feat was received with stunned surprise and shock by Americans and produced apprehension throughout the Free World. In this climate of near hysteria, many jumped to the conclusion that the Soviets had surpassed the United States in its science and technology and that they had achieved a guided missile capability that posed a fearful threat to our security. Some are too young to remember how psychologically vulnerable the American people were to this event. Edward Teller, in a television program, remarked that the United States had "lost a battle more important and greater than Pearl Harbor," by falling behind the USSR in scientific achievement. On another occasion, when queried about what might be found on the moon, he replied, "Russians." On its editorial page, *The New York Times* was gripped by the emergency and editorialized about national survival. Lyndon Johnson, Senate Minority Leader, spoke with flamboyant historical sweep, "The Roman Empire controlled the world because it could build roads. Later—when moved to sea—the British Empire was dominant because it had ships. In the Air Age, we were powerful because we had airplanes. Now the Communists have established a foothold in outer space. It is not very reassuring to be told that next year we will put a better satellite into the air. Perhaps it will even have chrome trim and automatic windshield wipers." And the *New Yorker* ran a cartoon which noted, in effect, that the Soviets had the ballistic missile and we had the Edsel.

Advice to Eisenhower

Among the actions taken by President Eisenhower as he sought to allay fears and reassure the American people was to summon scientists to advise him personally on our space and defense programs, on ways to insure the general health of American science and technology, and on improving the quality of our education in science. He called on the President of the National Academy of Sciences for advice, and he asked the Office of Defense Mobilization Science Advisory Committee to meet with him, the committee which he was later to reconstitute as a committee directly advisory to himself. On October 15, he met with the ODM Science Advisory Committee, then chaired by Dr. Isidor Rabi. The President afforded the group a full opportunity to air their views and to make proposals, which they did with frankness and vigor. He wanted to know whether American science was being outdistanced by Russia. The United States had great strength, said Rabi, but we must be aware that the Russians had gained impressive momentum and were effectively mobilized steadily to build their scientific and technological strength. They could possibly pass us, Rabi emphasized, if we were so inept as to permit it to happen. The United States had caught up with Europe, he noted, and then passed it.

Edwin Land then made one of his eloquent speeches in which he said that American science needed the help of the President. Better than anyone else, he, the President, could kindle among young people an essential enthusiasm for science and lead people to understand it as a joyous, creative, rewarding adventure. The President clearly was impressed by Land's plea that he could, through active intellectual leadership, seek to create a more widespread understanding of science. It is interesting to note that he undertook a series of speeches on science and defense that were partly inspired, I think, by this discussion with the Science Advisory Committee.

Rabi then made a specific proposal. There was no one around the President, he pointed out, who could help him be aware of any scientific component that might exist in the important policy matters coming before him. Science was not represented on his staff. He should have a full-time Science Adviser—a person he could live with easily. I then carried Rabi's proposal one step further and urged that there be a strong Science Advisory Committee reporting directly to the President who could back up his Adviser.

This is the story of the discussions which a few weeks later led the President to appoint a Special Assistant to the President for Science and Technology.

The TCP

Actually, the Eisenhower Science Advisory Committee was not the first to bear the presidential title. President Truman had appointed such a committee in April 1951 and the late Oliver Buckley became its chairman and, in effect, Science Adviser to Truman. It was not until the Eisenhower Administration, when Lee DuBridge was its chairman, that this Committee was presented with a really major opportunity worth its mettle. At a meeting with the President, the military problem of surprise attack was the principal item of discussion, and the President, in effect, challenged the committee to help him get a hold on this problem. As a result, a task force was appointed, known as the Technological Capabilities Panel, consisting of about forty scientists and engineers, and this group presented its conclusions at a full-dress, expanded session of the National Security Council in February 1955—a session which Robert Cutler was to describe in his memoirs as the high point in the deliberations of the Eisenhower NSC.

This study did much to re-establish confidence between the scientific community and the administration, a confidence which had been badly damaged by the Oppenheimer case and the tensions of the McCarthy period. It brought to the attention of the President a group of scientists and engineers who had fresh contributions to make to national policy.

The Von Neumann Missile Committee

There was still another scientific group, the brilliant and decisively influential von Neumann Missile Committee, of which both Doctors Wiesner and Kistiakowsky were members, that helped in establishing a relationship of confidence between scientists and top policymakers. These two panels, I am sure, were important factors in the confidence which the President came to have in what he called "his scientists."

There is not space enough to go into the work of the first years of PSAC in any detail. It established a panel which recommended that the old NACA be converted into NASA, and the whole advisory group resonated with the President in insisting that our space program be in civilian and not military hands. Working closely with the Bureau of the Budget, a Space Bill was introduced in the Congress in April, 1958, and became law in early fall. The Committee secured the acceptance of a proposal for the processing of scientific information in government that avoided the creation of a center where all scientific literature would be processed by a computerized behemoth. It brought to the President and the Secretary of State information and recommendations which led them to reopen discussions with the Soviets on the limitation of nuclear tests, and this led to the Geneva Conference of Experts. In fact the Eisenhower PSAC brought into government, views and analyses which led to more open-minded discussion of disarmament issues. PSAC strongly supported curriculum reform in education and it played a role in the formulation of the National Defense Education Bill. It presented to the Cabinet a proposal for establishment of a Federal Council for Science and Technology, which was promptly authorized by the President. These were all specific outcomes of PSAC studies and recommendations, but I think more important were the relationships of confidence and free discussion that PSAC had with the President and the President's associates. There was never any difficulty in seeing the President or bringing matters before him for decision. After his retirement from the presidency, he told a friend that some of the best experiences he had at the White House were the meetings that he had with PSAC. These meetings, in which there was free-for-all discussion, were memorable events for PSAC itself. They made it possible for a group of scientists to come to understand the President's problems, his views and goals, and to learn how to make themselves useful in the light of this understanding.

The Eisenhower PSAC felt strongly about the futility of trying to achieve additional security by the unlimited pursuit of weapons technology. They recognized the importance of advancing our weapons technology in order to prevent the United States from becoming a second-rate power and they felt a deep obligation to assist in the strengthening of our military position. In fact perhaps their most useful role with Eisenhower was to advise him on weapons systems and the military budget, but there was a preponderant view in the committee that the security of the country could best be served by moderating the arms race. There was also a preponderant view that we were enmeshed in too much secrecy and that every effort should be made to achieve a more open society as a way to a more open world.

It must be said that the committee served under highly advantageous peace-time conditions that were almost unique to the Eisenhower years. It had free access to a President who knew he needed their help. The military establishment had not matured in its use of science and technology, and there was, of course, no place where there had been developed great capacity in space technology. The National Security Council, because of its small staff, had little capacity for in-depth studies of weapons technology. Under these conditions it was inevitable that the President would look to PSAC for advice on both weapons and space technology. The President looked to his Science Advisory Committee as a source of objective advice that he felt to be not always available from all the government departments. Especially did he seek help in dealing with the competitive claims of the three military services.

"Too Conservative and Unimaginative"

While it was the source of numerous innovations, the PSAC organization was sometimes criticized as too conservative and unimaginative. Indeed it found itself constantly opposing blue-sky proposals that were popping up all over, when it felt they were sufficiently screwy or inadequately thought through. The committee, for example, felt compelled to ridicule some of the occasional wild-blue-yonder proposals by some Air Force officers for the exploitation of space for military purposes. In their ardor to support innovative ideas for their services, these officers, often more romantic than scientific, made proposals that indicated an extraordinary ignorance of Newtonian mechanics, and PSAC made clear to the President the silliness of these proposals.

A PSAC panel opposed the building of nuclear-propelled aircraft and were consequently lambasted by some Congressmen for their "lack of vision." Some of the command and control systems proposed by the Department of Defense were attacked by PSAC as technologically unsound and too expensive.

PSAC had another important characteristic. A majority of the members had no political ambitions and no career objectives in government. In giving advice, they sought to be nonpartisan, whatever their private political beliefs might have been. They sought never to embarrass the President by differing with him publicly. They would have rejected as repugnant and ridiculous any idea that they could appropriately be described as a "priesthood," a term which some political scientists have used to describe the scientific community. They were motivated primarily by a feeling of obligation to make their specialized learning and skills available to the government in time of need, and by a confident feeling that they had important contributions to make. This absence of political ambition made it possible for them to work with the elective, appointive, and career people in government in a way that did not arouse antagonism or fears of territorial aggression.

PSAC was fortunate in its relation to the National Security Council. As Special Assistant for National Security Affairs, Robert Cutler had played a key role in the appointment of the Special Assistant and in bringing PSAC in direct association with the President. He was cordial to PSAC and was responsive to its proposals and recommendations. This provided a coupling that has not always existed in successive administrations.

Finally, this group of science advisers had a deep sense of responsibility to science, along with an unshakable faith in its importance both to the individual and to the nation. They loved science and wanted others to share their enthusiasm for it and to discover its inner power to make men and women a little more creative, a little more civilized, and a little more humane. These convictions about the values of science brought their advisory work for government an additional meaning and zest that made the experience memorable.

Science and Technology: Government and Politics

Jerome B. Wiesner

The Presidential Science Advisory mechanism has been useful, is important, and may even be termed vital.

If one looks at the performance of Jimmy Carter—particularly the problems he has in the arms control field—one sees the difficulties that come from not having a sustained effort by people who are really responsible only to him and who have a strong allegiance to him as President. I think that President Carter has had to ad hoc his way through a lot of problems that he would have handled better if he had the assistance of a Science Advisory Committee. He has a very able Science Adviser who often uses the services of ad hoc committees of distinguished people to help him make specific decisions; but such a system is substantially different from one that provides continuity and integration as well as providing answers to specific questions. When it's operating effectively, the science advisory group should spend much of its time anticipating problems.

The PSAC under Doctors Killian and Kistiakowsky and myself not only dealt with crisis problems, but tried to be orderly and forethoughtful and anticipate Presidential needs, so that when important problems arose, we'd be prepared to work on them intensely. We had ongoing panels in many areas that were not in the limelight, so that if one of those matters became urgent, there was an advisory group or panel that had been thinking about it ahead of time. In the instances of Eisenhower and Kennedy the President had a close relationship, not only to the Science Adviser, but to the PSAC as well, so that they felt very comfortable in letting their hair down in very frank discussions. Both of them obviously enjoyed the give and take. Lyndon

Jerome B. Wiesner (b. 1915) was Science Adviser and chairman of the President's Science Advisory Committee (of which he had been a member since 1957) under President Kennedy and for a short time under President Johnson. He has been president of the Massachusetts Institute of Technology since 1971 and previously was a member of its faculty. He is an electrical engineer and during World War II served in the Radiation Laboratory at M.I.T., in the development of radar and radio communications. He has served on numerous government boards and commissions and has been active in arms control efforts and in science policy matters. He is a member of the National Academy of Sciences, the National Academy of Engineering, American Academy of Arts and Sciences, and the American Philosophical Society and is the author of Where Science and Politics Meet. This is an excerpt from an interview with Jerome B. Wiesner.

Johnson was a poor listener and didn't trust anyone he didn't fully control. Nixon didn't like give and take, didn't trust many people, and was particularly mistrustful of scientists who were, in his view, all liberal and against him. The general perception is that President Carter too is uncomfortable with PSAC-like groups and thus shies away from them, but in my one close contact with him—at the Camp David energy meeting—he ran the meeting effectively, obviously enjoyed it, and seemed to benefit from it. Unfortunately the meeting was not as helpful to him as it would have been if the participants had been able to prepare for it.

I am certain that Mr. Carter would have dealt more effectively with problems like that of energy policy that involves so many complex and interesting issues of technology, timing, economics, available resources, environmental problems, politics, etc., if he had a continuous White House-based group supporting him. Now, you can properly say that the Energy Department has been working on all of them, but the point is that the Energy Department is a vested interest, though it tries not to be. I remember President Eisenhower saying more than once, and President Kennedy, too, that PSAC was the only group that worked for him that had no other responsibilities and no conflicting interests. In fact, Kennedy once told a newspaper man that the science advisory mechanism was the only thing that kept the government from pushing all in one direction.

I think he was right. The Science Adviser by himself can do a lot, but if you have a good Science Advisory Committee, involving an able and distinguished group of people who have both the respect of the external community and the President's confidence, the mechanism can provide much more support for the President than one individual possibly can. So I think the whole apparatus is important, is vital, not just the Science Adviser.

Insuperable Impediments?

It has been repeatedly stated that the Federal Advisory Committee Act and the Freedom of Information Act constitute insuperable impediments to the functioning, and therefore the re-establishment of a PSAC. I don't agree with that view. I believe that the President could still have consultants and other people working with him within the White House who would be protected by the Presidential privilege. It might involve a showdown to prove it, but I believe that the President could structure an advisory group in such a way that closed meetings would be possible. It could be a Presidential Panel of Consultants or something similar, that would clearly be part of the President's staff. The members could be appointed as individual consultants. But it would only work if the President used it effectively.

I believe that President Carter could use a PSAC-like group for the remainder of his term to help revitalize his administration. He's being urged rather vigorously to do this by several groups, and I hope that he will consider doing so. Of course, one of the facts of life is that existing power groups within the government aren't particularly anxious to see PSAC re-created. The opposition, I'm sure, includes some members of the White House staff and of the National Security Council staff and of the Office of Management and Budget. And, of course, many people in the agencies would prefer not to have an independent judgment passed on what they're doing. But it seems to me it's an essential thing for the President. I don't think the President understood this or believed it at the start, because he and some of the people close to him felt that he had an advantage over previous Presidents because of his technical background. But the fact of the matter is that he doesn't have the time to spend studying an issue, concentrating on it. I think that's appreciated now.

When I was Science Adviser to President Kennedy, the PSAC was very supportive of him. It was certainly in no way a threat, nor did it dilute my authority as Science Adviser. Occasionally the PSAC members thought I acted without enough consultation with them. These were usually times when one had to act quickly. And there were other occasions when I acted in what you might call a political sense, where I made the judgment that there was an issue that was dominated by political considerations. For example, when I was a member of the Eisenhower Science Advisory Committee I wasn't particularly supportive of the manned lunar project. This was the consensus of the PSAC. We didn't think it was good science, though it obviously was a great adventure. And the President's Science Advisory Committee took a quite negative position, which I supported. In the Kennedy Administration, as Science Adviser, I argued that one shouldn't support it as science, but when the President decided he had to have an aggressive lunar program as a political matter, I supported his decision and I didn't offer PSAC an opportunity to argue against it with the President because I knew that he had decisive political pressures on him that would just result in a confrontation to no good purpose. Some members of PSAC did indeed resent the fact that they hadn't had an opportunity to argue with the President, but I knew that Kennedy was himself of two minds about the manned lunar program and had based his decision on an extensive examination of the domestic and international political problems. He finally decided that the US prestige was so coupled to it that he had no choice. It was not a scientifically-motivated decision.

People also had widely differing views of the science advisory mechanism's purpose and particularly the Science Adviser's role. They often regarded Killian and Kistiakowsky and myself as representatives of the scientific community. I never felt that that was my role. We functioned in that role when our perception told us it was in the national interest and when it was not inconsistent with our primary role as adviser to the President. But I never felt that I was there primarily as an advocate of science. I was often criticized—sometimes meanly—because I didn't always reflect what someone considered to be in the best interest of science if it was in conflict with what I considered to be in the President's best interest. But those conflicts were rare. I can remember several occasions when people said, "You're not representing science adequately," and their astonishment when I'd say, "That isn't my task." The fact of the matter is that we did support science and technology, as well as higher education generally, because of their importance on the national scene.

President Kennedy was pretty relaxed about what I did, perhaps because I'd had a long personal relationship with him. So, occasionally I'd do things which I thought were important to science with which he was not necessarily wholly in agreement. I used my judgment. As an example of this early in the Kennedy Administration, Jim Shannon wanted to support fertility research at NIH, something that was potentially politically embarrassing to the President. I told him that I was going to support Shannon, and the President said, "Fine. If we get into trouble, I'll blame it on you."

Another example involved presenting the Fermi Award to Robert Oppenheimer. Not everybody on the staff thought it was in Kennedy's interest for us to honor Oppenheimer with the award. But Kennedy thought it was, as did I, and he allowed it to be done. There were many other issues of this kind.

The Need to Restore PSAC

Returning to the need for restoring a President's Science Advisory Committee, it's true, as some people say, that Frank Press has been doing a good job without the help of one, using low-profile ad hoc panels and advisers, but he'd be able to do an even better job with a senior advisory group. He's gotten along very well with the White House staff and with OMB, and they probably don't want a PSAC, fearing it would intrude on their authority, so he may have a political problem now. But if PSAC were recreated, the various skeptics would accept and get along with it, and ultimately learn to work with it. Actually, President Truman's Budget Director and staff supported the original recommendation for its creation.

A restored PSAC would be severely inhibited if it were required to function in open meetings and meet all of the other requirements of the Federal Advisory Committee Act and the Freedom of Information Act. But even such a restricted PSAC would be better than none. The country is much the poorer for its absence, and so is the President. I would restore it even with the limitations, if they can't be eliminated. The President could attract members of outstanding stature, wisdom, and selflessness to serve on it. The prestige and sense of responsibility inherent in a Presidential appointment would insure this.

The Freedom of Information Act provides for exemption of National Security issues. And I think you could deal with problems like energy in open meetings. You would have to deal with them in a different way. You'd have to explore issues in a more general way and then the Science Adviser would have to synthesize conclusions from the general discussion. I've watched the public performance of the OTA Council and I think that on the whole we have not been inhibited. Of course, we were not dealing with classified material. We've dealt with a wide variety of things. There are usually a lot of reporters present and I find that we got used to them; we're not inhibited. So much goes by that the reporters don't know how to deal with it all. But it's also true that OTA is not as close to the seat of power as a PSAC would be, and I think an open PSAC would probably generate more press than an OTA. I think open meetings would impose a severe limitation, but would be an interesting thing to try.

Yes, on balance, if we cannot have a PSAC exempted from the Freedom of Information Act (except for classified subjects), we should have one even with those limitations. There would even be some offsetting advantages. One problem would be eliminated. PSAC got into serious trouble in the Nixon period because members got so frustrated they began to talk—some called it "leak"—outside the committee, and especially to the Congress. If meetings were open, that wouldn't be a problem. But, as I have already said, things would have to be treated in a much more general way and much more broadly. Individuals might find themselves constrained but my suspicion is that they'd get over it. On the other hand, I think it would be so much better to be a private institution—that is, with privileged discussions—that I'd try very hard to structure it that way.

The most important aspect of PSAC service was the feeling that you were helping the President of the United States and in fact that the President was listening. He didn't always agree, nobody expected him to. PSAC members became frustrated only when they no longer had access to the President, not when the President didn't accept what they said. I never felt that the President had to accept my advice. I tried to give him the best advice I could, but I always expected it to be one input of many.

Broader Range of Oversight

Turning to matters of scope, I think that the President's Science Adviser ought to have a broader range of oversight than he now does, so that he includes responsibilities for the field of national defense. The biggest set of problems the President still faces are related to the national defense issues. The original view in the Carter administration was that Harold Brown and the President knew each other so well and thought so similarly that there was no need for a Science Adviser to be involved. But the important point, I think, is that no matter what Harold Brown thinks, he has an obligation and a need to be responsive to his Department of Defense constituency.

The most important thing I used to do for the President, and, for that matter, for Bob McNamara as Secretary of Defense, was to broaden their range of options for decision-making; for example, I would sometimes take the position that something proposed—something that the Joint Chiefs or the Air Force wanted to do—was unnecessary (as was often the case), and this position made it possible for the President and Mr. McNamara to find a posture that was comfortable somewhere in between. The President can't do that today; the President doesn't have the time to arrive at an independent judgement. And so, no matter what he or the Secretary of Defense thinks they are constrained to accept the demands of the professionals by the lack of a reviewing process that's independent of the Pentagon, one that allows them the flexibility to come down between what the strongest military advocates want and what a group of independent advisers think is necessary for national security. Now he probably has to compromise between his own position and the professional establishment's position, without the buffering provided by an independent, respected, authoritative group. I think that's very costly, both in dollars and in policy restraints for the nation.

When I was Science Adviser, I attended meetings of the National Security Council. I was not an actual member (no Science Adviser has ever formally been a member, as I recall it) but I believe my letter of appointment essentially authorized me to participate in NSC meetings. I certainly participated as though I were a member of the NSC; I was not just a spectator. No one in the group ever pulled rank on anyone else. I wasn't always comfortable, though, because I was frequently in a minority position. But I was an accepted member of the NSC group, and I think that Dr. Kistiakowsky and Dr. Killian were also. Perhaps they may have felt a little less free to participate, I don't know; but Kennedy made it clear that I was to act as a member, and I did.

The optimal size of PSAC is a good question. In my time we always tried to hold the group to between 16 and 18, and I think that was probably a little large. But we wanted to have it representative enough. There are many pressures for growth. You want enough people in a given area so that you're not dependent on one person's views, yet not so many that the group becomes unmanageable.

Then there is the matter of scope. Initially PSAC concentrated on military and space issues, but it became involved when national health issues emerged as a major problem and when environmental issues came to the fore, and so on. It became necessary to change the composition of the group, and this in fact, I think, impaired PSAC's effectiveness. Earlier its membership came primarily from the physical sciences and engineering, with a concentration in military and space matters. I think you'd have difficulty with a group of only, say, eight or 10 people. That would not allow for adequate coverage. But anything bigger than 15 to 18 becomes unmanageable. If it gets too big, nobody feels responsible. Fifteen was not an unmanageable number, but it is close to the outer limit.

So, with the growing concern at the Presidential level with societal issues as well as national defense issues—with disarmament, the environment, energy, health care, international development, and others—PSAC could not have specialists in every field. It has been suggested that, therefore, it should be composed of generalists, or of specialists of such stature and experience that they can function as generalists.

PSAC might consist of an overall committee of generalists and a sub-organization of specialized standing committees—let's say, a biological or health-related area, a military group, an international group, one concerned with industrial technology, an environmental/energy group, and several more. Perhaps two or three members of the parent committee could participate in each of the panels, possibly with some rotation. The bulk of the standing committee people would be outstanding specialists, but not PSAC members.

So I think there are ways to deal with this organizational and personnel problem. In the past PSAC panels were almost always chaired by a committee member. That arrangement might be broadened so that there would be two or three or four PSAC members on the important national security panel. I wouldn't call it a military panel because the area of arms control and related issues should be included. The national security panel might have a sub-group that dealt with naval matters, another that dealt with arms control matters, and so on. I realize that this would involve an extra layer of people; nonetheless, I think that it should be considered.

There was always debate about whether the Science Adviser should also be chairman of the PSAC. It always seemed on balance to be a good idea to have one adviser reporting to the President rather than to have two, although people, particularly I. I. Rabi, made a strong case for an independent chairman of PSAC. If you had the structure outlined above, you could have two or three deputies assisting the Science Adviser, and the deputies could chair some of the subgroups. In fact, I had a deputy for military matters, a deputy for space matters, a deputy for basic science and a deputy for health issues. They generally did not chair the panels but rather helped them. Or you could bring in more prestigious part-time people and let them in fact be managers of these areas—the panel chairmen.

The Current Crises

If PSAC existed now, it would probably be concentrating on the energy problem, and the arms control situation. Those are the two major issues the President is deeply involved in at this time where he obviously needs all the help he can get. And there is the matter of health care. That is more of a management and fiscal problem than a scientific problem, but the science issues are there and cannot be untangled. There are always issues of Federal support of education and Federal support of specialized education, such as medical, science, and engineering. In my time, the Science Adviser tended to be the person who worried about higher education by default. He did so, mainly because there was no one else to do it. The President allowed me to do it. I even meddled in elementary and secondary education, but not to the same degree, and I did that on the basis of the need to do research in education and good education is prerequisite to a career in science and technology.

A question has been raised as to how the scientific and engineering communities, including those in the academic world and those in industry, can best provide inputs of knowledge and opinion to the Presidential level. The President has only a limited number of hours in the day. The practical channels for such inputs, in addition to the Department and Agency levels, are a reconstituted PSAC and the Science Adviser.

Advice to the Congress

Finally, I will comment on science advice for Congress. They need it, of course. The Office of Technology Assessment is one source. But that's not really what OTA was set up to do. Unfortunately, one of the big strains inherent in OTA is the degree to which it responds to Congressional need. There was a vacuum there, and here was a technically competent group and they began getting all kinds of requests for immediate help, and they provided a lot of it. The other Congressional groups, service groups, such as the Library of Congress and General Accounting Office, provide that kind of help too, but they're not staffed adequately to do it. The OTA's panel system, which was not unlike the PSAC panel system, did help Congressional committees and is helping them. But with regard to technical assistance for the Congress, I don't think the present system is adequate. The committees of the Congress tend to run their own consultative groups, and sometimes they work effectively and sometimes they don't. I believe that there is a serious problem in the Congress. And it's not easily fixed because the individual committees are almost completely autonomous. There's no integrating body for the committee structure. The committees tend to be jealous of their prerogatives, although many of them do use the OTA studies. The OTA does studies, generally on request, and to a small degree it provides an integrative mechanism that didn't exist before. I don't know how to deal with the Congressional need adequately.

OTA has been in trouble of one kind or another more or less continuously, partially because of the natural forces in the Congress, including the one I described already, namely, that there is no responsible authority. Each Congressman and Senator is highly independent. OTA also has to live with the House-Senate conflict, the Republican-Democratic split, liberal-conservative polarization; so that on almost every issue there tends to be disagreement and the director has to walk a tightrope in ten-dimensional space. OTA was in trouble too because people were suspicious of Senator Kennedy's motives. Kennedy really had no hidden motives except to make OTA work, but nobody ever believed that. Kennedy has now more or less withdrawn and that particular problem has disappeared; but there is no one as strong as Kennedy now to support it, and the overall position of OTA may be weakened as a consequence. I really don't know whether OTA will survive. It will be too bad if it doesn't, because it is important and in fact has been making a very useful contribution to the operation of the Congress. Some of its studies were really outstanding.

In speaking of the Congress, one should consider the dual role of the President's Science Adviser in reporting to the Congress, as he now does by statute, as well as to the President. There are hazards in this, but it's no worse, really, than those that confront the Director of the Office of Management and Budget or the head of a Department or Agency. One just has to insist on separating the statutory responsibilities and issues which can be discussed from those personal, privileged things which are done for the President which cannot be disclosed. Frank Press seems to be getting along well in this situation. There's a great deal of respect and understanding in the Congress for the problems of such an official if he handles himself well. However, the present Congress is much more obstreperous than the one I faced when I was Science Adviser. It seems to be much less disciplined, and some members think there should be no Presidential privilege.

In conclusion, I believe that the need for a Presidential Science Adviser is greater than ever, and growing, and that his responsibilities should be broadened; and that the President needs a President's Science Advisory Committee as well. The President should reestablish it, preferably free of the requirement of public meetings but better with that constraint than not at all.

The President's Need for Science Advice: Past and Future

Donald F. Hornig

Introduction

It is no longer necessary to argue that an understanding of science and the technology derived from science is essential to the wise conduct of national affairs. That is obvious from many issues today. Weapons of the future may or may not include laser or particle beams. Science may or may not make it possible to verify a SALT treaty by "national technical means." The viability of nuclear power and our willingness to use it depends, among other things, on our confidence in the means we choose to store nuclear wastes and our estimate of their long term safety. Our willingness to utilize coal as a primary energy source is conditioned by its probable impact on the environment and by the possibility of a runaway "greenhouse" effect brought about by the inevitable build-up of carbon dioxide in the atmosphere. Balancing a growing world population against a food supply which depends on fertilizers and pesticides derived from petroleum poses not only world problems but policy problems for the US. At a more mundane level, imagine the economic problems which would be induced in Central and South America by the development of a good, cheap synthetic coffee and the foreign policy questions that chemical achievement would raise.

Scientific judgements affect energy policy, defense policy, economic policy, health policy, transportation policy, and environmental policy. In fact, science and technology even touch the post office and urban affairs and, if there were reason to believe important new gold deposits could be located or much lower grade ores refined, I suppose that these could be factors in monetary policy.

As a consequence of these things, there are scientific and technical officers and organizations in every major department of the Federal government, except the Treasury, and most departments and agencies support research and development

Donald F. Hornig (b. 1920) was Science Adviser to President Johnson (1964-69) and chairman of the President's Science Advisory Committee (member since 1959). He is now Professor of Chemistry and Director of Interdisciplinary Programs in Health at the Harvard School of Public Health. During World War II he was a staff scientist at the Underwater Explosives Research Laboratory (Woods Hole) and at Los Alamos. He is a member of the National Academy of Sciences and of a number of industrial, academic, and governmental boards. He was President of Brown University (1970-76).

efforts on a substantial scale in areas relating to their mission. Currently over $25 billion, or 6% of the Federal budget, is expended on research and development. Despite the magnitude of this effort there has never been an overall science policy to guide it, nor has there been a systematic overall mechanism for coupling scientific analyses or judgements into policy making. However, many thoughtful observers think this is as it should be. Since research and development is carried out in relation to many national goals, it should not be thought of in isolation, and pluralism in its planning and conduct is a virtue.

In most countries the support of science and the development of a policy for research and scientific manpower is in the hands of a Ministry of Science and Education. In the British government there is also a Ministry of Technology. France has gone a step further with the creation of the post of Secretary of State for Scientific Research and Atomic and Space Questions plus a staff organization, the Délégation Générale à la Recherche Scientifique et Technique, which cooperates with the Commissariat Générale du Plan in developing overall policies and plans. In contrast, the US structure has developed pragmatically. The scientific and technical component of each agency has grown to meet felt needs, and, with the exception of the National Science Foundation, is defined in terms of the agencies' programs and goals.

The question is, if matters of defense policy are handled through the Secretary of Defense, the stimulation of the economy by the Secretary of Commerce and so on, what is the need for a staff office for science in either the White House office or the Executive Office of the President? Or if there is a need for such an office, how big should it be? How should it relate to the agencies or other staff offices such as OMB? What should be expected of it? How should it relate to the Congress and the public? All of these questions have been addressed in a growing literature as well as in the several years of hearings which preceded the passage of The National Science and Technology Policy, Organization and Priorities Act of 1976.[1]

What tends to be ignored in many abstract and idealized discussions of the science advisory process is the necessarily finite capacity of the President and his staff. Advice and pressure for action flow in continuously from many quarters, and a constant question is how to deal with the many demands. Most matters must therefore be handled outside the White House or Executive Office; only the Office of Management and Budget even attempts to keep up with the entire range of programs and activities. Consequently, any approach to science advice must be related to the interests, priorities and assimilative capacity of the President and those around him. The entire staff mechanism must be educated but its attention can only be gained in relation to what is or might be on its agenda for action. Perhaps what is required more than advice is a communication channel between the world of science and technology and that of politics. Because of these considerations it is interesting to contemplate the evolution of the role of the Science Adviser and his relation to the President and the presidency.

The Eisenhower Years

The office of Science Adviser and the President's Science Advisory Committee were established in the White House by President Eisenhower in the form approved, but not activated, by President Truman in 1951. The immediate reasons for doing so were straightforward. The successful Soviet launch of the first orbiting satellite, a satellite much larger than was even being contemplated in the US program, produced a tremendous public reaction, both in the US and abroad. The launch was seen both as an immediate threat, insofar as it demonstrated long range Soviet missile capability, and as a long-term challenge. American technological superiority had long been taken for granted, but overnight the discussion turned to whether the US was about to be surpassed, if that had not already occurred. Even the quality of American scientific research and higher education was called into question. The degree to which the press and the public reacted was totally unexpected and showed an unanticipated understanding of the relation between scientific and technological stature and world power.

The Creation of PSAC

Under these circumstances the President felt compelled to act, both to restore the US position and to demonstrate his own leadership. One of his moves was the appointment of Dr. James Killian, President of MIT, as Special Assistant for Science and Technology, and the creation of the President's Science Advisory Committee (PSAC). Thus began a unique experiment in modern government. The Science Adviser was appointed to serve the immediate needs of the President in areas which were of critical importance to him and to the country. Dr. Killian was invited to sit in on cabinet meetings and National Security Council meetings and became privy to the secrets of technical intelligence such as the U-2 reconnaissance aircraft. There was little theoretical debate over the functions of the Science Adviser other than disappointment in some quarters that the Special Assistant had no powers of command or action. He was not a missile czar. His role was simply to assist the President to do his job as political leader, as the head of an enormous bureaucracy and as commander-in-chief of the armed forces, by mobilizing the best scientific brain power in the country behind him.

The Eisenhower years have been well chronicled in Killian's Memoir[2] and Kistiakowsky's diary.[3] The science advisory system (the Special Assistant, PSAC and a variety of special panels) remained close to President Eisenhower and he referred to them as "my scientists." They oversaw the development of NASA and the space program. They studied missile characteristics and helped bring order out of the chaos attending the parallel development of missile programs in each of the armed services. They became involved in the development of satellite reconnaissance and of nuclear weapons. Perhaps more significantly for the future, they looked into nuclear testing and the possibilities of monitoring a comprehensive nuclear test ban. Given the importance of these problems, the high stakes of money and power involved in coping with them, and the difficulty the President had in obtaining dispassionate advice he could trust, it is not surprising that this dedicated group which set high standards of analysis and good judgment should have become an important part of the Presidency. The Science Adviser and the President's Science Advisory Committee helped shape policy on major matters in significant ways.

Even in the Eisenhower years the Science Adviser began to look beyond the immediate problems which first brought the office into existence. From the first, the health of basic research, particularly in universities, along with the improvement of advanced education in the sciences, was a central concern. Two of PSAC's early public documents, "Education for the Age of Science" and "Scientific Progress, the Universities, and the Federal Government," attempted to communicate an understanding of these matters to the public and the Congress.

One of my own first impressions of PSAC was its effort on the great cranberry crisis in 1959, when traces of a herbicide, aminotriazine, which had been shown to be carcinogenic in mice, were found on cranberries. Almost all of the crop was withdrawn from the market that Thanksgiving. The Science Adviser and PSAC urged a "rule of reason" and launched a study of pesticide use culminating in a public report, "The Use Of Pesticides," in 1963 which gave scientific support to the concerns expressed by Rachel Carson.

Attention was also given to strengthening the institutional base for science and technology and it was after a PSAC recommendation that President Eisenhower established the Federal Council for Science and Technology. Its members were the top ranking scientific and technical officers of the departments and agencies which had major research and development programs, and it was chaired by the Special Assistant. The idea was to create a body which could coordinate the Federal research and development program but more importantly could focus Federal scientific thinking on major problems. It was to be the internal equivalent of PSAC. It is significant that the plans for FCST were presented to the cabinet by a PSAC member, Dr. Emanuel Piore. Unfortunately, the FCST has not lived up to the high hopes initially held out for it.

The first three years represented a first stage of Presidential science advice which focused sharply on the Presidency and the major issues facing it. Its philosophical focus was clearly delineated by Killian at the outset. Its purpose was to put the best scientific brains at the service of the President and his associates, either to be sure that he had the best available facts and analyses at his disposal or to look ahead and call his attention to problems and opportunities which might not otherwise come to his attention. This led to a style in which a small group of dedicated people worked very intensely on a small number of major problems.

The Birth of OST

The Science Adviser to President Kennedy, Jerome Wiesner, had a somewhat different conception of his role. To be sure, he carried on the functions which had been initiated by Killian. As an electrical engineer and communications expert, he helped to improve White House communications. He continued to oversee the space program via PSAC Space Science and Booster panels. However, he was not a partner, apparently, to the Presidential decision to undertake the Apollo program to land a man on the moon within the decade. He worked closely with Dr. Harold Brown, then Director of Defense Research and Engineering, on defense problems and had a key role in the cancellation of the Skybolt air launched strategic missile. He and PSAC worked very hard on the nuclear test ban and can claim considerable credit for the Atmospheric Test Ban Treaty. Above all, Wiesner was close to President Kennedy and aside from formal memos and reports was able to convey informally some of the spirit of simultaneous rationality and dreams for the future which characterize science.

The Role of the Adviser

But aside from efforts focused on the Presidency, Wiesner saw the Science Adviser as a focus for what is now called science policy. He saw the office as the science center for the entire government, a source of scientific and technical input and leadership wherever or whenever it was appropriate. This led him, together with PSAC, to initiate studies of scientific manpower, energy research and development, scientific information, the impact of new technologies on the life sciences, and strengthening the behavioral sciences. He helped to establish the office of Assistant Secretary for Science and Technology in the Department of Commerce and collaborated with the Commerce Technical Advisory Board in studies of innovation in civilian technology such as housing. In another direction he supported the Office of International Scientific Affairs in the Department of State and the role of our own and foreign science attachés. Above all, with few exceptions the office of the Science Adviser became the White House contact point for the entire governmental science apparatus. However, both James Webb, Administrator of NASA, and Dr. Glenn Seaborg, Chairman of the AEC, maintained independent channels to the White House and health research remained largely outside the orbit of the Special Assistant and PSAC.

It is not the place of this article to review the history of science advice or to describe events and activities in any detail.[4] What is important, because it reflects on our judgment of future possibilities, is that the initial rather clear concept of the offices evolved into something much more complex. Some of the new functions were either incompatible with or got in the way of the old ones. Others basically changed the nature of the office.

Such a change came with the institutionalization of the Office of Science and Technology in the Reorganization Plan No. 2 of 1962. As a result of the expansion of interests and activities the Special Assistant required a larger staff than the President was willing or able to maintain on the White House budget. The solution was to create the Office of Science and Technology in the Executive Office of the President with its own budget, its own staff and a director who might or might not be the Special Assistant to the President.

PSAC and Dr. Wiesner were aware of the implications of this change from being part of the White House staff to being part of the Executive Office, and were troubled by them and discussed them at length. Having a statutory base provided stability since the office did not potentially terminate at the end of each President's term, although in 1973 the change did not protect the office from being effectively termimated by another Reorganization Plan. But the change also generated a new climate within the office and a new relation to the White House and the government.

Before 1962 the Special Assistant was a confidential adviser to the President. What was expected of him was determined by the President and he reported only to the President. He was not expected to testify before the Congress and never did. After 1962, that was changed. Rather than being part of the White House Staff, OST was an institution like the Office of Management and Budget. The Director's nomination was subject to confirmation by the Senate. He was charged with advising the President in respect to:

- "major policies, plans and programs of science and technology of the various agencies of the Federal government, giving appropriate emphasis to the relationship of science and technology to national security and foreign policy, and measures for furthering science and technology in the nation;
- assessment of selected scientific and technical developments and programs in relation to their impact on national policies;
- review, integration and coordination of major Federal activities in science and technology, giving due consideration to the effects of such activities on non-Federal resources and institutions;
- assuring that good and close relations exist with the nation's scientific and engineering communities so as to further in every appropriate way their participation in strengthening science and technology in the United States and the Free World;
- such other matters consonant with law as may be assigned by the President to the Office."

A Life of Its Own

Finally, although not stated in the Act, he would have to defend his activities and his budget before appropriations committees of the Congress. Moreover, he was expected to testify on numerous scientific and technological matters and to explain the administration position when science or technology was involved.

After 1962 OST became an institution with a life of its own. Its staff grew, but not commensurate with the bewildering array of its possible responsibilities such as the review, assessment and coordination of billions of dollars of research and development. Impossible expectations were generated and that situation has been augmented in the National Science and Technology Policy, Organization and Priorities Act of 1976. That Act significantly expands the preceding list of responsibilities and, in addition, calls for the preparation each year of a five-year outlook and the submission of an annual Science and Technology Report to the Congress.

As we look at the present and the future, the question is what the experience thus far teaches about how the science office and the Science Adviser can be most effective. Certainly, in comparison with other governments, the unique feature of U.S. structure has been direct personal access to the President. Presumably the special things the Science Adviser can do should be related to that, but how? Should his office be the analytical branch of government and concentrate on objective reports? Should it concentrate on long range forecasting? Should it be a scientific OMB to assess, evaluate and coordinate Federal research and development budgets and programs? Should it help to inform the public and the Congress on critical issues? Or should it focus on the immediate problems facing the Administration? All of the preceding experience suggests that the answers depend on the President and his relation to the Science Adviser as well as the historical situation and the other resources open to the President.

Johnson and the Great Society

It may be of some interest to examine the Johnson period, 1964-68, which encompasses my own time in the White House. These were years in which the Office of Science and Technology reached a certain maturity, at least in terms of size of staff and number of consultants. A new President, Lyndon B. Johnson—entirely different in style and outlook from his predecessor—had been thrust into the Presidency without preparation. I had no previous aquaintance with President Johnson, having been asked to serve as Special Assistant by President Kennedy just a week before his assassination (and asked again two months later by President Johnson). It was an entirely new situation. I had little feeling for the strong, dominant personality who saw everything in political terms, and President Johnson had little feeling for academicians and scientists, although he always held them in great respect.

As a consequence, the relationship to the President can be described as friendly but arms-length. My wife and I were invited to White House dinners and social occasions and we occasionally utilized the Presidential box at the symphony, but were never a part of, or even near to, the inner circle. During the first two years, particularly, I saw him frequently in the Oval Office and never dealt with him through an intermediary, but our discussions were usually fairly formal and in the nature of reports and questions. Only a few times did he call me on the telephone to raise a problem on which he needed immediate help. The President used the talents of PSAC, the OST staff, and the Science Adviser and was happy to hear from them, but one never had the feeling that he depended on them to shape his views. I should add that it was my impression that most of the White House staff, with the exception of McGeorge Bundy, Horace Busby, Douglas Cater, and sometimes Bill Moyers, weren't sure why there was a Science Adviser or where OST should fit in their scheme of things. Nonetheless, when I thought, after about two years, that it was time to leave, he was very firm and persuasive in arguing that I must stay on. Among other things he argued that if I left he would make OST a part of the Bureau of the Budget. Whether he really meant it I do not know, but I took it very seriously.

During the Johnson years there were about 20 professionals on the OST staff, each responsible for one or more areas of activity. Most of them were responsible for PSAC or Federal Council for Science and Technology (FCST) panels, and two or three of them usually chaired FCST panels. PSAC continued to play a major role in foreseeing and defining problems and addressing its expertise, judgment and experience to their study and analysis. However, the trend toward dealing with a much broader range of problems put more and more of the work in the hands of the panels. Moreover, as the backgrounds of PSAC members became more diverse, the internal coherence and ability to achieve a broad consensus gradually broke down. In the latter years the problems of PSAC were accentuated by opposition to the Anti Ballistic Missile (ABM) program and the Vietnam War and tension over the permissible public posture of PSAC members. It should be added that although the problems never erupted as they did later under President Nixon, the White House sensed the opposition, and PSAC's position of trust eroded. To bring in the range of expertise required, OST employed some two or three hundred consultants, usually on an ad hoc basis for a limited time, some as members of PSAC panels and others participating in a variety of working groups.

An Enormous Range

The activities of OST and PSAC spanned an enormous range. They have been catalogued in some detail in the history of OST in the Johnson years which is on deposit in the Johnson library, but, unfortunately, much of the material is still classified.[5]

The activities, broadly speaking, fell into two basic categories which tended to compete with each other:

a) Assisting the President and the administration to take proper account of scientific or technological factors which affect a large variety of national policies concerning national security, the economy, international affairs, and so on.

b) Working with the Bureau of the Budget, and the agencies in managing the Research and Development program and such other programs as are affected by scientific and technical considerations. These activities involved the review and assessment of programs, the coordination of programs among the various agencies involved and, above all, the establishment of links both with the BOB and the agencies. Their end product was usually the addition or subtraction of funds.

It would be tempting to design a bureaucracy to address these problems systematically. However, in my view the urgent need is not for another broad oversight organization. It is for one which will meet the needs of the President and by doing so maintain access to him. This includes more than providing advice when asked, because a busy President receives advice from so many quarters that he rarely has time to ask. Instead, by being in touch with the President and those around him, the Science Adviser must be aware of the issues which are arising and present the scientific facts and judgements which might otherwise be ignored or misunderstood. Furthermore, he must look ahead and make the President aware of upcoming opportunities and problems. Lastly, and possibly most importantly, he must help a President who usually has little feeling for science achieve a sense of its relation to the future and of the importance of maintaining a strong national base, particularly in fundamental research.

Interactions with the President are complex and take many forms. For example, as part of the development of his legislative program in 1964, a PSAC task force under John Tukey developed recommendations which found their way into legislation, Presidential messages and Executive Orders. PSAC also carried out a series of long-range studies which were made public and to which the President contributed statements, such as:

> Restoring the Quality of the Environment, 1965; The Space Program in the Post-Apollo Period, 1965; Effective Use of the Sea, 1966; The World Food Study, 1967; and Computers in Education, 1967.

The contributions of other PSAC panels, such as: Anti-Submarine Warfare, Naval Warfare, Ground Warfare, Strategic Nuclear Warfare, Military Aircraft, and Biological and Chemical Warfare, and the work of PSAC itself on Vietnam and AMB never produced formal reports and usually did not go to the President at all. They were the subject of discussions and memos addressed to such people as the Secretary of Defense or the Chairman of the Joint Chiefs of Staff. Most particularly they were dealt with at biweekly lunches with the Secretary of Defense, the Undersecretary, and the Director of Defense Research and Engineering.

In the course of five years at least 113 memoranda on an extraordinary variety of topics went to the President. Many of them were paralleled or followed by discussions. Only a few of them responded to particular personal interests such as the possibilities offered by water desalting, especially by gigantic nuclear plants; the technological gap between Europe and the United States; and the Northeast power blackout and electric power reliability.

Many others sought to inform him or interest him in issues which might involve him or in which he ought to become involved, such as: jet aircraft noise at airports; the SST and sonic boom; environmental quality; strengthening academic science and basic research; telecommunications organization in the U.S. government; technology for urban problems; high energy physics and the large accelerator; and science and international problems.

Substantial Impact

An analysis carried out for the History[5] by the OST staff shows that a surprising number of the efforts had a substantial impact which can be traced in budgets, programs and messages.[6]

Communications from the President were of many sorts, and he assigned tasks in many ways, sometimes unconventional. For example, in early 1965 he telephoned me (in my kitchen) to complain of the preparations for the visit the next morning by the Prime Minister of Japan. He wanted a good idea to present. After all night telephone discussions across the country I presented him with a proposal while we waited for the Prime Minister to arrive. One hour later the US-Japan Medical Program was born. Perhaps because of that success I received a similar call the afternoon before his meeting with the President of Korea which led to the dispatch of a delegation to Korea and the eventual founding of the Korea Institute of Science and Technology. It became clear to me that being benevolent and progressive, science is a marvelous lubricant in international affairs.

More commonly assignments were made by memo or in messages. In 1965, he designated me by memo to assist him in coordinating water resources research and in 1965 he requested me by letter to develop a long-range water resources research program. These requests resulted in a report, ''A Ten-Year Program of Federal Water Resources Research.'' At the signing ceremony for the 1965 Water Resources Planning Act, he directed me to coordinate all Federal efforts on desalting and in a memo to the Heads of Departments and Agencies, ''Strengthening Academic Capability for Science Throughout the Nation,'' I was directed to supply periodic reports on our progress. Some requests, such as that in his message of February 7,

1968, to Congress, were more unusual, in that case for the Director of OST to study new nonlethal chemicals and weapons which would not cause permanent injury for use in crime control.

A final item should be mentioned although we failed completely at it. At one time the President was persuaded that, with so much going on in the world of science he should be able to announce and discuss important and interesting developments and maintain the attention of the American people. As he put it "for $18 billion per year there ought to be something to say at least once a week." I hope future Science Advisers succeed better than I did.

In sum, under President Johnson, OST and the Science Adviser dealt with a bewildering variety of matters. It is hard to ascertain any central themes. Geared to the political decision process, its priorities and agenda reflected the concerns of the President, the Executive Office, the Congress and the people. At the same time, it expanded its efforts to identify emerging problem areas and new opportunities arising from scientific and technological areas.

While a central feature of the White House science structure has been its ability to mobilize some of the most qualified members of the scientific and engineering communities to serve as consultants on an array of panels, the existence of a PSAC to act as intellectual glue and to provide a body which under ideal circumstances regarded itself as part of the President's staff was indispensable. However, the proliferation of areas of interest made PSAC less coherent and moved it further away from direct contribution to the concerns of the President.

In terms of evolution of the White House science structure during the Johnson Administration, the greatest change that occurred was the growing influence and importance of the OST staff. The staff was able to infuse technical advice into the framing of government policies and decisions (a) through the development of personal relations with the White House and Executive Office Staffs and with policy-making officials in the Executive departments and agencies; (b) through participation in high level, interagency task forces and committees; and (c) through continuing contact with first rate consultant groups. However, while this development added to the influence of OST, it also diluted it. Individual staff members frequently found greater satisfaction in personal impacts within their areas of expertise than in contributing to shaping major issues. The temptation was great to move from advice to direction.

During the same period the hope that the Federal Council for Science and Technology would evolve as a decision-making mechanism declined, and it has served chiefly as a means for administrative coordination. In part this can be attributed to OST emphasis on pushing decisions back to the responsible agencies and in part to the increasing reliance on staff and lead agencies rather than interagency committees for executive leadership and action. Whatever the reason, the FCST has not been effective in any administration except as a means of communication.

Later Years

The pattern of the Johnson years seems reasonably typical. The relations with the President were not close but the interaction was adequate and he continued to involve the director of OST until the poison of Vietnam introduced questions of loyalty and trust. Even then it was possible to discuss such questions as the effectiveness of bombing in North Vietnam, but all in all it was hard to attract the President's attention in the last year.

The basic pattern of activity described here was continued under Nixon but difficulties developed from the start. The Nixon staff early inserted itself between the Science Adviser and the President. Political tests and criteria were introduced, something which never happened under previous Presidents. PSAC members spoke out publicly and testified before Congressional committees in opposition to the administration. A process of gradual estrangement ended in the abolition of the Office of Science and Technology in the Executive Office of the President. The lesson to be learned is that, however well the task is performed formally, the office and the Science Adviser derive their utility from their relation to the President and his other advisers.

The Future

To me, this is the main lesson for the future. The Office of Science and Technology Policy and its director (and PSAC, if in some form it is re-established, as it should be) must focus on those issues which are appropriate to the Executive office of the President. There are no rules by which those items can be defined; some are big and some are small. Some relate to current crises and others are preparation for the future. If that is kept in mind, and the focus is on not just the President, but the Presidency, a large number of studies, analyses, projections and forecasts are in order, but they should be in support of main themes. The staff, therefore, must be organized to support that objective.

The relation of the Science Adviser to the President is very important. It is not to be expected that many will be personal friends but the Science Adviser must be understood, respected and trusted, both by the President and by his other advisers.

One final point should be made. The problem of the Science Adviser in bringing scientific and technological understanding to bear on Federal policies is replicated throughout the government. One of the major tasks of the Office of Science and Technology Policy is to support and encourage the scientific voices and to help ensure that they are heard, not only in their own departments and agencies but, when appropriate, in the highest councils of government.

References

1. "A Legislative History of the National Science and Technology Policy, Organization and Priorities Act of 1976." US Senate, Committee on Commerce, Science and Transportation; Committee on Human Resources (April 1977).
2. James R. Killian, Jr., "Sputnik, Scientists and Eisenhower: A Memoir of the First Special Assistant for Science and Technology" (Cambridge: MIT Press, 1977).
3. George B. Kistiakowsky, "A Scientist at the White House: The Private Diary of President Eisenhower's Special Assistant for Science and Technology" (Cambridge: Harvard University Press, 1976).
4. See, for example, David Z. Beckler, "The Precarious Life of Science in the White House," *Daedalus*, Summer 1974, *103*, pp. 115-134.

5. "The Office of Science and Technology During the Administration of President Lyndon B. Johnson—November 1963-January 1969." Vol. II, Documentary Supplement. In the Johnson Library, University of Texas, Austin, Texas.

6. See Annex B, Ref. 5, "Policy and Program Changes Influenced by OST, PSAC and FCST (1964-68).

Current State of White House Science Advising

Edward E. David, Jr.

There is a surprising unease in Washington about the current state of White House science advising. One source of this feeling is the Congress and, in particular, some members of the House Science and Technology Committee and the Senate Subcommittee on Science, Technology and Space. Their feeling is that the White House Science Office is not fulfilling the role laid out for it in its enabling legislation. That legislation was passed in 1976 and was modified significantly by a reorganization plan in 1977 submitted by the Carter Administration. The changes removed some of the broader functions envisioned by the Congress. The statutory reports required by the legislation have been prepared by the National Science Foundation rather than the White House, and have been disappointing. Also, the size and structure of the White House Science Office have been eroded.

Beyond the Congress, some Washington observers have joined the chorus of criticism. There have been accolades for Frank Press (*National-Journal*, 1/6/79), but there is the feeling that somehow the larger contributions of science and technology are not falling into place in the Administration's programs and policies. President Carter's Congressional Message on Science and Technology gave lip service to the importance of scientific activities on many fronts, but their effective utilization in programs and policies leaves something to be desired, according to these observers. The complaints seem to echo the ''lack of leadership'' charge which has been leveled at the President himself. In this time when science and technology are being looked to as major resources, it is important to evaluate such comments and to propose improvements should they turn out to have substance. The remainder of this article is focused on such an analysis and some suggestions for improvement.

The history of White House science advice indicates that technology was more often the subject than science itself. This is particularly true if a broad definition of

Edward E. David, Jr. (b. 1925), was Science Adviser to President Nixon from 1970 to 1973. Now President of Exxon Research & Engineering Company, he was formerly Executive Vice President of Gould Inc. and Executive Director of Research, Communication Principles Division, of Bell Telephone Laboratories. A member of the National Academy of Sciences and of the National Academy of Engineering, he is Chairman of the Board of the American Association for the Advancement of Science and of the Aerospace Corporation and a member of the Executive Committee of the M.I.T. Corporation.

technology is accepted. This broader definition would include much more than merely engineering hardware; for example, computer software, system design techniques, and modeling to aid policy formulations would all be included. However, even without these additions, the White House Science Office has dealt heavily with technologies associated with military hardware, energy production, transportation vehicles, health care apparatus, and the like. It is, of course, true that the White House scientists have also recommended science-related activities, for example, programs of high energy physics research and increased computer education for the next generation of scientists. However, technology has been a major focus for the activity from its inception.

The Washington critics believe that the White House office should be doing more to infuse technology in the broadest sense into the operations of the White House. Science and technology particularly should be aimed at producing major benefits for the nation. These observers have been watching for major technical strands in such policies as energy, health care, economic growth, inflation control, and so on. They believe that what is needed is a rethinking of science and technology policy in the framework of making them responsive to national problems. Other observers believe that the White House Science Office should be looking for ways of turning technological development toward the public interest, so-called "appropriate technology" for example. Ways should be found to encourage public participation in governing technology development and in determining the science programs which tend to generate new technologies. Science and technical considerations should become more influential in national decision-making. These views do not exhaust the desires of the observers but they typify the attitudes. We may now ask about the realism of such criticisms, and particularly if the science apparatus can become the "conscience of the White House."

A Systems Approach

It has become a shibboleth that the formulation of national policies and programs which involve science and technology call for a systems approach in which economics, social factors, and politics are all interwoven. The science office is not in a good position to deal with these broader policy issues, even though they may involve some elements within the office's competence. The dynamics of the White House staff and its operations are hardly susceptible to control by the White House Science Office since the former are politically animated. Politics is seldom compatible with science and technology, and scientists and engineers are not adept at the rough and tumble of affairs. Adequate overall programs in the areas of great national concern depend increasingly on swings in the social milieu. Again, the science office is not in a good position to anticipate such swings. Even economic predictions fall to the Council of Economic Advisers. The Scientific Adviser and his staff should not be expected to lead in the formulation of national policy except where science and technology are paramount.

This limitation was very clearly seen in the early days of White House science activities. In the fall of 1970, I met with all of the former Science Advisers and some other long time contributors to White House activities. The consensus of the group included two important points. First, it was essential for the Science Adviser to help the President. That means supporting his programs, thinking of new ways of accomplishing his objectives, and bringing science and technology to serve those objectives. Secondly, the office should concentrate as much as possible on the physical sciences and engineering. These were the areas in which the office had been recognizably successful in the past and had made major contributions to the national welfare. The two cases in point were the defense programs of the 1960s and space exploration. The group felt uncomfortable with broadening science activities beyond these areas and into the very complex, politically volatile issues which are now being proposed for involvement by the office.

More particularly, the advisers believed that the social sciences could lead the office into political conflicts where considerations other than scientific were controlling. The subject of racial balance in education and its effects on student achievement is an example, though the office did not address this topic specifically. Even involvement in health and biomedical research policy was viewed as risky because of its avid lay following, both in and out of the Congress. It was with regard to biomedical research that we first heard (in President Lyndon Johnson's day) the opinion that administration and planning of research were too important to be left to scientists.

If one accepts such views, does it mean that science advice is restricted to narrow scientific and research fields? One possibility beyond this is that science advice should become advocacy oriented. There are indeed new style scientists and engineers in the land who are described as single-issue oriented. Might the next Science Adviser be such a person? Under these circumstances, the science office would assume a role similar to that of the White House Council of Environmental Quality and its advocacy chairman, J. Gustave Speth. Another possibility is a person following the "technological imperative" tradition. It may be that the science office is destined for such a future. However, the political dynamics indicate the office would not likely have a long tenure under such conditions.

Advocacies for Universities

As a matter of fact, the White House Science Office has had a reputation in the past of playing an advocacy role for the research universities. This reputation was particularly strong during the first term of the Nixon Administration but antedated that time as well. Regardless of the accuracy of this reputation, it was one of the principal factors in the abolishment of the old Office of Science and Technology. It was one of, if not the, principal reason for the initial doubts of the Carter Reorganization Task Force which was tempted to recommend abolishment of the office just after it had been reestablished during the Ford Administration. From a purely operational viewpoint, offices with an advocacy role and an outside constituency to serve can exist within the White House structure only as long as the cause is popular as perceived by the political arm of the White House. So, while it is conceivable that the science office could adopt advocacy for various causes, that role would not be stable for the long term. In my view, the Science Advisor's post must be filled by a statesman rather than an advocate, if it is to remain viable.

If we finesse the advocacy role as impractical for the longer term, we can come to some other possible roles for the White House Office. One of these is to provide extensive scientific information to the Congress, the public, and of course to the White House staff and the President. However, to political activists in the Administration, such activities seem peripheral and assume the appearance of "busy work." This is certainly one reason why the one- and five-year reports mandated in the OTSP legislation have been downgraded among the office's concerns. It is this rationale which has led me over the years to believe that the only stable role for the office is that of serving the President's interests directly. Serving the President is clearly incompatible with a responsibility to serve the Congress and the scientific community at the same time. Yet this multiple constituency is inherent in the enabling OSTP legislation. This conflict can be managed, however, through careful attention to the diverse interests of the three constituencies, giving primacy to the President and his staff. The current Science Adviser has carried off this balancing act well.

Despite this definition of Science Adviser-OSTP function and the feasibility of performing the function, there still appears to be among observers in Washington an unease about the situation. This may have to do more with the substance of the advice and reports emerging from the office than from the methodology and mechanisms employed. It has been said, for example, that the office should be looking for ways to *reduce* expenditures on R&D rather than increasing them, perhaps by shifting more of the R&D burden to industry. Then, too, the office's contributions to the energy, defense, health care, and educational programs have been oblique, if not non-existent. The "turf protection" game has been a barrier. These are extreme views, but observers on the scene in Washington are mouthing them.

Basic Contradiction

All of this tends to say that it is unlikely that the Science Adviser and the OSTP can play the broader role that is envisioned by the critics. The Congressional dissatisfaction with the office illustrates the basic contradiction. Perhaps a more realistic role is for the office to provide the basic underpinnings for imaginative step-out policies which address some of the major problems of our times. In this I am looking backward at some of the achievements of past Science Advisers; for example before the missile crisis of the 1960s, the Gaither panel, a precursor of the present White House science advice mechanism, predicted the strategic significance of ICBMs. It was about at that time too that the Seaborg* Report showing the necessary coupling between research and graduate education appeared and animated Federal research programs in the universities for a number of years before the Mansfield Amendment crippled them. The primacy of quality in research was clearly established by a PSAC study under E. R. Piore and W. O. Baker entitled "Strengthening American Science, 1960." The study creating NASA and the space exploration policies that have served well through the ups and downs of funding was chaired by E. M. Purcell. Indeed, these examples typify the role envisioned by the Killian Committee in 1974 when they studied the situation in

---

* "Scientific Progress, The Universities and The Federal Government," PSAC, 1960.

depth. This experienced, distinguished group produced a report** which bears reading today as we re-examine the functions and role of the White House office.

If, indeed, there are elements missing from the current White House science advice performance, it is in this function of providing paths to the future based upon new technological possibilities and basic relationships in the research and development process. It is indeed unrealistic to expect this kind of performance from the White House OSTP staff itself. Traditionally this staff has provided the framework and the coherence through which such statesmanlike contributions could be made. There is no reason to believe that the current OSTP staff is any less capable of this function than those who have preceded it. Indeed, the missing piece appears to be the broad and deep contributions by dedicated members of the scientific and engineering communities willing to spend time and effort on behalf of the Administration. In past times, this effort was supplied by the President's Scientific Advisory Committee and its panels.

The current trend is toward an insular situation. Since the reestablishment of the White House Science Office, it has assumed an inward look in conformance with the attitudes of the White House Staff generally. Much of the planning and committee work formerly done by distinguished contributors from outside the government is now performed by interagency task forces. For example, in the Domestic Policy Review on innovation, recommendations by outside task forces were filtered and re-cast by an interagency task force on whose recommendations the final actions were based. It is not that members of interagency task forces are less capable than people in general. It is rather that such task forces inevitably are bound by the bureaucratic interests of their agencies. Their results tend to assume a bureaucratic least common denominator. Contributions from outside the government, through balanced committees, can produce more novel and imaginative approaches to problem-solving.

It is interesting to contrast the Federal science task force which visited Peking in 1978 with the contingent from the American Academy for the Advancement of Science which visited later. The Federal group consisted entirely of government appointees and though many of them are outstanding scientists in their own right, they represent a much narrower slice of the disciplinary spectrum than the AAAS contingent. The latter included representatives from a very wide variety of disciplines such as sociology, industrial technology, physical and life sciences, library sciences, economics, and included several distinguished government people as well. These people were all elected members of the AAAS Board and so had standing in the broad science community. The reports that were published and the speeches that were given as a result of this trip showed a depth of penetration into the Chinese situation which would have been difficult to achieve with a different, narrower delegation.

The inward-looking attitude of the Administration and consequently the science office is leading to a serious erosion of that office's influence on broad policy matters. We see other elements taking over the leadership which was previously exercised by that office. The National Academy of Sciences, the AAAS, the Office

---

** Report of ad hoc Committee on Science and Technology, "Science and Technology in Presidential Policymaking," National Academy of Sciences, Washington, D. C., June 1974.

of Technology Assessment, the Industrial Research Institute, the American Chemical Society, and many others are assuming leadership in science and technology matters which the White House Science Office, through its previous mechanisms, once dominated. Indeed, we may very well be witnessing the demise of the White House Science Office in its traditional form and in a way that is more fundamental and serious than its clear abolishment in 1973. However, the Office can regain its proper function if it reverts to its traditional mechanisms, emphasizing excellence and scientific integrity.

Earlier I pointed out that I have consistently taken the position that the proper role of the White House Science Office and the Science Adviser is to aid the President and not to represent the science and engineering constituency. I still believe strongly that that is true, and, in addition, that the science office cannot be a creature of the administration and Congress simultaneously. It must owe its principal allegiance to the former and serve the latter within that limitation. To this clear charter, I would add a corollary.

The Total Wisdom of S&T

In serving the administration, a principal technique would be to bring the total wisdom of the scientific and engineering community to bear for purposes where science and technology can make a paramount contribution. The original Science Adviser arrangement, together with the President's Science Advisory Committee and its panels, was aimed exactly in this direction. The voices of people of accomplishment and experience are needed at the White House level and with access to the President and his principal staff. At the present time this function appears to be largely missing. Yet, it is historically an effective and well worked mechanism.

A particularly important case in which such distinguished voices were influential was through the President's Foreign Intelligence Advisory Board. This group, too, has been abolished. PFIAB was an important balance wheel in the intelligence picture and aided the White House measurably in bringing the intelligence community into the service of the President using the most advanced technological capabilities. PFIAB deserves substantial credit for bringing into being the advanced satellite sensors which are the mainstays of intelligence collection and treaty verification today. This feat was done with the cooperation of the White House Science Office and special panels which were set up to monitor and advise on what was then the most exotic technology.

These sorts of arrangements should be re-established, for they can provide a reservoir of wisdom on behalf of the national good working through the very top levels of government. The 1976 Act establishing OSTP contained provision for such a mechanism if it were invoked by the Administration. The time is auspicious for such a restoration.*** It was through such mechanisms that the great science-based accomplishments of the past two decades have come about. If leadership in

---

*** The issue of confidentiality has been raised in view of the Freedom of Information Act, the Ethics in Government Act, and the Federal Advisory Committee Act. Where confidentiality is required, proceedings can be protected and contributors can avoid conflicts of interest, according to legal advice, if care is exercised.

integrating science and technology policy with national and international goals is wanted by the White House, it can be found among the people of accomplishment and promise in the technical community. Today old, worn proposals for approaching the nation's needs are all too frequent. New fresh approaches are possible, but they will not emerge unless the principal role for the Science Adviser and the science office becomes rallying the wise, both young and old, to the cause. In my view, this is the aspect which is missing on the present scene.

Science Advice—Out Of And Back Into The White House

H. Guyford Stever

On August 13, 1974, five days after he had been sworn in as President of the United States, Gerald R. Ford spoke to me at a reception at the White House about the science-advising job for the Presidency. He said that he would like to get together with me soon, to talk over his ideas about science-advising, and he asked me to arrange through his staff in the White House a meeting in the near future. Though most proponents of the return of the Science Adviser to the White House staff had given up any hope of having it happen during President Nixon's second term, this hope returned as it became clear that Mr. Nixon would not last a full four years, and there was talk that the return of the Science Adviser to the White House could be accomplished in the period of Mr. Nixon's successor. Mr. Ford's move to set up a meeting on the subject dispelled any doubt in my mind that he wanted to move, although the details of his ideas were still not clear.

Although I previously had spoken to then-Vice President Ford about science budgets and science programs, we had not talked much about his attitude toward having the Science Adviser in the White House, simply because, as Vice President, he was committed to supporting the programs of President Nixon, and I, in turn, had been assigned by President Nixon to make the science-advising system work with the roles of Science Adviser and Director of the National Science Foundation combined. I did know, of course, that several groups of interested scientists and engineers had spoken to then-Vice President Ford and that they as representatives of the scientific and engineering communities outside of the government had in fact come away from the meeting believing that he was quite receptive to the idea of the return of the science-advising mechanism to the White House.

H. Guyford Stever (b. 1916) was Science Adviser to President Ford and the first Director of the White House Office of Science and Technology Policy established in 1976. Previously he was Director of the National Science Foundation (1972-76); president of Carnegie-Mellon University (1965-72); and Professor of Aeronautics and Astronautics and head of two engineering departments at the Massachusetts Institute of Technology. He served in the OSRD throughout World War II and was Chief Scientist of the U.S. Air Force (1955-6). He is a member of the National Academies of Sciences and of Engineering and has been a member of numerous governmental boards and panels.

Although it was clear from the beginning that President Ford wanted a change, it took longer than expected. Most important, President Ford's primary job was to restore the morale of the people of the country following Watergate and the resignation of President Nixon, and to tackle the severe economic problems which the country faced. In the last six months of the Watergate era, the Gross National Product growth had fallen and inflation had taken hold. Stagflation was the buzz-word of the time.

President Ford held a White House conference on the economy in September 1974 and the Science Adviser, still operating from the base in the National Science Foundation, with the Science and Technology Policy Office and the Energy Policy Office as the staff offices, was invited to participate. Since the primary aim of the conference was to get the economy changed round in a relatively short period, our contribution was not particularly central to the subject at hand. We did remind everyone that the contribution of science and technology is primarily on a long-term basis and we reminded ourselves that our science and technology policies ought to be aimed in that direction—to keep up, maintain, and strengthen our capabilities in those fields. (As an aside, those of us who had participated in the marshaling of the strength of science and technology in World War II and those who had participated in the rapid assembling of our strength in space did know that strong policies and programs can get very great results from the science and technology community in emergencies and on a time scale of a few years; but a time scale of a few months is too short, except for isolated or relatively small issues.)

The White House conference did serve its purpose for science and technology policy though for, at the reception following the conference, President Ford again brought up the subject of our getting together to discuss what should be done, and this time he directed a member of his immediate appointments staff to set up a meeting within two weeks. Although the two weeks stretched from September 27, the date of the reception, to October 17, the date of the actual meeting, we had finally accomplished our aim of starting to talk.

From my standpoint, this was a very satisfactory meeting. President Ford showed several characteristics in all the meetings that I had with him. In the first place he listened and he asked questions which were knowledgeable and indicated that he knew a great deal about the subject and recognized several sides of the issues. He didn't rush the meeting through so that he could get on to something else. He made clear what his own wishes were and, most important, there was no question that he wanted to re-establish the White House science structure on somewhat the same lines it had followed before, although there were some variations that he wanted considered.

The Question of the PSAC

One question in Mr. Ford's mind had to do with the reestablishment of the President's Science Advisory Committee as part of the total package of the science-advising structure. I guess all Presidents who had experienced the operations of PSAC had some reservations about it because that body had had the right to make studies on any subject of its own determination and, when some of those studies were not approved for publication by the White House mechanisms, embarrassment would ensue, so much embarrassment that they usually got published with the expected further embarrassment. President Ford did not show any strong feeling one way or the other on PSAC, but we did discuss the newly-formed Office of Technology Assessment in the Congress. I made the point that I thought that the OTA and its advisory council and committees were becoming the panel of scientific advisers for the Congress; OTA was not confining its operations to technology assessment. Mr. Ford replied that he had never thought it to be anything else but a scientific advisory mechanism. (As an aside, the OTA did in fact move in the direction of scientific advice but is now confining itself more to its original purpose.)

The matter of the desirability of re-establishing the President's Science Advisory Committee was left open, but all other aspects of re-establishing the White House science advisory mechanism—including the use of scientists as consultants—were clearly indicated as desirable. At the end of the meeting the President directed the only other person in the meeting with the two of us, a member of the Domestic Council staff, to begin the effort to get together a Presidential position on the re-establishment of science advice in the White House.

At this time President Ford made what I consider to be a wise and important decision. He said that he wanted the re-establishment of science advice in the White House to be a matter of legislative action as opposed to a reorganization directive from the President. I believe that this stemmed in part from his knowledge and experience gained as a long-term member of the House of Representatives and in part from his knowledge of the dismay, confusion, and generally unsatisfactory result which comes from an abrupt Presidential directive. He knew that the legislative process, if successful, would give greater strength to the office. He also knew that the legislative process would give all concerned parties opportunity to express themselves on the form and the substance of the office, that there would be a feeling of participation by the affected community, and that the time taken to consider openly all of the issues would result in wiser decisions. Looking back on all that has happened during the next few years, I am even more impressed by the wisdom of that decision than I was at the time, though I agreed heartily then. As a follow-on to that decision, he also directed me to continue as both Science Adviser and Director of the National Science Foundation.

The Role of Rockefeller

In spite of this clear concept and direction from the President, it still took from October 1974 to June 1975 for the Administration to get its position publicly announced. The President wanted Vice President Rockefeller to take a leadership role in the affairs of science and technology in his administration. This was a good decision, for Vice President Rockefeller was interested in the field and had had some experiences working with scientists and engineers, and his enthusiasm was of great value throughout. His contact with the Congress through his role as President of the Senate was also invaluable. But it still took some time to get him involved, because he had to go through confirmation hearings and then he had an important role in trying to straighten out some activities in the intelligence area, which at the time had some deep problems. After he saw his way clear in the beginning of 1975 to put in some time on science and technology policy, he wanted to call on some of the people whom he trusted—whom he had worked with before in his role as a private citizen looking into courses of action in the employment of science and technology for progress—and they took time to become acquainted with the issues.

In addition to the delay of Mr. Rockefeller's participation, there were others, because each of the major staff units of the White House were going through transition in personnel and direction. Although the staff members who had been associated with the Watergate affair had long since left the White House, there were still many previous appointees who were on their way out. Furthermore, the phenomenon of protection of turf was evident. The Office of Management and Budget liked the arrangement of the Science Adviser being out of the White House because the Director of the National Science Foundation could actually put more resources in the form of money and manpower on some of the issues which the OMB felt were important; and on those which they didn't think important, they didn't have to contend with a White House Science Adviser who was in constant and daily contact with all of the other staff units and the President himself. The National Security Council was pleased to take care of national security without substantial participation by the White House Science Adviser. As you recall, President Nixon's Reorganization Plan, which was promulgated in January 1973 and which established science advice outside of the White House, had not given the Science Adviser any role of importance in national defense affairs, although there was a small loophole that he could participate when needed. The Domestic Council also did not call often on the Science Adviser, although Secretary of the Treasury George Shultz, who was also a Special Assistant to the President on matters of domestic affairs, was always a very good friend of science, technology, and the Science Adviser. Presumably these are normal staff unit reactions, which had to be overcome, and the result did turn out quite well.

The proposal by the President was introduced in the House of Representatives on June 11, 1975, by Mr. Teague and Mr. Mosher as an Executive Request, and the bill was referred to the Committee on Science and Technology, of which Mr. Teague was chairman and Mr. Mosher was minority leader. It was a relatively simple bill, establishing the Office of Science and Technology Policy within the Executive Office of the President; and it assigned duties and functions to the Director making him the chief policy adviser to the President with respect to scientific and technology matters and included in the areas for advice scientific and technology matters of major national policies, programs, and issues; the adequacy and effectiveness of Federal scientific and technological policies, programs, and plans for meeting national goals; the utilization of new ideas and discoveries in science and technology in addressing important national problems; the coordination of scientific and technical activities of the Federal Government; and such other matters as the President may direct. This set up the White House office and the role of the Science Adviser, but it did not address directly the President's Science Advisory Committee. It did provide, of course, for the use of consultants from the science and engineering world and it also provided the capability to let contracts and make grants for studies by the outside scientific community. It didn't specifically address the function of coordination in the form of the Federal Council on Science and Technology, but it did assign coordination to the Director of the Office of Science and Technology Policy. Most important, it was a clear signal that President Ford wanted to re-establish science advice in the White House.

Before going to the substance of science and technology advice in the White House during this period of transition, let us continue with the story of the form. In the spring of 1975 the focal point of the development of the public law entitled "The National Science and Technology Policy, Organization, and Priorities Act of 1976" shifted to the Congress—to both the House of Representatives and the Senate. This does not imply that both the House and the Senate had not been active in the matter before, but the signal coming from the White House in the spring of 1975 clearly encouraged both to pick up speed and come out with an agreed-upon position which could be gotten through the administration.

From the administration's side, Vice President Rockefeller now assumed an important role, that of communicating directly with the House Science and Technology Committee and with the three committees in the Senate which were concerned with science and technology and this whole matter: the Special Subcommittee on the National Science Foundation of the Committee on Labor and Public Welfare with the subcommittee chairman being Senator Kennedy; and the subcommittee on Science, Technology, and Commerce of the Committee on Commerce with the subcommittee chairman being Senator Tunney; and the Committee on Aeronautical and Space Sciences with the committee chairman being Senator Moss. The minority leader of Mr. Kennedy's subcommittee was Senator Laxalt, and the minority leader of Senator Tunney's subcommittee was Senator Beall. The principal leadership on the House side was clearly that of Mr. Teague and Mr. Mosher, both of whom played a very active part, and on the Senate side, Senator Kennedy with active interest from Senator Moss.

Differences of Opinion

The Senate and House actions from June of 1975 until the spring of 1976 are well documented. It is sufficient to say that there were great differences of opinion as to the direction of the bills, that the administrative representatives were often asked for their opinion about them, and that the House of Representatives' views were more complete and more solid than those of the Senate, though both houses contributed important ideas; and by the spring of 1976 the National Science and Technology Policy, Organization, and Priorities Act of 1976 had passed the House and the Senate and was signed into law by President Ford on May 11, 1976, at a pleasant ceremony in the garden of the White House. That still is the existing law governing the operations of the Science Adviser and the Office of Science and Technology Policy in the White House.

Still continuing the discussion of the form as opposed to the substance of science and technology advice, it is well to assess what was gained in this transitional period with science in the White House, out of the White House, and back in the White House again. It is my opinion that the period of the spring of 1973 and through to the summer of 1974, the Watergate period, was a good time for science not to be in the White House. Perhaps more important, the enlistment of a massive community effort, with many individuals and organizations in the science and technology area, and in related areas as well, expressing their views as to just what forms science and technology advice should take, how it should be organized, how large it should be, and who should participate, was a very important exercise. It reaffirmed some of the important features and, I believe, eliminated some features that were not so important. It showed the science community where its friends were in the Congress and the administration and in society in general. I think it pointed up a more realistic conception of what science could and could not do for the country—where it was important and where it wasn't. There was even a chance for descriptions of how science worked, which were more widely read and understood. Some would say that this was the hard way of getting all of these things done, but I am not so sure that they would have been done at all had it not been for these events which removed the science apparatus from the White House and then brought it back.

All of the discussion so far has concerned the activities of the administration under Mr. Ford and of the Congress—both the Senate and the House of Representatives—in re-establishing by law the White House science apparatus. It came after three years of very abundant discussion and negotiations by members of the science and technology community and by leaders in Government and in Congress. The form which governed the operation of the science apparatus when it was outside of the White House came about rather abruptly and with far, far less general discussion—in fact, none. Because of President Ford's decision not to re-establish the White House science apparatus by Executive Order but by legislation, there was a period from August 1974 to the summer of 1976 in which it operated under the same guidelines under President Ford as it had from January 1973 to August 1974 under President Nixon.

The events which led to the Reorganization Plan No. 1 of January 26, 1973, are not entirely well known to me, for as Director of the National Science Foundation in the year previous to that (though I had frequent contact with Edward David, the Science Adviser to the President then, and the National Science Foundation staff had frequent contacts with the staff of the Office of Science and Technology) we were not parties to the decision to change things. Following President Nixon's re-election in November 1972, I began to hear rumors of pending change in the White House science advisory structure, but when members of my staff told me of these rumors, I instructed them to stay out of that arena. Of course, I do not know if they did. When I returned from my first visit to Antarctica and the South Pole in the middle of December of 1972, the rumors were stronger and my instructions to the staff were stronger also. When I returned from a lengthy Christmas-New Year's vacation in January of 1973, it was very clear that the intent of President Nixon was to remove the science apparatus in the White House and divide its functions.

The Responsibilities of the NSF

Shortly after the above events, I was asked by George Shultz, Secretary of the Treasury and Special Assistant to the President, to join with Roy Ash, the Director of the Office of Management and Budget, and him in some discussions concerning what the National Science Foundation would do with some new responsibilities handed to it. I was not asked to invite the National Science Board, the governing board of the National Science Foundation, to participate at that time, and I was allowed to ask only Raymond L. Bisplinghoff, the Deputy Director of the National Science Foundation, to work with me on any ideas. This curious and quick action illustrated a very important dilemma for all Directors of the National Science Foundation, for they are Presidential appointees and are responsible to the President, and yet they serve as ex officio members of the National Science Board, the policy-governing board of the National Science Foundation and advisers to the board on policy matters. In any case no jurisdictional battle had time to evolve, for shortly thereafter the Reorganization Plan came out and surprised the science world. Some things in it were a surprise to me, too. I am not sure which staff people in the White House actually composed the portions in that Reorganization Plan which occurred under the heading "Streamlining the Federal Science Establishment," but I detected some expert input which must have come from the Office of Management and Budget and from the staff of the Office of Science and Technology itself, the latter apparently trying to preserve as much as possible of the previous functions of that office.

So the marching orders for the science-advising structure from the spring of 1973 to the summer of 1976 were all contained in these words:

When the National Science Foundation was established by an Act of the Congress in 1950, its statutory responsibilities included evaluation of the Government's scientific research programs and development of basic science policy. In the late 1950s, however, with the effectiveness of the U.S. science effort under serious scrutiny as a result of sputnik, the post of Science Adviser to the President was established. The White House became increasingly involved in the evaluation and coordination of research and development programs and in science policy matters, and that involvement was institutionalized in 1962 when a reorganization plan established the Office of Science and Technology within the Executive Office of the President, through transfer of authorities formerly vested in the National Science Foundation. With the advice and assistance from OST during the past decade, the scientific and technological capability of the Government has been markedly strengthened. This Administration is firmly committed to a sustained, broad-based national effort in science and technology, as I made plain last year in the first special message on the subject ever sent by a President to the Congress. The research and development capability of the various Executive departments and agencies, civilian as well as defense, has been upgraded. The National Science Foundation has broadened from its earlier concentration on basic research support to take on a significant role in applied research as well. It has matured in its ability to play a coordinating and evaluative role within the Government and between the public and private sectors.

I have therefore concluded that it is timely and appropriate to transfer to the Director of the National Science Foundation all functions presently vested in the Office of Science and Technology and to abolish that Office. Reorganization Plan No. 1 would effect these changes.

The multi-disciplinary staff resources of the Foundation will provide analytical capabilities for performance of the transferred functions. In addition, the Director of the Foundation will be able to draw on expertise from all of the Federal agencies, as well as from outside the Government, for assistance in carrying out his new responsibilities. It is also my intention, after the transfer of responsibilities is effected, to ask Dr. H. Guyford Stever, the current Director of the Foundation, to take on the additional post of Science Adviser. In this capacity, he would advise and assist the White House, Office of Management and Budget, Domestic Council, and other entities within the Executive Office of the President on matters where scientific and technological expertise is called for, and would act as the President's representative in selected cooperative programs in international scientific affairs, including chairing such bodies as the U.S.-U.S.S.R. Joint Commission on Scientific and Technical Cooperation.

In the case of national security, the Department of Defense has strong capabilities for assessing weapons needs and for undertaking new-weapons development, and the President will continue to draw primarily on this source for advice regarding military technology. The President in special situations also may seek independent studies or assessments concerning military technology from within or outside the Federal establishment, using the machinery of the National Security Council for this purpose, as well as the Science Adviser when appropriate.

In addition to those marching orders the Reorganization Plan included among other things the abolishment of the National Aeronautics and Space Council of which the Vice President was chairman and which was an in-the-White-House niche for space.

The best piece of advice given to the new Science Adviser was from David Beckler, long-time leader of the Staff of the Office of Science and Technology in the White House, who told me that when James R. Killian, Jr., first established the Science Adviser's office, he decided that his most effective channel for influencing the massive events in which science and technology are involved was through the budget process. He established good relationships with the head of the Office of Management and Budget and insisted that his staff work with the proper staff elements in the OMB on all the issues of science and technology which had anything to do with management or budgeting, particularly the latter. There is an old saying in government that "the budget is a policy statement," and all of my experience in government confirms this statement. So, with this advice, we chose budgeting as the strongest line of attack to make our influence felt, and I am very pleased for several reasons. One reason is that it showed the National Science Foundation's greatest immediate strength in science policy, for we had been working more and more on policy studies and budget studies with the Office of Management and Budget on a number of emerging issues. The Research Applied to National Needs program had fairly substantial budget allocations for these purposes, and they came in handy at this point. Furthermore, the OMB welcomed these studies because they were often hard-put to get the background studies for some of their budgeting decisions.

There was another reason why I was pleased that we chose the budget route as our strongest line of attack on the science-advising problems. As the Watergate crisis in President Nixon's administration grew, it became clear that the strongest force holding together the operation of the administration was the Office of Management and Budget. In fact, it was almost the only really strong centralizing force in running the government during the most critical times of the Watergate crisis—the last six months. This isn't quite true, because Secretary Kissinger ran a tight show in Foreign Affairs and the military affairs of the country were strongly led. But the fact that the OMB had not been involved in Watergate made it a much more pleasant and effective attachment to have.

But one can't just say that the most important line of attack was through the budget process, though that is very important, but one must pick some highlights in that. And my choice for an example is energy. Perhaps it would be better to say the three Big E's: energy, environment, and economy, for those three went together more and more. So our first effective efforts were in the budgeting process for energy. Using some of the important study money of the RANN program and some of the RANN staff in energy policy studies, together with some others that we added, we became involved in the growing energy crisis.

The OST and Energy

Even before the Oil Embargo of October, 1973, energy was a front-page subject in science and technology and in government. Here, away back in the middle 1960s, the Office of Science and Technology had triggered some important and very broad-scale studies of energy in the future, and their reports covered the field very well. They didn't deal solely with supply, although there was a lot on nuclear, coal, oil, geothermal, solar, and all of the sources. However, it was very important to note that they had conservation considerations as well. In 1973 and in 1974, as the White House struggled to get energy better organized in the administration, we lent a hand to several things. We were asked to chair and establish the Energy R&D Advisory Council, which had a strong influence on the budget prepared in 1973 and presented in the spring of 1974, though Dixie Lee Ray, then head of the Atomic Energy Commission, did much of the work before it was presented to the White House. In particular, we increased the accent on coal and on conservation in the budget as prepared by the agencies and other special units established by the White House. We were asked to seek the advice of the science and technology community and the industrial community and other communities on possible appointees for the administrators of the Energy Research and Development Agency (ERDA), which was to be formed soon. We made a number of energy policy studies, using many outside contracting agencies and using National Science Foundation study money.

Incidentally, the scientific-advising activities in the field of energy did conflict with the responsibilities of the Director of the National Science Foundation for the NSF had been assigned the field of solar energy and part of the field of geothermal energy and, as a consequence, was an advocate for those fields in each budget cycle. The much-feared conflict of interest was in fact noticeable in that area probably more than in any area, although others emerged later. When ERDA was formed, the National Science Foundation transferred its responsibilities in the development of solar energy and geothermal to ERDA, and to a considerable extent the policy conflict was eliminated.

While I believe that our work in the budget area as exemplified by our advice on energy policy matters and energy R&D budgets to the Office of Management and Budget was a successful and worthwhile contribution, it did illustrate another one of the often-quoted weaknesses of the science-advising mechanism outside of the White House. We were not regularly asked to submit our position on all White House issue papers. Occasionally the OMB or the Domestic Council or some other unit of the White House staff would ask for an independent submission, but for the most part we had to find out, through the legwork of staff, what the issues were, and if we felt strongly enough to submit an independent view, we could do so. But the fact that we did not have regular immediate contact with all White House events definitely gave us a handicap. I would say that both a strength and a weakness of the science-advising mechanism at the National Science Foundation was shown by this energy budget operation, the strength being that clearly the NSF's funds were far larger than funds given to the White House science-advisory mechanisms in the making of many particular studies. The Energy Policy Office in the NSF under the direction of Paul Donovan used this study-making power to the hilt.

Support from the STPO

The other unit of staff within the National Science Foundation which supported the Science Adviser role was the Science and Technology Police Office under the direction of Russell Drew. Its instructions were to cover all fields of science policy other than energy, and it entered several arenas. One of the most persistent, pervasive, and yet most difficult to get hold of was the influence of research and development on industrial and economic strength. While almost everybody agrees that our science and technology have important effects on our foreign trade, our productivity, our capability to innovate in all of the fields from consumer goods to health and medical care, from the production of food to the control of the environment, quantitative measurements of the relationships are very difficult to get, and changes in our science and technological programs to influence our industrial and economic strength are even harder to bring about. Nevertheless, in this period, we succeeded in strengthening the ties between the Federal government and those representatives of industry who were in charge of industrial research and development. The Industrial Research Institute, which had been in existence for some time and whose membership was made up principally of the vice presidents for technical programs and research and development of the principal corporations of the country began to take a renewed interest in government R&D. STPO aided and abetted this renewed interest by supporting conferences and studies and other activities of IRI, and especially in serving as a link between IRI and important units of the administration. I am pleased to say that this activity, this new relationship, is growing stronger today.

Another case where the STPO and another unit of the National Science Foundation, the International Programs, got together to carry on in a strong way the functioning of the former OST, was in the area of support of international science and technology exchanges. The Science and Technology Exchange Program with the Soviet Union is the prime example. In mid-1972, following the Nixon detente initiatives with the U.S.S.R., the first specific agreement to be signed was that in science and technology. Edward David had led a group of colleagues in science and technology to the Soviet Union and worked out the details of that agreement and had begun to put together the United States half of the Joint Commission on Scientific and Technical Cooperation with the U.S.S.R. We took over the chairmanship, the staff management, and a great deal of the funding for that science and technology exchange program.

In my estimation, this joint program with the Soviet Union was successful in many ways. It did help to lessen the tensions in some very difficult times for the United States; there was some reasonable exchange of worthwhile science and technology information and experiment between the United States and the USSR. It did not create a threatening leak in the vast science and technology information banks in this country, for we were careful to choose the fields of cooperation in areas where both sides had something to gain. On this latter point, I am sure that the Soviets were disappointed, because they had hoped that this exchange would open wide to them the flow of the science and technology of our private industry; there were often difficulties in the relationship on that point. Also, although there were cases of resistance by American scientists to participate in science and technology exchange with the Soviet Union until they showed a better record with respect to human rights, the program did provide and does today provide an entree —a porthole through which conversations on that subject can take place. We did conduct private discussions on this subject with our Russian counterparts, although I do have to say that the scientists with whom we dealt believed that human rights discussions should be separated from discussions on science and technology exchange, and they expressed themselves strongly in that vein.

A Better Climate

Although, as I mentioned before, we operated under President Ford for 1-3/4 years outside of the White House and under the same Reorganization Plan that President Nixon had put forward, things began to change very rapidly when Mr. Ford was there. In the first place, it was much nicer to be associated with the White House; it was a much pleasanter place. Secondly, Mr. Ford himself was receptive and interested, and although he did not spend a lot of time with the Science Adviser, he certainly did on some issues and it seemed very fruitful. More important, his attitude spread over to his staff, and the staff members of the Science and Technology Policy Office and the Energy Policy Office in the National Science Foundation could be inserted more and more into the many staff activities that went on in the White House structure. We had Vice President Rockefeller as an enthusiastic supporter and a helper. All the units of the White House staff seemed to be easier to work with, especially when it was clear that eventually the Science Adviser and the science-advising mechanism would move back into the White House in Mr. Ford's administration as soon as the legislation could be brought forward.

We used this better climate to get re-established the annual award of the National Medals of Science, and to improve the efficiency of the appointment of Presidential appointees to important jobs in science and technology posts. We tried to arrange as often as possible meetings with the President and top members of his staff, with important leaders from abroad and from this country in science, and we succeeded in many cases.

The S&T Budget

The success of which I am proudest in all of the period with President Ford has to do with the amount of the science and technology budget. In his two and a half years as President, he was responsible for the putting together of three Federal budgets, those for fiscal years '76, '77, and '78. (If these dates are confusing, the reader must recall that there is a considerable lead time in the preparation of Federal budgets, and an incoming President operates for almost a year and a half on budgets that were prepared by his predecessor.) As a help to the Director of the Office of Management and Budget in putting together the R&D budget for the Federal Government, and with increasing opportunities for independent expression of our views, we worked hard to get President Ford to recognize the threat of a Federal R&D budget which was slowly but surely drifting downward a small percent per year in real dollars, and had been since 1967. He did recognize this. His Budget Director and his Deputy Budget Director also recognized this, and we set about to repair this. We succeeded, and in all three budgets for which President Ford was responsible there was a real growth in the R&D budget of the Federal government. Naturally, this growth was not uniform in all fields. President Ford made clear statements as to his R&D priorities, and there were three: energy, national defense, and basic research. In his last two budget messages he specifically mentioned these and, from conversations with him, it was clear that he was proud of this accomplishment. There are fields of basic research which did not fare as well as others, health and medicine being one, but that portion of the budget was always so strongly helped by the Congressional add-ons in their budget-making that the administration did not have to worry about it. There were several new initiatives in basic research, for example, strengthening the basic research program of the Department of Agriculture in their grants program. All in all, that exercise was very satisfying.

Another very satisfying event took place in the award of the National Medal of Science. In the early days of the award of that medal, there was one American scientist, Linus Pauling, who from every scientific standpoint was a very deserving candidate but who was a thorn in the side of many Presidents and governmental leaders because he did not agree with some of their most important national programs. For several years, among the names submitted by the Award Committee to the White House, from which list the White House would pick the awardees (taking either the first recommendations of the Committee or substituting some of the recommended alternates) his name would always be removed by the White House. After a while, the Award Committee grew tired of a lack of success and no longer nominated him. In Mr. Ford's first year, it was the opinion of several of us, however, that we should again nominate Pauling, and we did. I had several calls from various White House staff members who indicated that their eyebrows were raised but who were quite willing to discuss the matter with me, and it was finally agreed that his name should be submitted to the President by the White House staff with a full description of the case. It was, and the list of approved candidates came back with the President's personal little check mark of approval next to Pauling's name. It was a quite characteristic check mark because it was a left-handed check mark. Mr. Ford had a broad understanding of science and technology, its benefits and its problems, how the work was done and by what people; that understanding made the relationship with him as Science Adviser a pleasant one, either from a base outside of the White House or within.

As 1976 was approaching, it became clear that the Senate and the House would probably make up their differences and get out a science and technology bill. There were still some very strong arguments there, and while we in the administration were called in often to give our opinion, the fight was really within the Congress itself. But, in expectation of a change, we began to prepare a shift of the responsibility back to the White House. Naturally, there was great discussion as to what things the National Science Foundation should keep. Fortunately, the Congress recognized that the National Science Foundation had a strong capability to support policy studies in science and technology and that capability was retained in the NSF when the bill came out. But all of the science-advising strength really was to be returned to the White House.

"FIXIT"

The Federal Council on Science and Technology, which had been continued during the "out" period, was to become the Federal Coordinating Council for Science, Engineering, and Technology, which immediately acquired the name "FIXIT." There was to be re-established something in parallel with the President's Science Advisory Committee called the President's Council on Science and Technology, PCST, which, however, was to have a very specific job of a complete analysis and audit of the science and technology programs of the country, including the new advisory structure, and to make a report on this, with an interim report after one year and a final report at some specified later date. Interestingly enough, the PCST was directed also to recommend whether or not it should be made permanent. By the time I became Science Adviser in the White House, I had decided and told Mr. Ford that I did not expect to serve after he was re-elected in 1976 and that I would leave at the beginning of his second term. I say this here because, among other things, I wasn't particularly interested in getting involved in that recommendation on the continuation of the PCST.

There were some other provisions of the bill that I was rather glad not to have to face, one of which was the annual report on the state of science and technology. I knew this to be a difficult one because some years before I had been a member of a PSAC panel on science policy, chaired by Patrick Haggerty of Texas Instruments which recommended to President Nixon that there should be an annual report on science and technology. Mr. Nixon accepted right on the spot and turned to Dr. David and said, "Put that into effect." There was a great struggle within the White House staff to bring out such a report, but it proved just too big a task for a staff of that type and of that size with all of the other things they had do. Fortunately, now the responsibility for this was delegated to the National Science

Foundation and the NSF gave a contract to the National Research Council, which is now in the process of writing a pretty good report.

By and large, however, the new legislation was something that one could act on. Philip Smith, Executive Assistant to the Director of NSF, helped me as we proceeded to set up all of the elements involved, which were the White House office, the Federal Coordinating Council on Science, Engineering, and Technology, the President's Council on Science and Technology, and the Intergovernmental Science, Engineering and Technology Advising Panel, composed of leaders of state and local governments and some scientists and engineers who worked on problems to improve the Federal Government's performance in helping other governmental units in science and technology. When OSTP was finally started in the White House, we secured the help of Donald Kennedy and William Nierenberg as Senior Associates, and they helped get things going, especially as we enlisted various advisory panels from the science and engineering community.

The Congress had been rather slow in responding to the request to get out the new legislation. In some quarters there developed the idea that perhaps it would be better to put off the legislation until a new administration took over. Many of the people who expressed this were clearly Democrats who hoped that a Democratic administration would take over and could have the honor and fun of establishing it. I believed that that was a wrong opinion, for either a Democrat or a Republican. It is too easy for government to drop a good idea right out of sight in the change of administrations.

I think that I was confirmed in the above view when I experienced the transition days from one administration to another in late 1976 and early 1977. There is an immense amount of confusion no matter what administration comes in. The Office of Science and Technology Policy was there, with offices, with staff, with very good equipment, with good relationships with the other continuing units of government (such as the Office of Management and Budget). A new Science Adviser could be appointed and new staff could be appointed, and they would get there early enough to feel an orginal part of a new administration. If they had to come along later, I am not sure they would have come along as well or even at all. There are just too many other things to do for a new administration. I was interested to note that during the transition, the OSTP staff (and many of the staff did not survive the transition later) carried on the work and people began to rely on them. The incoming administration recognized they had an ongoing operation that could help. I was very pleased that the timing of the return was when it was.

Conclusion

In conclusion, one can live and learn from all of these experiences, and, clearly, there are many, many lessons. Perhaps the most important one is that science and technology, in spite of ups and downs, does have a recognized place in the scheme of things at the level of the White House and at lots of other levels in our society. There certainly will be ups and downs in its relationships to society as well as to administration and legislative units. Society's attitude has changed in our lifetime towards science and technology. Right now there is a reasonably benign and appreciative attitude, although there are some strong anti-science and anti-technology foci. If the science advisers in the future can continue a record of service to their Administration and an even-handed approach to the many different science and technology factions over which they look, if the science community itself can be supportive and cooperative and understanding of the Science Adviser and his staff, if the science and technology communities in themselves can continue to be positive forces in our society, I have no basic worry about the long term importance and recognized position of science in the White House.

Advising Presidents on Science and Technology

Frank Press

Over the years a variety of recommendations has been made regarding the best ways to provide the President of the United States with timely and effective advice on matters relating to science and technology. From the Eisenhower years to the present administration many approaches have been tried, and with varying degrees of success. These have ranged initially from the appointment by the President of a single Adviser, to the legislative creation of an Office of Science and Technology Policy in the Executive Office of the President, with its Director serving as the President's Science and Technology Adviser, as is the case today.

During all these periods, however, one thing has become clear; the structure and use of science and technology advice in the White House has tended to reflect the background and outlook of the incumbent, and the attitudes and needs of his time. Each President has employed science and technology advice differently. Each has turned to his Adviser, and to the science and engineering communities in ways that were consistent with his personality, his methods of operating his office, and what the events of the day demanded. One result of this has been that each Science Adviser has had different experiences and served his President and the country in different ways.

This effect continues today. The current situation related to science and technology advising in the White House is enhanced by the fact that President Carter is a technically oriented person with an interest in the technical input bearing on national issues. The situation is also influenced by the fact that an increasing number of such issues are strongly related to science and technology—driven by their forces and possibly susceptible to change brought about by advances in our scientific knowledge and its applications. We see this in the important matters of energy, environment, health, and economic growth. We see it in matters of national security. And we see it in our international affairs, where both competition and cooperation in science and technology and technology transfer play such important roles today. Again, President Carter's interest in these matters, resulting,

Frank Press (b. 1924) is Science Adviser to President Carter and director of the Office of Science and Technology Policy, appointed in 1977. He is a member of the National Academy of Sciences and of the National Science Board. A geophysicist, his academic career has been at Columbia University, California Institute of Technology, and M.I.T. He has served on numerous governmental boards.

among other things, in our emphasis of science and technology in exchanges with the Peoples Republic of China and in the creation of an Institute for Scientific and Technological Cooperation to work with developing countries, indicates his receptiveness to timely scientific and technical information. This combination of a President's interest and the interests of the times, has given the OSTP opportunities to play a central role in the White House.

The fact that the President in reorganizing the science advisory apparatus decided not to make use of a formal outside group such as a President's Committee on Science and Technology has come under some criticism. Conditions for operating a PCST or PSAC are different today when one takes into account the Freedom of Information Act's requirement for open meetings, and the vast agenda beyond national security of concerns for the Science Adviser. However, the lack of these formal groups has not deterred the President, nor OSTP, from calling on the country's most knowledgeable people in all fields of science and engineering for broad or specific advice. Various panels have been convened analagous to the PSAC panels of earlier years. These have provided excellent counsel on a number of important national issues, including energy, defense, environment, technological innovation, government organization and relations with the developing world.

As an incumbent Science Adviser it would not be proper for the author to write in judgement of the current science advisory mechanism. Only history will shed some perspective on how successful it was as compared with others. However, I commend William T. Golden for his help in devoting this special issue of *Technology in Society* to the commentary of former Science Advisers, each of whom has played a different and unique role in our science and technology policy history. The recounting of their experiences and their perspectives on past policy-making in this field should make enlightening reading for everyone interested in learning more about how Presidents deal with the important issues of science and technology.

The Science Adviser's Role in the Reagan White House in 1986

John P. McTague

Science and technology issues in the Reagan White House were handled in the same manner as most other subjects. The President articulated a relatively small number of broad but clear general principles, and then expected his Cabinet and White House staff to effect specific actions accordingly. Lines of authority and responsibility depended not so much on formal job descriptions and coordination mechanisms as on individual initiative and *ad hoc* consensus processes. This led to a large degree of freedom of action, counterbalanced by the resultant infighting. The two major exceptions were national security and the budget process (or what in the private sector would be viewed as the controller's function). The National Security Council (NSC) and the Office of Management and Budget (OMB) provided relatively well-defined input, consensus, and resolution mechanisms, despite personalities and emphasis on individual initiative in even these offices.

Attempts were made to systematize the decision-making procedure, but none was wholly successful. The Economic Policy Council, whose members consisted of about half the Cabinet, the Chairman of the Council of Economic Advisers, and the Science Adviser, handled several science and technology issues, such as the emerging biotechnology regulations, but matters involving direct presidential input were treated separately and uniquely. The prime example was the reexamination of NASA's space mission following the Challenger tragedy, which occurred in the fourth week of my tenure as Science Adviser. Although there was a highly visible commission appointed to examine the causes of this event, the policy implications were handled by NSC meetings with the President, in which I fully participated.

The Strategic Defense Initiative had been created by the President in 1983, deliberately outside normal channels, and it continued to be handled on an *ad hoc*

John P. McTague (b. 1938) was acting Science Adviser to President Reagan from January through May 1986. Prior to that position, he was Deputy Director of the Office of Science and Technology Policy (1983–1985). During most of his research career, he was Professor of Chemistry and a member of the Institute of Geophysics and Planetary Physics at the University of California, Los Angeles (UCLA). He is presently Vice President–Technical Affairs at Ford Motor Company and serves on the President's Council of Advisors on Science and Technology (PCAST) and on the Secretary of Energy Advisory Board (SEAB).

basis with substantial resultant turmoil. Although most routine policy concerns were dealt with by normal interagency means, such as the fragile Federal Coordinating Council on Science, Engineering, and Technology (FCCSET), there were two areas I felt to be sufficiently important to raise directly with the President. The first topic concerned the attacks on freedom of communication of basic science, attacks which originated in segments of the national security community. When the arguments were laid out to President Reagan, he came down clearly on the side of continued free communication and issued a National Security Decision Directive to that effect. This, however, did not deter some individuals from continuing to undercut his stated position.

When the White House Science Council issued its major report on the Health of Universities in early 1986, I believed this topic also deserved Presidential attention. In a letter to the chairman of the panel, David Packard, the President endorsed the report's recommendation to strengthen the research base at universities and set a goal to double the National Science budget, a goal which has since become a perennial but elusive presidential initiative.

Science and Technology and Presidential Science and Technology Advice

Gerald R. Ford

The importance of science and technology to the well-being of our people as individuals and as a nation is quite evident to anyone of long service in the Congress. Perhaps those of us with such service see the impact of science and technology even more clearly than scientists and engineers, for we see both a dedicated constituency in science and technology, whose members find their fame, fortune, and inner sense of self-accomplishment, and also a concerned citizenry needing and desiring the fruits of science and technology, who at the same time want to make sure that its power is channeled in positive ways.

My long service in the House of Representatives acquainted me not only with the broad generalities of science and technology but also with a wide range of detailed issues, large and small, as a good proportion of the legislative actions on which I voted contained elements of science and technology. Its growing impact on our society could be sensed in the increase in our times in the number of major Federal agencies involved with science and technology, the growing need for rapid progress in government-funded R&D to solve critical economic problems, and the increasing Federal burden for regulation of the results of new science and technology.

When I came to the Presidency, I recognized that priority-setting was a principal need in R&D. I knew that the morale of the scientific community was somewhat low, partly because of the apparent downgrading due to rough handling in the White House by both President Johnson and President Nixon, but mainly because the Federal government through many actions in both the administration and the Congress had tightened its research and development funds over a period of almost a decade. Most important to me, though I believe both in returning science advice to the White House and also in strengthening the R&D budget, was the selection of priorities. My R&D top priorities were three in number: *energy*, a front-page issue of both immediate and long-range import; *national defense*, a matter of shoring up the long-range military R&D, which had badly lagged because of our

Gerald R. Ford (b. 1913) was the 38th President of the United States (1974-77). He is a lawyer, was a member of the 81st to 93rd Congresses from Michigan, and was minority leader in 1965, and U.S. Vice President in 1973-74. He served in the U.S. Navy in World War II.

Vietnam involvement, which we were terminating; and *basic research*, without which the United States cannot stay first-rate. I knew also that we must try to handle the myriad of other fields in which R&D is important: health, space, commerce, international relations, and many others. One of my proud accomplishments is the fact that we did stop the downtrend of R&D funding and turn the curve upward again.

On the more immediate topic of the structure of the White House scientific advisory apparatus, there was no doubt in my mind when I came to the Presidency that it should be restored. Many recognize that each President must have some freedom to organize his White House staff, though I believe that science advice was certainly one of the important components of any such organization. Still, I wanted to give the White House science advisory structure a firmer base than it formerly had. This could only be done through legislation which would emerge from the well-tested techniques used by the Congress of hearings on the many differing ideas which were then being proposed. Having spent a lifetime of service in the Legislative branch, I believed greatly in its strengths. Also, I wanted Vice President Nelson Rockefeller to play a leading role in carrying through my R&D programs in the Congress. This he did, and I am indeed grateful.

We accomplished what we set out to do in the restoration of science and technology advice to the White House. The new Office of Science and Technology Policy took its place with the other strong units of Presidential staff—the Office of Management and Budget, the Domestic Council, the National Security Council, and all the others—and played its role in helping me on the tasks which had any science and technology components. It is my belief that the Office of Science and Technology Policy can continue to serve Presidents in the future from the strong base which has been created.

Reconciling the Science Advisory Role with Tensions Inherent in the Presidency

Elmer B. Staats

As both the Presidency and the science advisory function have increased in importance and complexity, so have the context and design for institutionalizing the science advisory mechanism* as a major force within the Executive Office of the President. Experience during the past 40 years with various formal structures and ad hoc arrangements has revealed inherent conflicts and tensions which must be recognized and dealt with to insure Presidential recognition of the need for and use of science and technology advice as a major component in policy decisions and executive leadership.

World War II dramatized the importance of science and technology to national security. The advent of the Atomic Energy Commission, the National Science Foundation (NSF), and the National Aeronautics and Space Administration, followed by the restructuring and increased emphasis on science and technology in other civil agencies, led to a pluralistic system for supporting research and development (R&D) which is a fundamental characteristic of the US science and technology enterprise. From this pluralism, in which each agency sponsors R&D in support of its own mission requirements, the need for stronger central oversight and coordination emerged.

---

* "Science advisory mechanism" and "science advisory office" are general terms for any office offering science and technology advice and analysis within the Executive Office of the President (presently the Office of Science and Technology Policy).

Elmer B. Staats (b. 1914) is Comptroller General of the United States by President Johnson's appointment in 1966. Previously he was Deputy Director of the Bureau of the Budget under Presidents Johnson, Kennedy, Eisenhower, and Truman. He participated in development of legislation establishing the Atomic Energy Commission and the National Science Foundation. In 1962 as Deputy Director of the Budget Bureau, he helped draft and supported the reorganization plan establishing the Office of Science and Technology; and later, as Comptroller General, testified in support of the establishment of the present Office of Science and Technology Policy.

Federal support of R&D policies affecting science and technological innovation in both public and private sectors involves major issues which transcend individual executive departments and agencies and cannot be dealt with adequately below the Presidential level. In recent years, Congress also has become increasingly concerned about the roles of science and technology in the economy, quality of life, and international affairs, and has demanded more explicit disclosure of Presidential strategy.

Thus, there are three essential functions or dimensions of the science advisory role in the Presidency: oversight and coordination of Federal R&D resources, advice to the President and executive agencies on short-term policies and long-term national issues having significant science and technology components, and presenting Presidential strategy for science and technology to the Congress and the public.

How can the science advisory role best be institutionalized in both structure and process as an integral part of the Presidency while preserving adequate flexibility to suit the needs and style of each incumbent? Experience has shown that there is no single best way to organize the Presidency to assure that the major issues involving science and technology are dealt with satisfactorily. Inherent conflicts and tensions, both within the Presidency and in the nature of the science advisory role, make this an extremely difficult problem.

This article focuses on these constraints to establish a context for examining the structure and process of the Presidential science advisory mechanism and to consider means for improving that mechanism. Although frequently mentioned in the plethora of literature concerned with the Presidential science advisory role, these conflicts and tensions deserve special attention as major determinants influencing the effectiveness of integrating science and technology into Presidential leadership. Some of these stresses are inherent in the Presidency. These are revealed by a brief description of the Presidency and the structure and operation of the Executive Office as it exists today. Others are associated directly with the three essential advisory functions which I believe must be performed at the Presidential level. These three functions are discussed in detail with selected historical notes to illustrate the nature of the conflicts and to identify some lessons learned about what is and is not workable. From this analysis several important factors are noted which I believe should be considered in assessing alternatives for improving the structure and process of science advice in the Presidency. I conclude with some personal observations about the existing situation and how it can be improved.

Nature of the Presidency

In discussing issues pertaining to science advice and the Presidency, one must recognize that this is part of a set of broader issues centered on how the Presidency is structured and how it operates at any specific time. Two definitions are commonly ascribed to the word "Presidency." One means the institutional role of only the Chief Executive spanning the terms of individual Presidents. The other definition includes the structure and functions of the Executive Office which is made up of over 1,500 employees. In this article, the latter definition is used. It is important to distinguish between the *President* as an individual and the *Presidency* as an institution. Two key factors that need to be accommodated are the personal leadership style of the individual President and the institutional structure of the Executive Office, established to a large extent by the Congress.

The most important characteristic of the Presidency is its status as a political institution. It is the single most visible office in government. The public perceives the individual President as the head of government, responsible for most major government action or inaction. Indeed all major governmental functions either are initiated by or are reviewed within the Presidency—from broad new Federal policies or regulations to annual agency budget submissions.

This agenda, as well as the stances taken, are politically influenced. Prompt decisive leadership to solve urgent problems and establish a visible record of achievements in a four-year term that will assure reelection requires political expediency designed to gain and maintain support from the Congress and the public. Weighed against this requirement for political expediency is the need for an apolitical perspective when making major decisions affecting our nation's and the world's future, as well as each President's desire to leave a legacy of historical importance. This tension is inherent in the Presidency.

Subordinate to the political context provided by the President and his appointed Cabinet members and senior staff are several tiers of advice, analysis, and management. Closest and perhaps most important to the President are the many offices located in the Executive Office of the President. This office has evolved from a few staff members under President Franklin D. Roosevelt to an institution of over 1,500 employees. Included in the Executive Office are a variety of specialized offices, such as the Office of Management and Budget (OMB), the Council of Economic Advisers, the National Security Council, the Domestic Policy Staff, and the present Office of Science and Technology Policy (OSTP). In many respects, the Executive Office serves to focus and analyze the information from the rest of the executive branch for the President. Additionally, the office serves the President in implementing the administration's agenda by managing and coordinating the actions of the executive branch. Thus, as the functions of the Presidency have expanded and grown more complex, the Executive Office has become more important due to its vital role as a conduit between the President and the executive agencies, and for its analytical support.

Involvement of Congress

Concurrent with this increasing complexity and importance of the Presidency, the Congress has become more aggressively involved in national policies, the Federal budget process, government programs, and other Presidential initiatives. This has enhanced the conflict between congressional insistence on timely disclosure of complete information and the President's desire to preserve the protocol of "executive privilege." This tension is compounded by the fragmented structure of Congress which makes it difficult to satisfy the needs of Congress as a whole, as well as those of individual committees with special interests.

Other tensions inherent in the Executive Office of the President relate to the stature and rapport among the various senior staff members in the Executive Office and the Cabinet members; e.g., competition for power and influence among the several elements of the Executive Office and the desire of Cabinet members for independent executive authority to run their own agencies without Executive Office interference. These tensions are enhanced or diminished largely by the degree to which the President clearly articulates and demonstrates the relative stature of each element in the hierarchy.

National needs dictate that certain responsibilities of the *Presidency* transcend the views of any individual *President*. Yet each President comes into office with his own conceptions of the Presidency and his own ideas about organizing it. The preferences of individual Presidents may or may not recognize the importance of science and technology in national policy. Thus, a dynamic tension can result from the imposition of science and technology advisory functions established by the Congress on a President who may not recognize the need for them. The challenge is how to balance each President's relatively short-term political orientation and his need for flexibility with the longer-term institutional requirement for a science and technology advisory mechanism in the Presidency.

The transition from the Office of Scientific Research and Development (OSRD), which was dismantled after World War II, to the current Presidential science advisory mechanism was initiated by President Truman in 1950. Since then, the vicissitudes of the Science Adviser and his office have included expansion and contraction of responsibilities, shifts in orientation, and changes in structure. Early emphasis was on strategy for mobilizing scientific and technological resources for war and strengthening national security. Then came concern with the space race, and eventually—but somewhat sporadically—policy emphasis shifted to domestic affairs related to the economy and quality of life. More recently, the need for greater attention to science and technology in international affairs is being recognized.

Organizational Alternatives

Following the untimely demise of the science advisory function during the Nixon Administration, Congressional hearings in 1974 and 1975 explored three organizational alternatives. One called for retaining the arrangement whereby the Director of NSF would continue to serve as science policy adviser to the President. Another alternative suggested creating a Cabinet-level Department of Science and Technology. The third, endorsed by the National Academy of Sciences, proposed the creation of a science advisory mechanism similar to the former OST but headed by a council of three Science Advisers who would function in a way similar to the Council of Economic Advisers. During these hearings, I testified in support of legislation to reestablish a science advisory function in the Executive Office but indicated my preference for a single Science Adviser (assisted by a small staff) appointed by the President and confirmed by the Senate. Thus, congressional initiative, in consultation with the Ford Administration, in May 1976, statutorily established the Office of Science and Technology Policy in the Executive Office of the President. Public Law 94-282 was designed to institutionalize the three essential Presidential science advisory functions, which were alluded to in the introduction, while preserving organizational flexibility to suit the preferences of individual Presidents.

Patterns traced by these historical developments reveal tensions in science advisory mechanisms similar to those inherent in the Presidency itself. For example, there always has been pressure to give adequate attention to long-range strategic planning while simultaneously solving urgent short-term problems and addressing cyclic concerns such as evaluation of ongoing programs and the Federal R&D budget. Congress' desire to institutionalize the science advisory function so that it will have continuity and accountability, and thus transcend the term of any individual President, competes with the Presidential pressure for short-term expediency and flexible style. As within the Presidency, various entities within the science advisory mechanism have experienced disagreements and struggles among groups with vested interests.

Although the three science advisory functions essential to the Presidency are closely interconnected, in some respects they are incompatible, which makes it difficult to place all of these responsibilities in one office. Because of the importance and peculiar needs of each of these functions, they are discussed individually in the following sections in the context of selected historical events which illustrate the dynamic tensions involved.

Oversight and Coordination of Federal R&D Resources

Oversight and coordination of Federal science and technology resources are fraught with special tensions because of the attempt to balance departmental individuality with the need for an integrated government program in any given area. Since each department is responsible directly to the President and Congress for carrying out its own specific missions, the roles of science and technology are therefore different in each agency. The present science advisory office, OSTP, and its appendaged committees assist the President in providing coherent direction for the multitude of interrelated Federal science and technology programs.

Oversight and coordination have many components: information exchange among agencies on program content and management practices; interagency sharing of scientific and technological resources; developing "crosscutting" Federal policies and administrative procedures such as patent policies and personnel procedures; study of how program objectives relate to present and anticipated national needs; assessment of program overlap; how programs can best meet national goals; and monitoring progress and evaluating existing programs, budget planning, and assignment of program priorities among agencies. These components often overlap and are not easily distinguishable from one another.

Various organizational entities have been attempted in the past 25 years to attend to the function of oversight and coordination. They basically divide into two categories: interagency committees—the Interdepartmental Committee on Scientific Research and Development (ICSRD), Federal Council on Science and Technology (FCST), and Federal Coordinating Council for Science, Engineering, and Technology (FCCSET)—and centralized science advisory offices serving the

Presidency—NSF, the Office of Science and Technology (OST), and OSTP. The committees have been involved primarily with coordination and the offices were primarily responsible for evaluation and policy efforts.

The Role of ICSRD

ICSRD was the first interagency committee to be formed after World War II. Its aim was to fill the gap in the planning and central direction of Federal government science resources caused by the dismantling of the Office of Scientific Research and Development (OSRD). The "Steelman Report" of 1947 recommended:

> that a Federal Committee be established, composed of the Directors of the principal Federal research establishments, to assist in the coordination and development of the Government's own research and development programs.[1]

Responding to this recommendation, in December 1947, President Truman established ICSRD and instructed it to study "and report upon current Federal policies and Federal administrative practices relating to Federal support for research."[2]

The institutional base for ICSRD—NSF—was established in May 1950 with complementary evaluation and policy responsibilities. Part of NSF's charter stated that the foundation would be responsible:

> (a) to develop and encourage the pursuit of a national policy for the promotion of basic research and education in the sciences [and] (b) to evaluate scientific research programs of the agencies of the Federal government.[3]

This first link between NSF and ICSRD as a means of providing coordination and evaluation proved relatively ineffective. There was confusion in differentiating the roles of the interagency committee and the foundation. The foundation shied away from evaluating the research programs of peer agencies and did not provide ICSRD with any additional clout.

The outbreak of the Korean War in mid-1950 accelerated attempts by the Executive Office to coordinate government funding of science and technology research. Based on a study he conducted in 1950 while serving as a Special Consultant to the Director, Bureau of the Budget (BOB), William T. Golden prepared a memorandum for the President in which he recommended the appointment of a Presidential Science Adviser and a science advisory committee. This dual appointment, in Golden's view, would "assure continuous alertness and a flow of information and advice on matters of science and technology to the highest policy levels of government." In April 1951, President Truman named a science advisory committee comprised of 11 scientists and appointed Dr. Oliver E. Buckley as chairman. This committee was subsumed in the organizational structure of the Office of Defense Mobilization in the Executive Office, but retained the option of direct communication with the President. Thus, for the first time, the role of science adviser to the President was institutionalized.

In 1957, President Eisenhower announced the creation of the Office of Special

Assistant to the President for Science and Technology and then appointed Dr. James Killian to this office. Two weeks later President Eisenhower reconstituted the science advisory committee as the President's Science Advisory Committee (PSAC) in the Executive Office of the President.

An assessment of ICSRD's work appeared in the 1958 PSAC report *Strengthening American Science*. PSAC reported that the membership of the committee was not sufficiently high level to perform coordination activities at the policy level. The report stated that ICSRD:

> has constituted a useful mechanism for the exchange of information among research and development agencies and has been a source of policy recommendations dealing principally with scientific and technical personnel problems and the administration of Federal laboratories. But these attempts have had limited objectives and the fundamental problem remains unsolved. Each agency and department continues to formulate its own policies in science and technology with insufficient reference to the policies of others.[4]

The successor to ICSRD—FCST—was established in 1959 at a higher policy level. In signing the Executive Order establishing the council, President Eisenhower stated:

> I expect the new Council to consider and evaluate these opportunities [in scientific fields such as meteorology] and to encourage all Government agencies further to increase the quality of their efforts in these fields. By fostering greater cooperation among Federal agencies in planning their research and development programs, by facilitating the resolution of common problems, and by reviewing the impact of government policies on the programs of non-governmental institutions, the Council should be able to contribute greatly to the development and advancement of our national programs in these important and critical areas.[5]

FCST was formally linked at this point to the Special Assistant to the President for Science and Technology in the White House and to PSAC. Thus, the council was linked more closely to the policy role of the science adviser and PSAC than to the dormant evaluation and science policy planning responsibilities of NSF.

A Congressional study of 1961,[6] stated that the President's Special Assistant, PSAC, and FCST, all supported by only a small full-time staff, could not effectively advise the President on coordinating existing science and technology programs or future science planning. At that time, the author had responsibility in BOB, working with White House staff and the Science Adviser, to develop organizational alternatives for President Kennedy in the area of science and technology. Reorganization Plan No. 2 of 1962, which resulted from this inquiry, established OST within the Executive Office and transferred to it NSF's evaluation and coordination responsibilities. The re-organization recognized congressional and BOB perceptions that NSF, as a small mission-oriented agency, could not successfully coordinate and evaluate scientific and technical activities over the wide spectrum of government. Neither the first Director of NSF, Alan Waterman, nor his successors had accepted this mandate. The Reorganization Plan of 1962 formally linked PSAC and FCST to OST.

Learning from Experience

The science advisory organization set up in 1962 for coordination and evaluation is very similar to that established in 1976 by legislation and amended in 1978 by Executive Order. Since OSTP and FCCSET operate under many similar constraints, the lessons gleaned from the experiences of OST and FCST during the last two decades are applicable today.

Throughout its history, the Federal Council on Science and Technology was most effective in gathering information and developing program inventories. Ostensibly this proved the least threatening to the agencies and, thus, they were highly cooperative.

The council also was relatively effective in the determination and coordination of administrative policies and practices for R&D in the agencies. The Congressional Research Service's history of interagency R&D coordination[7] lists 20 policy statements that FCST issued in the 17 years of its existence, some of which were incorporated into BOB circulars. These statements cover topics such as patent policy, Federal payment of publication costs for research, and various international agreements for research.

There are mixed reviews of the council in more substantive areas such as materials R&D, oceanography, and natural resources. The work of the council and its committees covered a large range of issues, from promoting communication among the agencies to issuing many descriptive reports through the President to Congress. However, it is generally recognized that, in coordinating programs in substantive areas, an interagency committee has many constraints. When there were disagreements among agency positions or when vested interests were strong and the stakes were perceived to be high, agencies frequently bypassed FCST. Without a clear mandate from an authoritative source, the council found it difficult to generate strong recommendations and to have them implemented.

Thus, to be effective, an interagency coordinating council must be linked with an institutional base which gives it leverage and staff support. The most logical base is the science advisory office responsible for evaluating Federal R&D programs. The council supports its sponsor in assessing national needs in program areas requiring multiagency participation and in coordinating the implementation and processes necessary to carry out interagency programs. This link proved most successful when the base organization itself tied into the R&D budget planning process, and thus had clout to utilize the recommendations of the council.

OSTP reports that it is intensely involved in determining agency R&D budget strategy with OMB. For example, in the past two summers, the directors of OSTP and OMB have cosigned a letter to agencies stressing the importance of basic research, and asserting that it should have a special budget priority. OSTP relies on interagency committees, both informally and as part of FCCSET, to study particular areas of national need involving science and technology (e.g., disposal of radio-active waste, dam safety, and earthquake prediction) and to make recommendations about how programs should be structured and funded in these areas.

OSTP favors task-oriented ad hoc committees for coordinating and evaluating specific areas rather than the appointment of permanent standing groups with broad charters. OSTP claims that useful products are not forthcoming from the latter approach.[8] This ad hoc, task-oriented view also tends to focus and maintain the interests of OSTP and the relevant agencies and the recommendations of these groups, thus enhancing their effectiveness. In addition, OSTP has used this ad hoc arrangement to evaluate basic research programs of the mission agencies. In this case, advisory groups were selected primarily from outside of government, and have assisted OSTP in evaluations of the Department of Defense and the Department of Energy.

Policy Advice

The policy advisory function for science and technology is similar to policy advice in other areas (i.e., social, economic) in that it can be described as a continuum ranging from short-term, daily considerations to the analytical assessment of long-term problems. The short-term part of the scale relates to the quick-response requests for advice on issues such as drug research and the technologies of energy production. Longer-term analysis of present issues (such as the potential for solar energy or a survey of Federal dam safety) fall in the middle of the scale. The part of the continuum relating to the longest-term analysis covers strategic planning—the systematic identification of potential future problems and the examination of options that would try to avoid or minimize these problems.

Capability for policy advice to the Presidency on science and technology exists in three different groups: the staff of the science advisory office, supported by other agencies, principally NSF and the National Academies of Sciences and Engineering; interagency committees; and outside institutions and experts serving on panels or as individual consultants. The science advisory office in the Executive Office provides the focus, giving out assignments and shepherding projects among the groups.

The tensions pertaining to policy advice may be grouped into three interrelated categories. First, there is the balance to be sought in the mix of short-term versus long-term analysis. Second, advice and analysis from the science and technology perspective on national issues involving science and technology must frequently compete with advice offered from other staff offices within the Presidency as well as from outside sources. The third tension is associated with offering well-informed, thoughtful advice on scientific matters without what has been perceived as advocating the vested interests of the scientific community.

The first category of tension for the policy advice function is particularly germane to the nature of scientific research and the length of time before any results are visible. To avoid crisis management, it is imperative that long-term, strategic planning be incorporated into Presidential policy-making for science and technology. The mix of short-term versus long-term analysis, which becomes part of the policy-making process, depends to a great degree on the sensitivity of the President and his advisers to longer-term issues. Long-term analysis also requires extraordinary discipline to prevent resources assigned to strategic planning from being absorbed into short-term ''firefighting'' work. Without a strong counterforce to champion the notion of strategic planning as vital to Presidential policy-making, and to insulate such efforts from political pressures, long-term analysis has little chance of serious consideration at these levels. To be effective, strategic planning requires insulation without isolation.

To make good use of support groups such as ad hoc panels, committees, and staff within NSF or other agencies, the science advisory office should maintain an interest in each assigned responsibility. Without almost constant attention, the long-term perspective offered by these groups will not greatly influence the work of the office. The science advisory office is also an institutional client for the work of these groups and guides them on the timing and packaging of the final output or report.

The second category of tensions stem from the policy arena in which science and technology policy advice must compete. For any given issue, there may be competition among several elements in the Executive Office to reach the ear of the President and his inner circle of senior advisers. Each President encourages this competition to a varying extent. Thus, science and technology policy advice must be understood in the political context. Scientific and technical relevance are not sufficient. Social, economic, and political considerations must be incorporated or the advice will be minimally effective.

The problem of political context seems to have plagued PSAC throughout its 14-year history. During the 1960s, the range of issues addressed by PSAC reports seemed unusually prescient. PSAC studies ran the gamut from energy and the environment to civilian technology and international trade—issues of considerable concern to today's policymakers. But, according to Harvey Brooks, a former PSAC member,

> [I]n almost every case we failed to get the attention of top policy-makers sufficiently to raise the issue to the necessary level of political visibility to generate concern and action.[9]

This failure, says Brooks, resulted from PSAC's inability to fully translate analysis into politically intelligible terms.

The demise of PSAC has been linked to the individual stances of members and former members on the supersonic transport, antiballistic missile, and Vietnam War issues. Many scientists vigorously opposed the Johnson and Nixon Administrations' positions, and these conflicts were made public. The resulting gulf between scientists and the Nixon Administration is an important lesson for those offering policy advice in any field. By taking public stances opposing the President, advisers lose a large degree of their influence and credibility. Scientific and technical advisers must be aware of the complex factors and other perspectives that affect the policy deliberations of decisionmakers.

Scientific credibility can also be eroded by the perception, right or wrong, that Science Advisers are primarily advocates for the interests of the scientific community. Unlike other advisory units in the Executive Office, such as the Council of Economic Advisers or the Domestic Policy Staff, the science advisory office is unique in that it assists the President in setting policy for a clearly defined community of interests—the scientific community. Each set of Science Advisers has been judged by the extent to which they conceived their advisory role as singularly objective and nonpartisan. The perceptions of past Presidents that scientific advisers were biased in favor of the scientific community led to diversification of membership on advisory panels.

Each of these categories of tensions—the short-term/long-term mix, the competition with other analyses, and the intrinsic tensions of the support groups—have been dealt with to some extent by the present OSTP. The office appears to recognize the political, social, and economic context of its advice and frames its studies and recommendations accordingly. Additionally, the office draws on a large range of individual consultants, ad hoc panels, and interagency committees to provide in-depth analysis.

The most often heard criticism of OSTP pertains to the mix of short-term and longer-term studies. For the most part the OSTP staff responds to short-term, almost daily requests. OSTP uses outside sources to help formulate advice on mid-term and long-term issue-specific policies. However, critics argue that the OSTP staff is not appropriately sensitive to the need for a close tie-in with long-term strategic planning. OSTP has been involved in Presidential Domestic Policy Reviews (e.g., industrial innovation and nonfuel minerals policy) which addressed urgent current issues relative to future strategy. However, a major strategic assignment, the five-year outlook report, was farmed out to the Director of NSF. Although it is impractical for the small OSTP staff to prepare this report, critics question whether this assignment has been removed too far from OSTP. Having the study performed by special resources detached from the exigencies of daily OSTP operations is entirely appropriate subject to two provisos. OSTP should guide the course of the study and prepare an overview to be included in the final package presented through the President to the Congress.

Communicating with the Congress

The third science advisory function, presenting Presidential strategy to the Congress, involves tensions rooted in the confidentiality afforded by the President's need for executive privilege and the Congress' desire for information. Since the late 1950s, executive attempts to set "manageable parameters" on the advisory system have clashed with legislative efforts to participate more fully in the policy process. The genesis of the Office of Technology Assessment (OTA), established in 1974, was shaped largely by a Congressional desire to marshal countervailing expertise to match the capabilities of the Executive branch in coping with issues having strong science and technology components. In the past decade, the Congress has also mapped out a new and demanding role for itself through increased exercise of the oversight function.

First delineated in the Legislative Reorganization Act of 1946, the principle of oversight was formally integrated into Congressional responsibilities with the 1973 adoption of a recommendation that all standing committees in the House incorporate the oversight function within their regular activities and that each establish an oversight committee. In their quest for more and better information on national science policy, Congressional committees with oversight responsibilities for scientific and technical issues rely heavily on the head of OSTP for testimony revealing executive strategies in these areas. Today, the Science Adviser must carefully balance his task as confidential adviser to the President with his statutory position as Director of OSTP, which requires him to testify frequently before congressional committees on scientific and technical issues.

Serving Two Masters

The dilemma of serving two masters—executive and legislative—has characterized the history of the science advisory mechanism since the late 1950s. Early arrangements for science advice did not provide an administration spokesman to testify before Congressional committees on science and technology issues. Congressional complaints over this inhibition peaked in the years preceding President Kennedy's Administration. In 1959 hearings, Hubert Humphrey outlined the constraints which executive privilege placed on Congressional overseers. Said Senator Humphrey:

> [T]he reports of those councils [the Federal Council for Science and Technology] are executive privilege reports. Members of Congress never see them. We see the reports as they are filtered, strained, restrained and constrained. . .after they have been given the working over by BOB and everybody else. We finally get that housebroken, tamed report so that we never really get the gory details.[10]

In the Executive Office study leading to Reorganization Plan No. 2 of 1962, the author recommended clarification of the role which the Executive Office of the President would play, and particularly, the availability of a spokesman to appear before congressional committees to discuss its policies in the areas of science and technology. The Reorganization Plan of 1962, creating OST, included the provision that the Director of OST be appointed by the President and confirmed by the Senate. In this capacity, OST's Director could testify before Congressional committees.

Although OST's establishment removed the inhibition on testimony, the role of the Executive Office was not yet completely clarified. In 1963, the author testified on the need for improved, systematic communications between the executive branch and the Congress on R&D programs,[11] noting that a number of piecemeal and uncoordinated studies on science and technology, then emanating from a variety of agencies and private institutions, should instead be conducted by an entity with a central focus. The Executive Office, the speech indicated, should initiate holistic studies of major issues, analyses of indicators and statistics, identification and extrapolation of trends, and undertake "profound conceptual thinking." The author suggested that a presentation developed by BOB in conjunction with OST could perform the following services for the Congress: explain the major trends and changes in R&D, indicate emerging R&D investment opportunities, comment on the balance among fields of supported research, and furnish measures of the impact of Federal R&D funding on our supply of manpower, our industries, and our universities.* It was the author's conviction in 1963 that more adequate reporting procedures by the executive branch could facilitate legislative review of R&D.

At the time of OST's establishment, other advisory units in the Executive Office of the President, such as the Council of Economic Advisers, were already required by statute to prepare annual reports reviewing national policies and problems. Since a reporting requirement was absent from OST's enabling legislation, several parties began to recommend that the Executive Office of the President issue a report on the "substance, organization, costs, goals, problems, and progress of science and technology."[12] Such a report, said William Carey in 1965, could "supply in great part a conspicuously missing element. . .a framework for considering the state of science and technology in the larger economy of our public policy."[13] From 1965 to 1974, other proposals for preparing such a report to the Congress began to surface from many sources, including two Congressional subcommittees, a panel of the National Academy of Sciences, and individuals such as Nelson Rockefeller, who described the need for such a report in his 1968 bid for Republican candidacy in the Presidential race. And a 1970 compilation of Congressional hearings on national science policy pointed out that a report prepared periodically,

> would be especially useful to the Congress which could well use more leadtime than it usually gets in considering requests for support or regulations regarding scientific and technological ventures. If submitted at the same time as the Federal Budget, the utility of the reports would be particularly strong.[14]

The recommendations briefly sketched above achieved statutory authority in 1976 when President Ford signed Public Law 94-282. In this statute, OSTP was assigned responsibility to prepare a five-year outlook and an annual science and technology report. A legislative history of Public Law 94-282 indicates that the Congress intended to build a long-range policy capacity into OSTP, which would offset its tendency toward dealing solely with short-term issues. However, in 1977, OSTP's reporting responsibilities were transferred to NSF. The rationale for this decision was that "the Executive Office of the President exists to serve the President and should be structured to meet his needs."[15] In testimony, Science Adviser Frank Press explained to a congressional subcommittee that:

> the President is concerned that his staff and his Cabinet office are overly tasked with reports that take time, effort, and money, and keep them from doing other chores.[16]

Tension Highlighted

The fate of the 1976 reporting requirements highlights the tension between the Congressional desire for information and the Executive need for flexibility. Some

---

* Other recommendations of interest included a suggestion for more adequate reports on agency research programs for the information of the Congress and the performance of special studies and reports, from time to time, on selected cross-cutting problems in the administration of R&D.

argue that OSTP as an office with limited resources and pressured by short-term demands on the Office of the President is ill-suited to conduct long-range studies or periodic reports, as envisioned in the 1976 legislation. Executive reluctance to prepare such materials for the benefit of the Congress and the public, however, detracts from what should be an ongoing dialogue between the executive and legislative branches on matters of scientific policy.

The first annual report was transmitted to the Congress more than seven months after the legislatively set date of February 15, 1978.[17] With limited input from OSTP, the report failed to fully disclose the administration's plans and strategies for implementing priority decisions in science and technology policy. This was partially rectified by the President's Special Message on Science and Technology given in March 1979. Assuming that the next annual report on science and technology will be released in February 1980, I believe that, at least in part, its contents should be planned specifically to supplement the Special Analyses of R&D in the Federal budget.

Since the mid-1950s, the Special Analyses section of the budget has provided the only formal discussion of Federal science-related activities from the executive perspective. But the Special Analyses of R&D activities has not been adequate to serve the needs of the Congress since it is not designed to provide a comprehensive picture of the administration's strategies for science and technology. If the annual report and the Special Analyses of R&D were presented as companions, however, a more thorough treatment of the R&D budget, with particular emphasis on the broad oversight and related policy issues, could then be made available to the Congress. The five-year outlook report, which has not yet appeared, may indicate alternatives for administration strategies in science policy. Through such reporting mechanisms, both the executive and legislative branches can receive three valuable documents facilitating fruitful discussion and review.

There are a number of difficulties attendant upon the current delegation of the dual reporting requirements to NSF. The annual report, for example, should treat crosscutting and interagency issues involving the evaluation of programs or policies of other Federal agencies. But as an affected agency, NSF is not the appropriate organization to provide the administration's rationale for its R&D budget proposals. Due to these constraints OSTP is the logical agency to enunciate the administration's position in these areas. Although it may be necessary to continue delegating much of the staff work for these reports, OSTP should assume responsibility for those portions of the annual report and the five-year outlook that present the Administration's rationale for its overall R&D strategy and budget priorities. This would provide a more comprehensive and authoritative basis for congressional oversight of R&D policy and budget issues. R&D proposals could then be more fully assessed through hearings and public comments. Such hearings would also include testimony by OSTP which would further support and clarify the administration's position.

History has shown that any delineation of policy responsibilities is susceptible to change. Each stage in the evolution of the science advisory mechanism represents yet another instance of the tension between Presidential attempts to establish useful "organizational arrangements" and the congressional desire "to receive, in an orderly fashion, a series of reviews,'' of national science and technology policy. On April 3, 1979, Congressman Don Fuqua stated that his committee would begin to review the R&D budget as a whole to understand how it is fashioned, managed, monitored, and evaluated.[18] Congressional attempts to continue the conduct of US science and technology policy on a bipartisan basis, however, must be tempered by a well-aged caveat. In 1963, the author told the Select Committee on Government Research that no information disclosure system can ever be a panacea. The search for at least an approximation of a panacea has dominated the history of the science advisory mechanism and can be seen today as Congress continues to closely question the form and substance of the annual and five-year reports which it mandated in 1976.

Conclusion

In justifying his Reorganization Plan of 1977, President Carter stated that the "Executive Office of the President exists to serve the President and should be structured to meet his needs." In this statement, as well as in subsequent actions which removed from OSTP some of the functions assigned to it by Public Law 94-282, President Carter has shown too little recognition of the institutional responsibilities of OSTP.

Notwithstanding the President's commendable desire to keep the Executive Offices small, a science advisory office within the Presidency should provide a central institutional core with primary responsibility for the three essential functions—oversight and coordination, policy advice on both current and long-term issues, and communication with the Congress. Such a core must strive to integrate these functions into a coherent *modus operandi* with an agenda that is visible and accountable to the Congress.

When necessary to delegate tasks associated with these three functions, the head of the science advisory office should retain responsibility for ensuring that any reports issued to the Congress contain the administration's interpretation and rationale for budgetary actions or other policy initiatives to which the reports are related.

Unquestionably, more emphasis is needed on "horizon-scanning." Through the exercise of strong central responsibility at the Presidential level, we can begin to formulate science and technology policies that transcend immediate concerns and to design a prospectus for the future.

Title III of Public Law 94-282 established a President's Committee on Science and Technology to perform a major review of science, engineering, and technology, including both examination of current issues and "long-range study, analysis and planning in regard to the application of science and technology to major national problems or concerns." In abolishing this committee, President Carter may have lost an opportunity for gaining extraordinary advice and assistance from a body of distinguished citizens with broad perspectives and a degree of detachment from government operations. I agree with those who believe that reconstituting such a group of advisers, similar to the former PSAC, with subcommittees and ad hoc panels for special studies, could provide strong support for, and add prestige to, OSTP. Such a committee could serve as both a sounding board for considering alternatives for Presiential initiatives on current issues and for identifying long-range policy issues for study related to national and international problems. To cope with the broad spectrum of current and future issues involving science and technology, such a committee should have broader membership than the former PSAC, which addressed narrower issues dealing primarily with national security and the space race.

During the past half-century, the responsibilities and complexity of the Presidency have increased greatly. The modern Presidency is the focal point in our government for initiatives in policy, international leadership, and executive management of the bureaucracy. Concurrent to this expanding role of the Presidency has been an increasing recognition of the importance of science and technology in society. Today there are few, if any, national policy issues that do not involve science and technology as major components.

It is important that we continually reexamine the structure and process of integrating advice from the perspective of science and technology into Presidential leadership. Recognizing the challenge to reconcile the science advisory role with tensions inherent in the Presidency, here are some key issues:

- How can the science advisory role be institutionalized in the Presidency to preserve continuity and accountability needed for congressional oversight while assuring flexibility to serve each President's needs?
- How can we strengthen the government's capacity for early alert of impending problems related to science and technology and for developing long-range strategy to avoid crises or at least soften their impact?
- How can we best determine whether our science base is sufficiently strong and viable to assure our continued international leadership and competitive position, national security, quality of life, and a healthy economy?
- What is the best structure and framework for dealing with decision dilemmas that involve establishing science priorities?
- How can we develop a national strategy and investment plan for research?

There seems to be little doubt that the all-pervasive impact of science and technology on national security, quality of life, the economy, and international relations is so important that Presidential decisions regarding these issues and strategies for resolving them must have the benefit of the best advice available. Objective, thoughtful, and imaginative advice from the science community is vital in matters such as arms control and international safeguards; national security and defense posture; foreign relations and sharing of technological resources with other nations; potential critical shortages of energy, materials, and food; environmental protection; and the economy.

Federal science leadership   the Presidential level requires special attributes. First, it must be statesmanlike, acceptable to, trusted by, and have direct access to the President. Second, it must be respected by the community of scientists and engineers. It must not be an advocate of science *per se*, but should serve as an interpreter and adviser concerning all matters with a science component.

In general, it seems to me that the most important need of these times is for the Federal government to insure an opportunity for those with important responsibilities and good ideas to make their views known. This should be followed by a process of testing the ideas by evaluating those areas and projects which have yielded high returns and those which have not succeeded. In this process, a thorough effort should be made to understand the requirements for administrative and political success as well as scientific and engineering progress. Know-how in systems management and in large-scale governmental administration, as well as in science and technology, should be an essential ingredient of governmental R&D advice and decisionmaking.

References

1. The actual title of the "Steelman Report" was *Science and Public Policy*. It was submitted to President Truman on August 27, 1947, by the President's Scientific Research Board, which was chaired by John Steelman.
2. Penick, J.L., C.W. Pursell, M.B. Sherwood, and D.C. Swain, editors; *The Politics of American Science: 1939 to Present*, (M.I.T. Press, Cambridge, Massachusetts), p. 229.
3. Testimony by Elmer B. Staats, Assistant Director, BOB, before the House Committee on Appropriations. Supplemental Appropriations Bill for 1951, 81st Congress, 2nd Session, 1950.
4. US House of Representatives, Committee on Science and Technology, Subcommittee on Domestic and International Scientific Planning and Analysis, *Interagency Coordination of Federal Scientific Research and Development: The Federal Council for Science and Technology*, undertaken by the Science Policy Research Division, Congressional Research Service, July 1976, p. 52.
5. *Ibid*, p. 64.
6. US Senate, Committee on Government Operations, Subcommittee on National Policy Machinery, *Organizing for National Security: Science Organization and the President's Office*, 87th Congress, 1st Session (Washington, D. C.: US Government Printing Office, 1961).
7. CRS *op cit.*, Appendix K, pp. 331-380
8. Letter from Phillip Smith, Associate Director, OSTP, to Harry Havens, Director, Program Analysis Division, U.S. General Accounting Office, as published in *Better Information Management Policies Needed: A Study of Scientific and Technical Bibliographic Services*, PAD-79-62, August 6, 1979, p. 43.
9. Letter from Harvey Brooks to David Beckler February 19, 1974.
10. US Senate, Committee on Government Operations, Subcommittee on Reorganization and International Organizations, *Create a Department of Science and Technology*, 86th Congress, 1st Session (Washington, D. C.: US Government Printing Office, 1959), Part 2, pp. 127-128.
11. Testimony by Elmer B. Staats, Deputy Director, BOB, before the Select Committee on Government Research, US House of Representatives, on the "Research and Development Programs of the Federal Government," November 22, 1963.
12. Carey, William D., "A Proposal for a Yearly Presidential Report on Science," *Saturday Review*, Vol. 48, November 6, 1965, pp. 57-58.
13. *Ibid*.
14. US House of Representatives, Committee on Science and Astronautics, Subcommittee on Science, Research and Development, *Toward a Science Policy for the United States*, 91st Congress, 2nd Session (Washington, D. C.: US Government Printing Office, 1970) Committee Print, p. 12.
15. Weekly Compilation of Presidential Documents, Vol. 13, July 15, 1977, p. 1015.
16. US Senate, Committee on Appropriations, *Department of Housing and Urban Development and Certain Independent Agencies Appropriations, Fiscal Year 1979*. Hearings, 95th Congress, 2nd Session, Part 1, pp. 930-931 (Washington, D. C.: US Government Printing Office, 1977).

17. Public Law 94-282, Section 209(a), *The National Science and Technology Policy, Organization, and Priorities Act of 1976.*
18. Opening Statement by Congressman Don Fuqua. Chairman of the Committee on Science and Technology. US House of Representatives, April 3, 1979, hearings on the Federal R&D Budget.

The Pleasures of Advising

William D. Carey

On the day following John F. Kennedy's assassination I encountered Jerome Wiesner in the Executive Offices where we both worked, he as Science Adviser to the President and I as an Assistant Director of the Budget Bureau. We stood and talked for a couple of minutes, unable to shake off the trauma that had descended on the whole White House staff. "I'll be leaving, of course," Wiesner remarked, as we parted. "I've been on Johnson's wrong side too often." Later, according to Wiesner, Johnson tried hard to persuade him to stay on.

It was a small but classic glimpse of the truth that an adviser who doesn't share the mind-set of the advisee has his days numbered. The rites of objectivity cut no ice when the adviser serves at the whim of his master as distinct from having a recognized institutional base like that of a Budget Director who is accepted as a professional (and convenient) nuisance.

This fragility in the role of the Presidential Science Adviser still marks the office after 20 years of high and low tides. Each new Science Adviser and each President determine the quality of the relationship. The Wiesner-Kennedy model, which some of us recall with nostalgia, flourished in large part because of the perception of Wiesner as "the President's man" who had been entrusted with a piece of Kennedy's brief. But Wiesner was not especially anointed in that respect. The White House was run that way, as staff aides of both the upstairs and downstairs variety all freely spoke for the President, with or without his leave. It was a lively, contentious, and effective fraternity while it lasted.

Objectivity is the boast of science, and the working assumption is that the Presidential Science Adviser comes to the White House without the stain of original sin. He is an innocent in fast company, a paragon of virtue, St. Francis among the pigeons. But this scarcely represents real life. A Science Adviser who parades those virtues will be in for a rude surprise. His phone will seldom ring, he will have a tough time forcing his way into the first circle of decision analysis, and he will eat lonely meals in the White House mess.

If a Science Adviser is going to count, he must be a foot-soldier marching to the program of the President, not the company chaplain. The Science Adviser is re-

William D. Carey (b. 1916), Executive Officer of the American Association for the Advancement of Science since 1975 and publisher of Science *and* Science 80, *served for 26 years under five US Presidents, in the Bureau of the Budget, of which he was Assistant Director at the time of his departure in January 1969 to become vice president of Arthur D. Little, Inc. His entire government experience was in Presidential staff work.*

cruited into a policy management system that is committed to a political agenda. His job, basically, is to use his wits and expertise to inform the choices of a President and to carry his share of the President's burden. It is a hard world, and the Science Adviser often has to be the bearer of unwelcome advice and exasperating uncertainty. He can do this in a straightforward way as long as he is seen to be supporting the main directions of the administration and not obstructing them. It is not a clear line of sight, and it is not unheard of for a willful President or one of his purely political henchmen to grind his teeth over the conscientious staff advice he is given, and to murmur, "Who will rid me of this troublesome priest?"

A Complicated Relationship

But the relationship is even more complicated. A Science Adviser in the White House does not suspend his own nature when the oath of office is being administered. He brings to the job a lifetime of experience and a personal stock of values, both conscious and unconscious. While he will assuredly look upon life differently after some months of wide-eyed White House service, and even undergo some mutation in his values, he is unlikely to be fully reborn. If his baggage of experience and values is not reasonably commensurate with those of the President for whom he is working, a qualitative discordance may intrude into the stream of advice and judgment that he transmits, and sooner or later it will cause trouble. This is seldom appreciated when a new White House team is being recruited by talent searchers, none of whom are likely to know the first thing about science, scientific advice, or particular human creatures who practice science. They rely chiefly on the opinions of sources they consider reliable and well-disposed towards the incoming administration, and the only safeguard against a disastrous mismatch is the brief interview between the President-elect and the nervous prospective appointee. This may be no worse than the method we follow in recruiting Presidents themselves, but we cannot say much more for it.

The cultural mind-set of the adviser is no small matter. He may bring with him some of the strong ethical convictions by which he lives, such as a healthy distrust of anything smacking of authoritarianism on the one hand, or a compulsive faith in the inherent goodness of all men, on the other. He may see government as a flawed and unreliable ally of science, or as a pump needing only a better faucet. He may have deep ethical conflicts about the strategic weapons race as compared with funding development assistance, or he may come with a built-in hostility towards extremes of environmental regulations. He may be conditioned more than he knows for or against the blessings and contradictions of the market economy. There are countless possibilities, and they are embedded in the tissues of the man. Who is to tell when, and how much, they condition the shaping and texture of advice? The Science Adviser is the sum of all his parts, strengths and frailities included. It may be a source of comfort (or discomfort) to remember that his advice is absorbed into a larger stream to which a score or more of other White House staffers have contributed their share. In the end, what counts is the mind-set and cultural luggage of the advisee. And if *he* behaves with transparent objectivity, he is likely to be crucified for indecisiveness and inability to lead.

The accountability of the staff adviser, scientist or not, is another hard question. His responsibility is to the President. His *accountability* is rather more confused. He is the President's man. As a statutory officer subject to Senate confirmation, his performance is also subject to Congressional oversight. So we have a dual account- ability, though the scales carry more weight on the Presidential side. Beyond this, he is in a sense accountable to his profession, and a President would be hard put to retain a Science Adviser who had lost the confidence of his peers. That gives us triple accountability, which is rather more than enough. Putting it differently, the Science Adviser is expected to satisfy three kinds of clients and still get his work done. The conventional wisdom argues that if the adviser has the full backing of the President he will ride out flank attacks from the other two quarters, but there is history to the contrary, without naming names. A staff officer, however beloved and cherished by the President, is on his way out when he becomes a political burden. Or, as in the case of Wiesner, when he *looks* as if he will be too inde- pendent for the presidential ego. Or still more, as with Edward David, when his constituency abandons its docility and decides to take the President on.

Accountability and Visibility

Accountability is also linked with visibility. There was a time, now long past, when the theologians of Presidential government taught that White House assistants should have a passion for anonymity. The very last attribute now granted a White House officer is anonymity. Visibility is the command of the hour, and a vigilant watch is kept on the behavior, the doings, and the sayings of the President's men. No private communication between a staff member and the President is off limits, and leaked memoranda constitute prime scoops for enter- prising media reporters. The one-time notion that a Presidential assistant could not be compelled to appear before a Congressional committee to explain and defend his advice to the President is in shambles. Visibility is everything, and a President today flees to the fenced compound at Camp David for confidential meetings with his staff. There, muffled by the crack of billiard balls and the baying of perimeter dogs, the Presidential dialogues are conducted with a semblance of security. Later, the disenchanted staff aide consults his diary and prepares to write his revelations. *Very* considerably later, at least one or two presidencies removed, the Science Adviser writes his. This is discretion, and legitimate history.

Advocacy is not unknown to Presidential advising, though it is considered un- becoming to the scientist-in-residence. We are not very clear as to why this should be so. It is permissible for one presidential assistant to stump the country for consumer programs, another for women's causes, and still another for the Panama Canal treaty. But the Science Adviser gives scandal if he is an advocate for the scientific community. He is "different." He is there on forbearance and on his good behavior. Advocacy *for* the President is good, but not advocacy for science. The eyebrows of the Office of Management and Budget would rise clear to its receding hairlines if the Science Adviser forgot himself so far as to plead for science. On the other hand, it is quite all right for a Science Adviser to plug for things that are dear to the President's heart and his political goals, and to try to talk the scientific community into supporting them. That behavior earns chips for the science adviser which he can cash in later on for the benefit of science policy. As Alice might have said, "It all depends."

With all this, the science adviser is by no means a contemplative figure shrouded in scholarly studies or oblivious to the general pandemonium of the White House. He is perpetually on the run, fielding the hot-liners that land all around him. They come from his master, from OMB, from the Domestic Policy staff, from the agencies, from Congress, and from his professional constituency. In one sense this is proof of the fact that he is in the mainstream, where he must be if he is to matter at all. The problem is one of selection. He can't do everything well, nor should he attempt to. One of his burdens is that everybody thinks he has the President's ear, and wants him to be a messenger for all sorts of "advice." Another is that he is expected to turn up as the main attraction at scientific and academic socials, or to receive and smile upon the hundreds of delegations of foreign scientists that might alight briefly in Washington. A third difficulty is that Congress wants him to have answers to everything. The really tough problem is that of sifting policy issues to select the few that can be vetted thoroughly and qualify as "Presidential" in significance. That is a judgment call that he would be mistaken to make in strictly "science or technology" terms. It has to be made, as well, with awareness of the political economy as it exists at a point in time, and here he is only one of a family of Presidential staffers. It is his judgment plus that of the NSC and OMB and perhaps the State Department or the Trade Agreements Negotiator. For this, he has to be wired into the staff network and the thinking of more expert political warriors than himself.

Hunch and Judgment

It is very hard for the outsider to grasp this process of issue selection, and almost as hard for the Science Adviser to explain it. Like much of the art of Presidential staff work, it rests on hunch and judgment along with a sharp eye for random clues. There is, in addition, a subjective factor which takes us back to the fragile character of the advisory office. One individual who served as a White House Science Adviser once remarked to me that he would not fight through an issue unless he thought his chances of winning were "better than even." Losers do not last in Washington, certainly not around the White House. For a Science Adviser to have learned that the life he loses may be his own is to advance in political wisdom.

Although it has been said before, the Science Adviser's best hope is for normalized relations with OMB. That durable center of Presidential power regards itself, with justice, as the Swiss Guards of the Presidency. Its good will is a pearl beyond price in the jousting of the staff system, while a whisper of its displeasure can be traumatizing. After 60 years OMB believes that it has heard everything at least once, which is very nearly the truth. For a Science Adviser not to do his staff homework before taking on the OMB is a disaster in the making, exceeded only by the mistake of attempting an end run around OMB. It isn't that OMB is against science. Quite to the contrary (although OMB would blush to own up to it) that office has contested effectively with one President after another to protect science from the Presidential wrath. But OMB very much dislikes to be wrong. Flawed staff work, or its appearance, is a capital offense in the eyes of the White House and OMB. What this means is that the Science Adviser has to have a staff of his own that matches that of OMB. Otherwise the Science Adviser will be frozen out of the OMB process, which was what happened at one stage in the Nixon administration.

The Science Adviser is not the President's sole source of White House advice on science and technology policies. OMB has plenty to say on these matters, all of which have a bearing—often a major one—on budgetary trends and future obligations. OMB's vote will go beyond costs, however, and will reflect its judgments of scientific need, feasibility, and utility. These challenges are not directed, like Senator Proxmire's prize awards, to specific research projects, but at proposed new commitments or systems. The choice of an approach to the breeder reactor, for example, will get OMB's detailed attention. The merits and performance characteristics of the third or fourth generation LANDSAT will be argued intensely. Alternative basing plans for the MX missile concealment strategy will see OMB burning the lights far into the night. Urgent schemes for a crash Federal program to stimulate research and development on synthetic fuels will see OMB in the center of the arguments, questioning government's role versus that of industry and pointing out long-term residual financial risks.

When some of us were young professionals in the White House Budget Office, the same kinds of issues were tackled mainly by generalists. When it came to marginal judgments on scientific and technical factors, they were outgunned by experts in the office of the Science Adviser or panels of the now-defunct Science Advisory Committee. But OMB has come a long way. Its Energy and Science Division alone has a staff of 21 professionals, more than half with doctoral degrees. Among them we find a Ph.D. nuclear engineer, two doctoral physicists assigned to nuclear programs, an astrophysicist and an engineering physicist working on space programs, a mining engineer, two more physicists, a geologist, and several others with equal advanced credentials in other disciplines. The point is not that they are competing with the Science Adviser for the President's ear, but that the two staffs together give the Executive Office an institutional strength in science and technology that is decidedly impressive. When their advice coincides, its weight is potent. When it takes different courses, the President can have confidence in the quality of both streams of judgment.

The Way It Was

To be sure, Presidential advising takes a variety of forms. It can be highly structured, as a collective and even deliberate exercise, notably when the National Security Council and the Cabinet meet with the President. Some Presidents, especially Eisenhower and Nixon, carried this kind of advising to an extreme with elaborate arrangements for formally planned agendas and special secretariats charged with writing up minutes and chasing after high officials to be sure they carried out assignments given to them. In those days, the penchant for managing advice went rather far. To knit together the foreign affairs, intelligence, and propaganda arms of the national security structure there was an Operations Coordinating Board of undersecretaries, flanked by a large and deeply staffed NSC Planning Board run by the President's National Security Adviser. Those were the years when

American confidence and power were at a peak, and before the disasters of Vietnam and Watergate made it popular to preach against the "imperial presidency" and call for the dismantling of the White House staff system. There may have been more than a trace of pretentiousness and elitism in the Eisenhower and Nixon approaches to collective advice and policy management, but there were fewer complaints about hesitancy, vacillation, or a vacuum of leadership. The power train was engineered to work.

A very different kind of White House advising, which seldom meets the public eye, is the tutorial style. At its apogee, it saw Walter Heller coaching President Kennedy in the mysteries of the new economics. Kennedy was a liberal, handicapped by a fiscal conscience, and he came to power with a bias for balanced budgets and fiscal restraint. It took Heller, as Chairman of the Council of Economic Advisers, to work Kennedy over and persuade him that government should exercise economic leverage to correct for the failings of the market economy. The result was a Presidential speech at Yale in which a grinning Kennedy abandoned his upbringing and declared, "Now I am a Keynesian."

To go back to the prior administration, one would find Arthur Burns in a comparable role with Eisenhower, though travelling in quite a different policy direction. More currently, we have observed President Carter studying the dismal science with Charles Schultze as resident tutor. It is also very likely that we would find the Science Adviser explaining to his engineer-leader some of the more arcane aspects of genetic experimentation involving DNA, or the uncertainties relating to the consequences of CO_2 concentrations in the biosphere if a government program for synthetic fuels should be undertaken.

The role of the speechwriter as a White House adviser may come as another surprise. A Budget Director in the Johnson administration once remarked, at the end of a hard day, that the worst threats to the budget came not from free spenders in Congress but from the President's speechwriters. Typically, a President will commit himself to address a crowd whose political support he needs, then gives his chief ghostwriters a general outline of what he will talk about. The writers then disappear to prepare drafts. As the date for the speech nears, drafts are shared among the political counsellors to the President, who call for more inducements to applause and news headlines. By the time the budgeteers catch up with the writers, gashes appear along the whole fabric of the budget and heated telephone arguments ensue. The President himself may even be called upon to settle the fight. Nor is it unknown that a President is so carried away by his own fervor that he will improvise and invent new promises which have never been staffed out. But the role played by the Presidential speech in focusing choices and alternatives is a serious matter in the era of instant and universal communications technology. The speech is a decision-forcing device of the first importance, and the speechwriter is no longer merely a gifted wordsman. He can make or break a President, depending on the substance of the speech and how critical its timing and presentation are to the President's fortunes. Very fine judgments have to be made as to what to include or leave out, how to say what must be said, and how far the Presidential neck is to be exposed. Writing for a beleagured and unpopular President is considerably harder than writing for a President who seemingly can do no wrong.

A Personal Experience

One day in 1948, in the midst of Mr. Truman's apparently hopeless race against Governor Dewey, I had a call from the President's Special Counsel (read "Chief Speechwriter") asking me to work up a campaign speech on atomic energy. Dewey had made an incautious attack on Truman's policies for the development of atomic energy, calling for the removal of "the dead hand of government" on the industry. Truman was especially proud of his stewardship of the bomb, and he was furious with Dewey. I drafted a speech which was strictly within the bounds of existing public policies, and sent it over to the West Wing. The next day, the President's man was back on the telephone, complimenting me for a good speech, but quoting the President as saying that it wasn't tough enough on Dewey. I said that I thought this was someone else's job, not mine. That afternoon the Special Counsel called again to tell me that the President himself had "fixed" the speech, and that he was sending over a copy. The copy didn't arrive until the next day, and what I read in it made me lunge for the telephone.

What had happened was that Truman and his inside advisers had convened in the Fish Room to work my draft over, and the President started reminiscing about the first word he had heard of the successful drop of the Hiroshima bomb while he was on a battleship en route to his first wartime Summit. Now he turned to his press officer, Charles Ross, and told him to go and get his Potsdam diary. Then, diary in hand, he read aloud to the group his notes, including the briefing by General Marshall on the mixture of enriched uranium and plutonium in the first bomb. "There," Truman said, "we have the lead for the speech. We'll just start with this page from my diary." And that was what I saw when I looked at the new draft. It was a spectacular security leak, and it was to be delivered in a stadium before tens of thousands of the faithful. It had to be stopped, and I called Charles Murphy and told him so. "Where are the other copies?" I asked. "Well, the President has one with him on the train. ·Jonathan Daniels has another, but he's on his way to North Carolina. I guess there are a couple more." We finally hunted down all the copies, incinerated the disastrous details, and got the speech out. A copy came back to me, after the speech was delivered with "applause" noted at a dozen points, and the report that Dewey had been taken care of. It had been a very near thing, however.

Issue Orientation

The chemistry of staff advice in the White House is affected by still another variant, and it is here that professionalism and political intuition converge with results that are not always foreseeable. This variant is the intensity of *issue orientation* as it dominates the traffic of advice. It is this factor, indeed, that stamps communication between OMB and the President on the one hand, and the science adviser *cum* both the President and OMB. The whole lifestyle of OMB is one of issue analysis, and as the stakes increase relentlessly there is a diminishing degree of slack at the margin to allow for miscalculation or misinformation—much less than was once the case. The political advisers in the White House are dealing with the same issues, and the cross-traffic of political judgement and professional advice produces the issue-connected struggles within the advisory system which trap a President and seal off his exits. A useful contemporary example is provided by the energy crisis, where policy management has been brought low because of the standoff between pro-energy advisers on the one side and pro-environment advisers on the other. A President simultaneously committed in policy terms to both an "energy independence" goal and an "environmental protection" goal is in enough basic trouble to start with. His problem becomes intolerable when he organizes his advice-giving staff in such a way that the checks and balances produce a zero result, or stalemate. In the Carter White House, some elements in OMB have tended to side with the Secretary of Energy on policy options while others in OMB and in the Council on Environmental Quality as well as the West Wing have maintained an effective and deadly small arms fire at the Secretary.

Trite as it may sound, staff advice is built on trust. A lot of trouble has been caused, in many presidencies, when that has been interfered with. Despite Lyndon Johnson's earnest words to the shattered staff of the dead Kennedy, there was too little mutual trust to justify their staying on. With each incoming President, the career staff of OMB has to earn White House trust all over again, especially if there has been a change in parties. When Eisenhower was elected, an emissary came to me and warned me that no time would be lost in ridding the Budget Bureau of the "socialists" who infested it. As matters turned out, Eisenhower's Budget Director quickly came to the judgment that the staff was uncorrupted, and nobody left. Eight years later, Kennedy's emissary sought me out to tell me that *their* first order of business was to rid the Budget Bureau of the "reactionaries" who infested it. Again, the new Budget Director found no reason for a purge, and nobody left. In each case, the socialists and the reactionaries were the same people. But it did not mean that the Eisenhower and Kennedy appointees in the West Wing ever were reconciled to the loyalty of the Bureau's career staff, even though there was little that they could do about it.

Such is not the case where relationships between the President's political advisers and the executive departments are concerned. The latter are almost invariably viewed as a well-entrenched resistance movement aimed at frustrating the President and determined to put their professional priorities and interests ahead of his. The advice and the position papers that well up from the agencies are taken with more than a grain of salt and, when they argue against something that the White House staffers favor, the language heard inside and outside the President's office will not bear repeating. It is not merely a Republican distrust of an overwhelming Democratic career force, because the Democratic Presidencies tend towards even greater antipathy and even run candidates for the Presidency who crusade against the government they propose to lead. If the decision-making process, resting as it does on staff advice, turns out to be crippled and deformed, the cause is to be found in part not in a failure of national confidence but in self-inflicted mistrust. That infection is not remedied quickly.

"Advising Advisers"

A last word might be offered on the subject of "advising advisers" because this is the process through which advice progresses through a relaying and tempering series of stages to the breathless President. Advice advances by increments, undergoing mutations as it breeds with intersecting inputs. Advice from lower levels is affected by advice trickling down from above. In the expression used by Flannery O'Connor, though not with this in mind, "everything that rises must converge." Advisers advise other advisers. Government is a great marketplace in which advice is traded back and forth in countless tongues and to suit all preferences and needs, and it presents an aspect of governance that is not taught in schools of public administration. Seen in this light, advice is not cheap. But it can be viewed as an infinitely renewable resource.

Staff work in the White House is a moveable feast. It ranges from the pedestrian to the exhilarating, from the absurd to the sublime, from success to disappointment. When it is good, it is very good. When it is not, the less said the better.

Helping the President Manage the Federal Science and Technology Enterprise

Lewis M. Branscomb

The Fundamental Issue

Sunday evening, July 15, 1979, President Carter spoke through 60 million TV sets to the American people. In an emotional, forceful speech he discussed the two sides of his responsibilities as President: running the government and listening to the people. His elaboration on this theme emphasizes the dichotomy in the roles of many White House officials, including the Science Adviser. Thus the President implicitly framed the question I wish to explore in this paper:

What is the balance that should be struck between the President's need for technical experts to help him manage the science and technology activity within the Federal enterprise, versus staff leadership to gather support for national consensus in the technical community on broad issues of the day?

A President's personal preference will, of course, determine both the identity of the Science Adviser and the extent to which the adviser participates actively in key policy and management issues. This may be expected to vary considerably as different individuals occupy both jobs. Furthermore, there is little inherent conflict between the managerial and political roles from the perspective of the White House. Indeed, they are almost inseparable. But interestingly, the scientific community does sense a conflict. Many scientists prefer to characterize the adviser's role as a spokesman to and for the scientific community. While consistent with the political aspects of the job, this attempt to "capture" the Science Adviser for his professional constituency may weaken his ability to function as part of the President's managerial team. *It is my contention that the country needs a stronger staff function to deal with technology matters than is compatible with the scientific community's more passive view of the adviser's job.*

That Dr. Press serves in both capacities can clearly be seen from recent examples. He was an active and important participant in the construction of the rapprochement with the People's Republic of China, and actively intervened with a number

Lewis M. Branscomb (b. 1926), vice president and chief scientist of the IBM Corporation, is president of the American Physical Society and a member of the National Science Board. He was a member of the President's Science Advisory Committee under President Johnson and has been director of the National Bureau of Standards and a member of the board of the AAAS.

of agencies to lay the proper programs in place to support the diplomatic objective of normalization through the science and technology agreements. He took the initiative in efforts to create an Institute for Cooperation in Science and Technology within the new technical assistance structure.[1] He has been heavily involved in efforts to improve the way agencies manage their relationships with universities and their sponsorship of basic research. He was actively involved in key appointments of executives in scientific agencies.

On the other side of the coin, Press has actively supported the SALT II ratification effort by seeking understanding and support from the scientific community. He arranged for the President to address the National Academy of Science, and has sought support among technical people for the President's energy strategy. And while he has been an effective advocate for expanded Federal investments in basic scientific research, he has made sure the beneficiaries of these efforts in the scientific community gave credit where credit is due—to the President.

I do not mean to suggest that the political side of the Science Adviser's role is at all improper or that a friend of basic research is not needed in the White House. Partisanship is not involved, and the events that give the President opportunity to sell his program to scientists equally give scientists a chance to criticize his proposals and raise new issues. Nor is the "political" role a new one. Although its participants did not see it that way at the time, the President's Science Advisory Committee (PSAC) had many characteristics of a political device. Let me explain.

PSAC was comprised (with a few exceptions of which I was one) of private citizens in a voluntary role as advisers to the President and his Science Adviser. Its members, though both Democrats and Republicans, inevitably became identified with the President's efforts to deal with complex technical issues, adding at least some luster to these efforts in the process. Geographical distribution played a role in the selection process. Through its many panels PSAC reached a large number of influential people. And among those were many who became members of political campaign committees such as "Scientists and Engineers for. . .(Kennedy, Johnson or Nixon)."

A Political Tool

From the scientific community's point of view, PSAC was a political tool too, as indeed, many consider the National Science Board (NSB) to be also. PSAC was perceived as the "voice of science in the White House" by scientists, just as Science Advisers have sometimes described their job (if the switch in metaphor can be excused) as providing "a window on the world of science."

The notion of the Science Adviser as primarily concerned with scientific affairs for the benefit of science is a very respectable view. When President Ford asked Vice President Rockefeller to develop a plan for re-establishing the job of Science Adviser, the National Academy of Sciences (NAS) was asked to prepare a report on the proper role and function of the office. The NAS Council commissioned a distinguished group chaired by Dr. Killian (who had occupied the post of Science Adviser) to prepare the paper. When it was completed, a vigorous debate in the council brought out a minority view that the President would be ill served if he did not look to OSTP for help in managing the technological activities in the Federal enterprise. Unhappily, in my view, this aspect of Dr. Stever's office never received adequate acknowledgement from the Executive office of the President. Perhaps the time was too short and the election already in view at the time.

Two years later, the election and inauguration of Jimmy Carter came and went before there was any public mention of this office by the White House. The job of Science Adviser was not filled until March 1977. During this period of post-election uncertainty, the efforts of the scientific community to draw the administration's attention to this neglect of their interests only served to reinforce the impression that the scientific community saw the office as their representative in the power structure. At a time when some wondered if the President really thought he needed a Science Adviser, it was pointed out that the real question was not, "Does the President need a Science Adviser?" but, "Does the Science Adviser need a President?"

In any event, the OSTP was established with less than the Congressionally established staff and budget, and without the Presidentially-appointed Council of Advisers provided in the statute. While this council was established for a specific task to study the functioning and reorganization of the science and technology activities of the Executive branch, and was given two years to complete its work, the Congress built in an option for the President to transform it into a body like the PSAC which existed until 1973.

Thus, the management capabilities of the OSTP were restricted from the outset. The President's advisers on White House and Executive office organization were not inclined to accept the need for a strong OSTP. Nevertheless, Dr. Press has been able to overcome many of these attitudes and has established a substantial role for himself in appointments and policy-making in technical areas.

The Science Adviser should:

(a) provide linkage to the scientific and technical community in order to attract their cooperation and support, and benefit from their warnings and advice; and

(b) assist the president and Cabinet with insuring effective management, organization, policy, and program effectiveness in Federal science and technology activities.

The Science Adviser as Manager

There are six major elements in this second role.

1. Appointments

The Science Adviser should ensure that the President's nominees for departmental Research and Development (R&D), assistant secretaries, and heads of major R&D agencies (NASA, NSF, NIH, NBS. . .) are fully qualified technically and managerially. The specialized Assistant Secretary posts were established by President Kennedy through Science Adviser Jerome B. Wiesner's efforts. The Science Adviser should suggest appointees and should have right of review of their qualifications before the final staff recommendation is made to the President.

2. Organization

Many organizational issues require management experience in R&D as well as a keen sense of the art of the politically possible. Example of such issues are: How should research in support of regulatory objectives be managed, by the regulators or in an independent agency? When should Federal R&D agencies be integrated with departments whose broad purposes they support, and when as independent agencies? How can the very capable (and very costly) national laboratories be re-employed to meet shifting needs, particularly when their needs are not yet buttressed by clear definition of the problem and a political consensus to support it?

3. Policy Setting for Government Operations

Many elements of management policy have strong effects on the science and technology enterprise. Federal procurement policy, for example, provides for Independent Research and Development (IRAD) funds to permit industrial contractors to charge some of the cost of proposed R&D to procurement. New contract mechanisms, such as the Federal Grant and Cooperative Agreement Act of 1977, now established under statute Public Law 95-224,[2] permits new forms of Federal cooperation in R&D with state or private industry. In all such matters, the OSTP should be heavily involved—as indeed, Wiesner was when the Bell Commission on Procurement was organized in 1962.[3] Other examples of management policy include patent policy, regulatory processes as they relate to industrial technology, means for providing incentives for commercial innovation, and the like. The Science Adviser should be an active participating member of the President's team for addressing public policies for economic matters.

4. Evaluating Quality and Effectiveness of Federal R&D Programs

This was a major function of PSAC and, although by no means universally welcomed by the operating units, in government it probably represents the most consistently valuable contribution of PSAC in the past. The President has no other staff competent to know what he is getting for his investment at the substantive—not political—level. Even without PSAC this job can be done with ad hoc panels of experts, and by requesting the advice of the National Research Council. Evaluation clearly supports the next activity in this description.

5. Objectives and Priority Setting

Most of the attention of government is focused on priorities rather than objectives, since priorities determine financial input while objectives may influence what the public gets for its investment. Perhaps the largest focus of such efforts in the past was in defense. The OST and PSAC also struggled manfully and with considerable effect over the character of the space program. But the most valuable objective-setting role for OST and PSAC was, I believe, involved in bringing major new issues to national attention for the first time. A fine example was the World Food

Study of 1967, lead by John Tukey.[4] This early warning and issue-characterizing role has both its managerial and political dimensions.

6. Resolving Internal Conflicts on Technical Issues

Federal agencies often have very different perspectives on the same technical issues. Not only are responsibilities often dispersed among many agencies, but their diverse channels of accountability to Congress and to the public accentuate these difficulties. The Science Adviser should be in a unique position to try to resolve these differences insofar as they are amenable to objective analysis and forthright personal leadership. To the extent he succeeds, he reduces the burden on the President of conflict resolution. This role can engage the Science Adviser in many controversies that transcend concerns of the technical community.

These six managerial functions are quite analogous to the responsibilities of chief scientists in a number of corporations (for example, International Business Machines, General Electric, and Xerox). In many other companies the same functions are carried out by the vice president for R&D or Chief Engineer. Just as the top White House officials have political backgrounds, top corporate executives generally have marketing or finance as their primary experience. They need experienced professional management support for the scientific and technical functions in the enterprise.

R&D is more difficult to manage well than many other activities. The quality of results produced are highly variable and difficult to quantify; a lot of subjective judgement is involved. It is relatively easy to organize people and facilities to engage in research and development. It is very difficult to insure that the technical work, once accomplished, results in a timely, useful and cost-competitive product that can be sold, installed and serviced with customer satisfaction.

So, too, it is with Federal R&D investments, except that there is an additional management problem that makes the effective use of R&D even more difficult than it is in the private sector. Except in a few constitutionally-reserved areas, the benefits of Federal R&D must be conveyed to the public by private—usually commercial—enterprises. Even the performance of Federally supported R&D is primarily (about 70%) in private or local public institutions.[5] Thus the government's efforts to engage the national R&D enterprise with specific public purposes are unlikely to be effective unless the quality of its management, organization, policy, supervision and priorities are very high.

The Role of the Office of Management of the Budget (OMB)

Can the President expect his administration to be run in this way? The only staff organization in the Federal government that might be expected to look after such matters across the whole administration, other than OSTP, is the Office of Management and Budget (OMB). The "M" is intended to cover all such matters, and there are and have been a number of outstanding people in OMB who have dealt wisely and successfully with technical programs. But few have had personal experience with scientific research or technology management. In any case, just as the central finance, plans and controls staffs of corporations are primarily focused on resource allocation, not R&D effectiveness, so the OMB is very heavily focused on the budget process. Its evaluation of science and technology activities are oriented to exacting justifications for expenditures, not to the operational qualities of the work.

There is a natural and appropriate relationship between the staffs of OSTP and of OMB. When they work together voluntarily, the operating agencies are more forthcoming and receptive to OSTP review because of the undeniable "clout" of the budgetary authority of OMB. Reciprocally, the expert professional resources available to OSTP can give the budget examiner a superior quality of insight into the quality and effectiveness of Federal programs. Recent Science Advisers have understood this symbiosis and have built workable relationships with OMB.

Similar relationships with other staff operations could also strengthen the Science Adviser's effectiveness. Most important, in my view, would be a bridge to the Council of Economic Advisers and the President's Special Trade Representative, as well as to the Treasury. The effectiveness with which the nation's technological capabilities are nurtured and applied, substantially determines domestic economic well-being and foreign trade competitiveness. The fate of the Domestic Policy Review on Innovation, a major administration effort led by the Science Adviser and Secretary of Commerce, will test the importance that OSTP sponsorship of such economics-oriented activities can command in the President's economic priorities.

Should a future President feel the need for stronger staff support in technology and science, the question will arise: What about resurrecting the President's Science Advisory Committee (PSAC)? Legal requirements for public meetings of such a committee, and the problems of political accountability make it impractical for a part-time citizen's committee to address many of the issues debated by PSAC. It certainly cannot substitute for full-time government staff, or deal with current operating problems in government. Nevertheless, a new PSAC could still usefully address longer-term issues deserving public attention. The World Food Study, organized by John Tukey for PSAC in the 1960s, for example, could have been conducted under the new contraints without serious difficulty. PSAC can contribute to the agenda for Federal attention and the analysis of technical issues to insure that policy questions are properly posed. But PSAC is no substitute for competent management in the White House and the agencies.

Presidential Attitudes and Government Organization

In the absence of encouragement from the President, it is certainly difficult for a Science Adviser to take management initiatives of the kind I have described. Dr. Press does not suffer a lack of confidence on the part of the President. Indeed, he probably has more personal access to the Oval Office than any of his predecessors. But, at the beginning of the administration, the White House staff expected the OSTP to play a more modest role, and one oriented more to external scientific relations and the well-being of the scientific enterprise than was envisioned by the

Congress in the statute[6] that established OSTP in the Executive office. Even with the substantial responsibilities Dr. Press has earned during his tenure, the staff resources authorized by statute have not been made fully available.[7]

A strong base in the White House for management attention to technological issues seems particularly important in view of the fragmentation of the organizational structure of technology functions in the United States Government. These independent science and technology agencies are unrepresented at Cabinet level. Their administrators enjoy an important informal relationship to the Science Adviser, but do not have to take policy guidance from him. In any case, he must be the President's critic of their program, not their public advocate. Many have suggested a reorganization[8] which would bring a number of agencies together into a consolidated Department of Science and Technology. Most suggestions include the National Science Foundation (NSF), National Bureau of Standards (NBS), National Aeronautics and Space Administration (NASA), and National Oceanic and Atmospheric Administration (NOAA).

Such a consolidated agency would have a broadened outlook from that of the NSF into areas of technology, environment, and policy, and give the President a more powerful operational capability with which to address the major issues concerning the nation's science and technology infrastructure.

But a more difficult area to manage is the Science Adviser's involvement in reviewing the scientific and technological strategies of the Cabinet departments and the technology components of their policies. The unwillingness of the White House staff to give OSTP a strong role in the energy field early in the current administration was apparent and, in my view, unfortunate. How can this role be developed, given strong personalities in the positions of departmental secretaries?

The answer lies again with Presidential attitude. If the President feels the need for checks and balances within his own family to insure the soundness and objectivity of the policies recommended to him, he will use a qualified member of his own White House team to review, evaluate, and advise. James Webb was a strong leader, too, but President Kennedy gave Science Adviser Wiesner the job of staff review of NASA, and the Apollo program's success was to a considerable extent assured as a result. The space shuttle program still needs the same kind of skeptical monitoring today in my opinion, as do many other areas of agency activity.

Summary

The above discussion argues that the management of the science and technology functions is as important to the Executive office as are efforts to maintain a constituency for federal science policy among the members of the scientific community. This management function was underemphasized when the Carter administration established the OSTP, however. Progress could be made by increasing the resources and authority of the Science Adviser in policy, organization and personnel decisions, and by creating a cabinet-level department which includes the key scientific agencies. Neither objective can be achieved by the Science Adviser of the Congress without the active support of the President.

References

1. Testimony by Dr. Frank Press to the Senate Foreign Relations Committee, May 15, 1979, re: Institute for Cooperation in Science and Technology.
2. Public Law 95-224: Federal Grant and Cooperative Agreement Act of 1977.
3. Bell Report to the President on Government Contracting for Research and Development, sponsored by David E. Bell, Director of US Bureau of the Budget, April 30, 1962.
4. World Food Problem: Report of President's Science Advisory Committee Panel on World's Food Supply, Vols. 1 & 2, 1967.
5. Research & Development AAAS Report IV: Federal Budget: FY 1980, Industry International, by Willis H. Shapley and Don I. Phillips, 1979.
6. Public Law 94-282: National Science & Technology Policy, Organization & Priorities Act of 1976.
7. Testimony by Dr. Frank Press to the Senate Foreign Relations Committee, May 15, 1979, re: Institute for Cooperation in Science and Technology.
8. "Science in the White House" by Lewis M. Branscomb, *Science*, May 20, 1977. Vol. 196, pp. 848-852.

Presidential Science Advising

Richard L. Garwin

Introduction

The history of White House science advice to the President of the United States extends only from about 1957 until 1979, a period rich in examples of both success and failure. As the historical background has already been described by others, this paper presents the author's experience and views for the future.

The text describes some functions which a Presidential Science Adviser can perform uniquely well, points out the tension between the job of President's Science Adviser and that of other Presidential assistants, explains the role of the President's Science Advisory Committee (PSAC), and emphasizes the degree of tolerance which a President wanting to serve his country may need in order that he and the nation benefit from the President's Science Adviser and the President's Science Advisory Committee. Some examples from the last two decades of history in which the President's science advisory organization has worked very well (and others in which it has not been brought into play) support these prejudices with experience.

Part of the PSAC product is available in its published reports. A view of the more introspective side of the President's science advisory organization may be gleaned from the few letters reproduced or excerpted here to illustrate several points. Others may or may not share these views, but I think that it would be good now to begin a reasoned discussion of some of these ethical and procedural questions.

The Role of the President's Science Adviser

It must be said once more that the purpose of the President's Science Adviser is not to speak for science in the White House but rather to speak to the President of those things which science and the scientific community can say to him—identify problems, provide solutions, incorporate a firmly-based vision of the future. The

Richard L. Garwin (b. 1928) served two four-year terms on the President's Science Advisory Committee (1962-65 under Presidents Kennedy and Johnson and 1969-72 under President Nixon), chaired several of its standing and ad hoc panels, and is a long-time consultant and adviser to the Executive and the Legislative branches of the US government. He is an IBM Fellow and Science Adviser to the Director of Research at IBM's Thomas J. Watson Research Center.

President's Science Adviser and the President's Science Advisory Committee have on occasion been criticized (particularly by the White House staff during the Nixon years) as a special-interest group for science, but I find this criticism both contrived and unwarranted, in that the White House staff didn't know that the President's science advisory organization was needed, wanted to emasculate it, and tried to criticize it in the most telling way possible—hence accusations of self-dealing, and the like.

The isolation of even experienced and capable White House staff members is illustrated by the first "comment" in a letter written for discussion at the December 1969 PSAC meeting:

R.L. Garwin
Dec. 2, 1969

THE ROLE OF PSAC AND OST IN
NATIONAL SECURITY AFFAIRS
(for discussion at PSAC December 1969)

In the past PSAC has been an important source of advice to the President in substantive national security matters. It has also been an influence in the organization of the government in these affairs, having been responsible for the creation of the post of Director of Defense Research and Engineering (DDR&E), for the creation of the Arms Control and Disarmament Agency (ACDA), NASA, and for many major policy decisions in their field.

In substantive matters touching on national security, the influence of the Science Adviser and of PSAC has been both widespread and deep. Some examples come to mind:

1. One of the most recent and most important successes of the PSAC/OST complex was the work of the BW/CW (biological warfare/chemical warfare) panel this summer under Ivan Bennett and Vincent McRae. The report of this panel was clearly responsible for the present government position in BW/CW as suggested in the President's statement of November 25, 1969.

2. In 1960, the Special Assistant to the President for Science and Technology, observed that the U. S. IRBMs in Europe had inadequate safeguards against accidental or unauthorized launch, and that the U.S. nuclear weapons in Europe on airplanes piloted by foreign nationals were also under inadequate control. With the aid of members of PSAC and of the staff, an urgent program was initiated to invent, develop, and to fit all such nuclear weapons with the permissive action link (PAL), an effective control contributing not only to the security of our forces, but also to the effectiveness deriving from flexibility of deployment.

3. In 1964 and 1965 the PSAC Military Aircraft Panel, in the course of its work, reviewed the Air Force and Defense Department proposals to develop a heavy logistics support aircraft (CX-HLS) which has now become the C-5A. After intensive activity in this field, reviewing the feasibility of the aircraft and its proposed use, the panel wrote several reports and made presentations to PSAC concluding that the mission of the C-5A could be accomplished far more cheaply by the use of ships (now the fast deployment logistic ships, FDL) and that these ships should be built rather than the aircraft. The presentation to PSAC was attended by the DDR&E and the Comptroller of the Defense Department. The Defense Department eventually did authorize an FDL program, but Congress has not seen fit to fund it. Events have, of course, validated the panel's position.

4. From the beginning of the Vietnam War, the PSAC Military panels have been active in their two traditional roles: (a) to provide for the President an informed view of the nation's military capabilities, and (b) to aid the decision process, the Defense

Department, the Bureau of the Budget, etc., in maintaining and creating the most effective and efficient military force. PSAC members and panels were heavily involved in the provision of electronic warfare apparatus for the aircraft flying in Vietnam, in the initiation and conduct of the ''barrier'' to infiltration, and in studies of the means and possible effectiveness of various aspects of the bombing of North Vietnam and of a possible blockade of North Vietnam.

5. PSAC Military panels have been instrumental in bringing night vision capabilities to an operational status in Vietnam. In 1965, after 4 years of routine advice and urging, the PSAC/OST pointed out that capable night vision programs were being managed by the Army on a schedule which would equip the Army world-wide without regard to the early capabilities which could be realized by adding night vision capabilities to a relatively few heavily utilized Air Force aircraft and Army helicopters. DoD response to the PSAC advice was to appoint a czar of night vision programs, until he was removed for another crisis a few weeks later. Eventually, the Science Adviser brought this matter to the attention of the President, who intervened with the Secretary of Defense to hasten this much needed capability.

Comments

While PSAC/OST has a record of important successes, its involvement has been in many cases almost accidental. For instance, the PSAC Vietnam Panel last month heard from Dr. Chester Cooper of the IDA and formerly on the staff of McGeorge Bundy, the President's Special Assistant for National Security Affairs. Dr. Cooper discussed with us decision-making in regard to Vietnam, making clear the secrecy, the ignorance, and the confusion attending our involvement in and conduct of this non-war with a non-organization. At a time when the PSAC Naval Warfare Panel and the PSAC Military Aircraft Panel and the PSAC Vietnam Panel were all intensively involved in the war, Chet Cooper did not know of the existence of PSAC and thus had no idea of the informed support and resources available to him right in this building. In fact, he didn't learn of the existence or nature of PSAC until he went to work for the Institute for Defense Analyses where he is now Director of the International and Social Studies Division.

The PSAC was elevated to the present role of providing advice to the President only after Sputnik. There is no magic in the PSAC capabilities. PSAC is composed of individuals whose integrity and energy have been repeatedly certified by their successes in science. The PSAC panel system makes available the most expert talent on any given subject, and the fact that the panel chairman is a member of PSAC normally provides a level of responsibility unusual in reports of expert groups.

However, the PSAC cannot function in national security matters without access, and its access comes from the President's need for independent advice not only on development programs but also (and this is often missed by the Defense Department and by others in the Washington scene) in regard to a continuing evaluation of the capabilities of the U.S. forces and organizations in comparison with possible threats.

Events of the last few years have demonstrated that continued progress and survival of this country are not guaranteed. The need for efficient PSAC/OST involvement in national security affairs is as great as ever. Neither the Bureau of the Budget nor the NSC staff have adequate talent or continuity to do this job by themselves, and indeed the PSAC function of providing technical advice on present capabilities falls outside the responsibilities of the BoB and the NSC.

Questions

1. Does the President (and his Assistant for National Security Affairs) have an adequate appreciation of the need and PSAC capability for the provision of *independent* advice by technically informed personnel on major national security issues?

2. Should a concerted effort be made to inform the White House family once again of the history, nature, resources, and talents of PSAC/OST in national security matters?
3. Would it be desirable for the official membership of the National Security Council to include the Director of the Office of Science and Technology?

This discussion paper is reproduced in full to give the flavor of PSAC deliberations on its own role in support of the President. That role is certainly not primarily oriented towards expanding scientific research programs. To avoid the semblance of self-dealing, the President's science advisory organization must not be involved in the preparation of an annual or five-year report of the health of science, which important job should be assigned to some other part of the government, but the President's science advisory organization should indeed help the President to understand whether the bureaucracy has done a good job on such a report, just as it should help the President to understand whether the mission-oriented agencies and the departments of the government are using science and technology to solve their problems and to identify both hazards and opportunities.

Thus, although in practice the President's Science Adviser will have managerial and coordinating jobs, like that of chairing the Federal Council on Science and Technology, his principal contribution to the Presidency and to the nation arises from his staff role in advising the President—in being aware of the President's agenda and concerns, in identifying both opportunities and problems for that agenda, and in maintaining and exercising links to the technical community through personal contacts and a small staff, so that the competence and the power of the President's Science Adviser extends beyond that of a single specialty. The Science Adviser and his staff also help to identify competent effective people for high-level positions in the Federal government.

That the managerial and coordinating role is assigned to the President's Science Adviser is due not to his unique competence but, on the one hand, serves to provide a power base for the office of the President's Science Adviser and, on the other, avoids the competition and the conflict which could arise if the Federal Council were chaired by an individual who did not have the vision and the privileged access of the President's Science Adviser.

Because his office has lacked a large supporting organization with its bureaucratic pressures and imperatives (such as those of the Department of Defense or Health, Education, and Welfare), the Science Adviser has sometimes played an important, truly managerial role, either *ad hoc* or as a continuing responsibility. One mechanism which has been used in the past is the "executive committee" or ExCom, chaired by the Science Adviser and supported by a fraction of the efforts of one of his staff members (and by a continuing high-level panel attached to the President's Science Adviser in this role). This structure, for example, helped to guide the evolution of the national technical means of verification—photographic satellites, and the like—so important to the nation's security with or without a SALT agreement.

My views on the role of the President's Science Adviser are reflected in the excerpt of my letter of September 10, 1970, to Dr. Edward E. David, the President's Science Adviser:

I agree that PSAC represents an unparalleled national resource, but it needs leadership and encouragement for the individuals to do the most that they are capable of, and as efficiently as is required to make adequate progress against the tide. PSAC members must be selected, in my opinion, primarily on the basis of their competence, courage, honesty, and energy—geographical distribution and industry-academic split being of far lesser importance. We must keep in mind that PSAC has a dual role—first, to do what it can do in collaboration with the Defense Department, the Department of Transportation, etc.; second, to inform the President, whether or not PSAC can influence the situation. In this latter category are judgments as to our current military capabilities, evaluation of the health care system, the adequacy or inadequacy of Administration or Congressional organization for decision making, the length of the budgetary process in matters regarding development, etc. In many cases PSAC will have no solution for a disaster which it recognizes, particularly if political and social questions are involved. On the other hand, it is very important for the President to be aware of even those things over which he has no control.

It would be very good if the Science Adviser were once again the President's right-hand man, because, as is evident from the ABM and other questions, a properly-chosen technical person can see opportunities and dangers to which non-technical people are simply blind. If the Science Adviser has interesting, useful, and novel things to say to the President, which the President can use, then the two can grow close together. Even questions that are regarded as largely political, such as the level of the U.S. forces in NATO, the question of chemical and biological warfare, the provision of health care, or the income-maintenance program, do have a highly technical side. For instance, there is hardly a decision to be made which would not benefit from modeling and simulation. Here I am thinking particularly of the consequences of an income-maintenance program, the expected cost of a system of health care, or certain policy decisions in regard to government support of railroads. I hope that it is clear to the President that the President's Science Adviser has valuable judgments for him not only on science but on most decisions and concerns of the President, presumably with the exception of strictly-partisan political matters.

I believe that the Science Adviser could serve a further useful purpose by instigating studies in the Academy and elsewhere, either independently or through departments of the government, on such matters as improving the efficiency and reducing the cost of elections, reducing lead time and budgetary cycle, experimental implementation of important national programs, sampling and management techniques to reduce time lag involved in maintaining statistical indicators adequate for a particular purpose, etc.

Perhaps we should consider the creation of more OST panels, or the possibility of advisory panels to the individual department, whose work is distributed for comment to the PSAC, without, however, occupying large fractions of the meetings.

In any case, the question of what PSAC should be and what it should work on currently needs review.

Another problem is that the Bureau of the Budget is currently making important decisions in academic science with the use of budget officials who have no understanding of research or technology. I can give some examples of this problem, and I wonder whether a more formal working relationship between BoB and the OST might be in order in this field at least.

As for the public disenchantment with science and technology, where are the government-supported crash programs to meet the threat of pollution, of ignorance, of poor health? We have had science and technology pressed into the service of building military forces (but not of fighting recent wars!); we have had science and technology pressed into the service of the NASA Apollo demonstration program; we have had science and technology building nuclear reactors; but somehow we have applied a higher standard of thrift and a more critical eye to the creation of experimental hospital systems, experimental school districts, new incinerator design, and in general to development which could, but not necessarily would, be helpful in solving our urgent problems.

What PSAC could do, however, was limited at the time of the letter by the White House staff environment, by the well-developed territorial imperative of the President's national security adviser, and perhaps by a general reluctance to expose the administration to people of independent stature.

The President's Science Adviser, in the role of confidential adviser to the President, faces the dilemma of every high-level responsible staff person or manager—there is more to be done than can be done responsibly. A typical human response under such circumstances is to behave arbitrarily—to persuade oneself that one's decisions are as good as anyone else might make under the circumstances. The next step after recognition that decisions must always be made on the basis of incomplete information and with inadequate time for reflection, is to neglect to acquire or to heed information which may be readily available, and that way lies disaster. During the period of existence of the PSAC from 1957 through 1972, the committee helped to provide additional resources of competence and responsibility to the President's Science Adviser, aided in avoiding the easy arbitrariness which accompanies the overwhelming responsibility and burden of positions like that of the President's Science Adviser, and brought to the level of Presidential consciousness and action important problems before they would otherwise have arrived there.

One such example is the PSAC report on "Insecticides and Pesticides," the outgrowth of the panel activity initiated when a PSAC member brought to one of the monthly meetings copies of Rachel Carson's "Silent Spring" articles serialized in *The New Yorker*. Preliminary discussion and investigation showed that the effects of insecticides and pesticides were a matter of some controversy, that there was a Federal role in both usage and regulation, that the subject was of importance to the administration and to several departments (which had sufficient conflicts among them that they would have difficulty in arriving at a rational policy), and so a typical PSAC panel and report preparation activity was begun.

The Nature of the President's Science Advisory Committee

The four-year term of a PSAC member, with approximately four or five members replaced each year, provided continuity and growth for the member in responsibility and capability. Service on many *ad hoc* committees persuades me that there is considerable merit in this continuing committee approach, attached at a high-level and with sufficient freedom to do for its client (the President of the United States) whatever is within its capability. That it serves the President permits its membership to include the very best among scientists and technical people approached for the job. That it set its own agenda in large part, put upon it the responsibility for choice of the most significant problems on which progress could be made.

Previous service on PSAC panels permits evaluation of the members and provides pretty reliable indication that a member will work responsibly as well as effectively. The four-year staggered terms, and the two days per month of meetings ensured sufficient familiarity with the prior work of the committee and its panels that any effort undertaken by the individual members were subject to a standard for comparison of competence and responsibility. In this regard, the PSAC differed from *ad hoc* committees, from lower-level advisory committees, and from committees which spawned panels having a life of their own and reporting independently rather than by means of reports which were reviewed during the preparative process by the parent committee.

The PSAC panels had a major educationial role, not widely remarked in commentary on the President's Science Advisory organization. For instance, a typical two-day session of the PSAC Military Aircraft Panel, at the time I chaired it, would involve an agenda of perhaps 10 items, each a discussion with a different element of the Office of the Secretary of Defense, the Advanced Research Projects Agency, the Navy, Army, or Air Force, or one or another of the intelligence services. We encouraged "briefers" to come early and to stay late, sitting in one of 30-40 chairs around the walls of the room, and learning from previous or successive presentations and from the discussion.

This was not only a mechanism for exchange of information among the services and within the Defense Department; it was also good training ground for improving the quality of briefings, for indicating to the services what was required in substance and in presentation, and resulted in a substantial cross-fertilization which would not otherwise had taken place. Many program managers, outwardly grumbling about the effort required to provide a presentation to one of the panels, in fact welcomed the opportunity for an occasional visit because it provided the occasion to ask hard questions of lower echelons of management, and to put their house in order for a change.

In sum, the President's Science Advisory Committee was, in my opinion, an exceedingly effective and beneficial organization. Clearly for the committee to be effective it must have some access to the President, but for a reason different from that which might first appear. The power of the PSAC lay not in speaking to the President or in speaking *for* the President, but in acting on behalf of the President in gathering information for the preparation of its reports and the formulation of possible policy options. Where the information was publicly available as, for instance, in the preparation of the PSAC Report on Youth, it seemed to me a waste of precious committee time and effort (high "opportunity cost") to do what could have been done by a committee of the National Academy of Sciences, a university or institute research department, or the like. The comparative advantage of the PSAC lay in being able to demand discussion with government departments, in having access to national security information which was not available outside the government, or in being acquainted with the concerns of the President or the imminence of programmatic decisions in order to provide a report which could be not only correct but influential.

Why a PSAC?

A committee structure has natural strength and weaknesses. Responsible and careful emphasis of the strengths, while avoiding the disadvantages, made the PSAC an important influence at times past. The committee structure presents also some hazards, and the name and the form is not in itself a guarantee of performance.

In providing scientifically and technically informed advice for the President and the White House Staff, the President's science advisory organization and the PSAC have both individual and collegial roles. As is known from personal experience, the creative act, the imaginative proposal, the incisive key to the solution of a knotty problem, arise primarily in still-mysterious ways from the individual mind. However, the selection of those minds, and their preparation, care, and supervision are much more a collegial matter. Thus when the PSAC has worked well, it served as a means to identify, inform, educate, and evaluate individuals with a national view and a particularly effective mode of operation. Typically, such individuals would be selected for special expertise in some field, brought in to work intensively on an *ad hoc* panel for two days a month, extending over a year or so, and would then (if their performance was outstanding) be available to fill a vacancy on the PSAC itself. Thus the net could be cast quite wide in order to find such individuals, clear them for work in the Executive Office Building and for access to national security information, and provide them with an opportunity to help solve important, ill-structured problems, closely observed by their peers and by the more experienced PSAC members.

Another collegial role for the PSAC lay in its review and endorsement of PSAC panel activities and reports. Unlike many other organizations, PSAC not only determined which subjects were appropriate for more thorough study and helped to set up and staff panels for work in these areas, but reports of the PSAC itself or of its panels ordinarily were reviewed by the full committee at midterm and finally before release.

Such review, normal in style and tone in an individual scientific discipline, but brutal by comparison with the niceties of ordinary life, was intended to insure that the investigation had been organized in a reasonable fashion, that all reasonable questions were answered by the report, that, although technically correct and well-founded, the report would speak to generalists rather than to specialists, and that the product was not only valid but significiant.

Participation in these reviews at the PSAC meeting was helpful both to PSAC members and to panel members. It was a socializing experience, and it was particularly good that those who were on the interrogating side at one session might find themselves being interrogated at the next, as a panel activity in which they were participating came up for review.

Thus the PSAC as a body not only helped select and justify choices for concentration from the problems and opportunities suggested by the members and others, it set a continuing tone and style for scientifically-based advising at the highest levels of the US government.

Ironically, the Federal Advisory Committee Act of 1972, which requires meetings of committees like PSAC to be open to the public (with few substantive exceptions such as matters affecting individual personnel or classified material) *reduces* the degree to which such a committee can be aware of the President's concerns or uncertainty, while in no way giving the Congress or the public greater knowledge of activities of full-time administration officials, employees, or groups. With this Act, in order to remedy certain practices of some federal advisory committees, especially those in which representatives of regulated organizations met routinely with the regulators, Congress (and the Nixon Administration preemptively) seriously impaired both the effectiveness of a PSAC in doing its job of identifying and analyzing problems and opportunities properly of concern to the President, but also *reduced* the degree to which such a committee could be expected to be aware of administration activities and so serve as an additional safeguard against the covert misuse of power. The Federal Advisory Committee Act is appropriate for a committee dealing with the support of science but not entirely so for a committee advising the President on the use of science in a world of danger and competition.

The Problem of "Loyalty"

The Nixon White House staff apparently viewed the PSAC as disloyal to the President, as not supporting the President's programs, and as being less than helpful in the attempt to move those programs through the Congress. The White House staff certainly has a desire and duty to advance the President's programs, but the President himself (and even more the nation) has a strong interest also in formulating programs and in making sure that they are both sufficient and correct. The President's science advisory organization has an important role to play in formulating and challenging of programs; it can do little in the pushing of programs through the Congress, and would do that only at great prejudice to its ability to command respect and help from the scientific community in its other, unique role. It is the critics' lack of understanding of the necessity for analysis and policy information which leads most frequently to a view that the only or the most important function of a PSAC is to marshal scientific and scientists' support for the administration's program.

The question of treatment of privileged information is a knotty one, but in my opinion less troublesome than it at first appears. The letters excerpted here were sent to other PSAC members in continuation of a discussion on this question of confidentiality of information:

My own views about the propriety of Administration advisers testifying to Congress are contained in the attached letter, which I sent to the members of the OST Ad Hoc SST Review Committee. I believe that the communication in any administration, and in particular in the present one, is sufficiently poor, and the need for education of responsible officials so great, that it is unwise for high-level advisers on publicly-known issues to restrict their advice to one person.

Certainly, it is absolutely necessary to hold in confidence any information regarding the views of the advisee. On the other hand, if it is established that certain advisers have personal positions contrary to the action taken by the Administration, this can be the cause of some resentment, more or less rational.

Enclosures (RLG letter of 09/25/69 to coauthors of the OST SST Panel report. Code of Ethics for Government Service.)

September 25, 1969

Dear Colleague:

President Nixon announced on September 22 that the administration would seek funds to complete development of the SST. One of you has since called me to ask clarification of any limitation on your activities in regard to public statements or correspondence concerning the SST. This general question of limitations on the activities of advisers is one on which there has been considerable discussion at intervals in government circles. My judgment, interpretation, and advice is as follows:

For service on an *ad hoc* advisory panel, one tries to find those individuals who by experience and ability are most expert in a given field. That one portion of the government seeks advice on a particular question from such an individual is not adequate reason for denying the provision of advice to others in the administration or to the Congress and the public. The important parameter in this case, in my judgment, is the ratio of one's lifetime of experience and reputation to the short period normally spent in the *ad hoc* review. The situation is obviously otherwise for a full-time government employee charged primarily with the responsibility for such decisions, and it is somewhat ambiguous, for instance, for members of the President's Science Advisory Committee and of the Defense Science Board.

However, in your activities in regard to the supersonic transport, you should

1. make no mention of the fact that you served as a member of Dr. DuBridge's review panel on the supersonic transport,
2. make no use of privileged information obtained by the panel in the course of its discussion and visits (as contrasted with information publicly available or to which you have had access in your other capacities),
3. avoid any reference to the report of the panel, as well as any quotation from that report.

Aside from these common sense restrictions, I believe that you are free to advise or speak publicly either in support of or in opposition to the supersonic transport program. Indeed, you are among those in the country best able to lend wisdom to the deliberations of the Congress and of the public.

Once more, I want to thank you for your work on the review, to record my pleasure with our association, and to indicate my esteem for the fine report, many of the conclusions and recommendations of which still remain to be implemented.

Sincerely yours,

Richard L. Garwin

May 25, 1970
                              (PSAC Members & Consultants;
Dear Colleague: Some OST Staff)

In connection with the discussion at the PSAC meeting on the propriety of public statements and actions by PSAC members, I would like to emphasize my belief that such guidelines should be available in writing. I believe that it is all too easy in discussion to believe
that others acquiesce in proposals which they actually find inconsistent with the obligations
they assumed upon appointment.

I would like to predict that only two viable policies exist governing public statements of
PSAC members:

1.   No PSAC member (except the Chairman acting as Science Adviser or Director of
     OST) may speak on any issue of national interest, unless the PSAC is unanimous on
     that issue. If members were permitted or encouraged to speak in support of Adminis-
     tration policies, while dissenting members were forbidden to speak against, one of
     two inferences would be drawn—either PSAC would wrongly be assumed to be
     unanimous, or it would be presumed that those against were forbidden to speak. It
     might as well be made public knowledge that only those supporting the President's
     policy are allowed to speak, in which case I believe that new members, faced with this
     policy, would hesitate to join.
2.   It could be recognized that in the long run and for the security of our nation and of
     the democratic system, informed discussion by responsible individuals is in the
     national interest. PSAC members would then be permitted to speak on any subject in
     a manner consistent with the safeguarding of information given them in confidence,
     the preservation of a confidential relationship regarding contact between PSAC and
     the President, and the safeguarding of classified information. There is already a
     substantial literature on conflicts of interest, with which we are all presented upon
     entering government service. In fact, I just came across the "House Concurrent Res-
     olution 175, 85th Congress, 2d Session (presumably 1958) which I enclose. This
     code of ethics for government service was, I believe, adopted when President Nixon
     was President of the Senate.

An individual taking a public position on issues in which important groups are vitally interested assumes a substantial risk. His position may be misrepresented: he may lack the
resources to continue a dispute against powerful interests; his position may be found to be
in error. PSAC can stand above all this by indicating that the member is speaking as an
individual and not for PSAC.

On a more personal note, I want to make it very clear that there was nothing in my
communications with the Congress on the SST that I obtained from other than
public documents. The question of the use of confidential information has been repeatedly
discussed and embodied in the conflict-of-interest regulations. Although Congress is a
branch of the government, I complied with those regulations in the strictest sense, notwithstanding the fact that my testimony was directed toward helping the Congress to exercise its
Constitutional responsibility in this matter.

The Administration is often correct. But the Administration is sometimes wrong. A key
question in government or in any organization is the extent to which it is desirable
somewhat to hamper positive and presumably desirable actions in order to reduce the
probability of big errors.

I realize that there is a danger in all this to the free access by PSAC to departmental information. In most cases, this free access was long ago a casualty of the departmental fears of PSAC's access to the President and to the Bureau of the Budget. Access we have, but not very free. It may worsen, or it may improve, as the President prizes secrecy above understanding, or vice versa.

Sincerely yours,
Richard L. Garwin
enclosure (Code of Ethics for Government Service.)   5500.7 (Incl 2)
                                                                      Aug 8, 67

HOUSE CONCURRENT RESOLUTION 175
85TH CONGRESS, 2D SESSION
(presumably 1958)

Resolved by the House of Representatives (the Senate concurring), that it is the sense of the Congress that the following Code of Ethics should be adhered to by all Government employees, including officeholders:

CODE OF ETHICS FOR GOVERNMENT SERVICE

Any person in Government service should:

1. Put loyalty to the highest moral principles and to the country above loyalty to persons, party, or Government department.

2. Uphold the Constitution, laws, and legal regulations of the United States and of all governments therein and never to be a party to their evasion.

3. Give a full day's labor for a full day's pay; giving to the performance of his duties his earnest effort and best thought.

4. Seek to find and employ more efficient and economical ways of getting tasks accomplished.

5. Never discriminate unfairly by the dispensing of special favors or privileges to anyone, whether for remuneration or not; and never accept, for himself or his family, favors or benefits under circumstances which might be construed by reasonable persons as influencing the performance of his governmental duties.

6. Make no private promises of any kind binding upon the duties of office, since a Government employee has no private word which can be binding on public duty.

7. Engage in no business with the Government, either directly or indirectly, which is inconsistent with the conscientious performance of his governmental duties.

8. Never use any information coming to him confidentially in the performance of governmental duties as a means for making private profit.

9. Expose corruption wherever discovered.

10. Uphold these principles, ever conscious that public office is a public trust.

The most serious (and ultimately fatal) problem between PSAC and the White House staff arose over the Safeguard ABM (anti-ballistic missile) system and the SST (supersonic transport) program, both of which had been studied thoroughly by PSAC over the years, and on both of which PSAC had come to firm conclusions as to the effectiveness and promise of the systems. The question was not whether confidential information given the PSAC by the President or acquired by the PSAC on behalf of the President would in this way be transmitted to the public or to the Congress. It was whether respected and informed people who had been brought in by the Executive branch would be available to provide their expertise to the Legislative branch in the performance of its responsibility.

The question first arose when PSAC member (actually Consultant-at-Large) and former Science Adviser Jerome Wiesner, asked the President's Science Adviser, Lee DuBridge, to inquire of the President whether President Nixon wished Wiesner to resign before taking a public position and a position in testimony to the Congress in opposition to the Safeguard ABM system. DuBridge reported at a PSAC meeting that the President had said that Wiesner should not resign but should give his testimony while remaining a PSAC member. It was in an attempt to flesh out and to pursue the possible consequences of such a decision that the discussion took place within PSAC as to who might testify on what.

The SST Problem

Additional problems were created for the PSAC by the US program to develop a commercial supersonic transport aircraft. Over the early 1960s, various panels of the PSAC including the Military Aircraft Panel which I chaired, the Aircraft Panel, and others had reviewed for the President various aspects of the SST program—in particular whether there were defense benefits from the development of the commercial SST—and had written brief reports either specifically on the subject or in the form of progress reports of the panel. The new Nixon Administration was faced with a decision as to whether to proceed with development of the SST, and Lee DuBridge, the President's Science Adviser, formed an OST panel, which I chaired, to report to him on this question. Our report was submitted to DuBridge on March 30, 1969, after the usual review by PSAC. The OST panel and the PSAC itself had had their opportunity to provide advice to the administration on the SST program.

So far as I am concerned, had the administration been honest with the Congress, the matter would have stopped there. Unfortunately, administration spokesmen in formal testimony concealed relevant information and lied to Congressional committees which had the responsibility to authorize and to appropriate funds for the SST program.

The Code of Conduct for government employees reproduced above had been formally distributed to PSAC members and consultants (as it had presumably to all government employees and consultants in 1967). I felt that it was intolerable for the administration to subvert the democratic process by concealing and misrepresenting relevant facts and reports. Note that this was not a case in which national security was involved (as in the necessity to conceal certain kinds of covert operations). Accordingly, I tried to present information, always over my own signature, as Congressional testimony, or in publications, independent of the PSAC and of its panels. I believe that this action and this policy is in the interest of the United States, but the Nixon Administration felt otherwise and decided in January 1973 to abolish the PSAC and the Office of Science and Technology, with the announced purpose of saving money.

One knows more about the inner workings of the Nixon Administration than of some others—for instance, of its search to identify universities whose Federal research funds could be cut in retribution for expressed views of their faculty or administration on the ABM program. These tendencies are latent in most organizations, but were only controlled less well in that environment.

The President's Science Adviser and the members of the President's Science Advisory Committee, when it existed, signed oaths of office. On the record were the concurrent resolutions of the House and of the Senate prescribing a code of ethics for government employees and emphasizing that their fundamental responsibilities are to the people of the nation and to the Constitution rather than to the President and to the cause of reelection.

At most times, and in most cases, the purpose of the President and the purpose of the nation are coincident. When they are not, advisers and advisory committees are faced with a serious problem. When an adviser or such a committee can no longer carry out it function, it is conventional and appropriate for the individual to resign. When there is a serious threat to the nation (as in a President gone wrong and committing illegal acts day after day) the problem is not simply one of resignation but to ask in what way the individual, by fulfilling his oath of office, can serve the Constitution and the national good. Under those circumstances, one serves the nation best by not serving the criminal at all, but with clear conscience supporting the Constitution, giving one's best efforts in the performance of one's job, trying his best within the administration to return things to the right track while being vigilant in determining when the threat to the nation becomes critical. At that time the adviser must act, no longer as adviser but as citizen.

Of course the normal operation of government and of the President's Science Advisory organization need have little concern for these soul-wrenching possibilities.

Conclusion

Experience with the PSAC and its panels from 1957 through 1972, with the Defense Science Board and panels of the Air Force, and with groups of outside consultants in the years from 1952 to the present, persuade me of the great benefit to be obtained from a working PSAC. But this benefit must not be compromised by political appointments to the committee, by exaggerated emphasis on geographic distribution and representation of minorities to the detriment of finding those people most capable at any time to do the job, and the purview of PSAC might have to be limited at present to the national security area in order to allow its effective operation in view of the exigencies of the Federal Advisory Committee Act. The President's Science Advisory Committee would again be an important tool in marshaling scientific talent in support of opportunities and in the solution of problems; it would also be a considerable force to restore and maintain integrity in government.

Don't Segregate Science Advice—Integrate It!

Patrick E. Haggerty

Once again the question of how best to provide advice on science and technology to the President of the United States is being raised and discussed. Many of those scientists and engineers whose personal experience includes participation in providing such advice seem to prefer a committee or council which will advise the President directly on science and technology. Yet the situation has changed considerably since World War II and the subsequent formalization under President Harry S. Truman of a science advisory mechanism which included a Science Adviser, a President's Science Advisory Committee (PSAC), and an Office of Science and Technology (OST), staffed full-time. This particular advisory mechanism was used by all subsequent Presidents until substantially terminated by President Nixon in 1973.

In 1977 it was resurrected in part by President Carter with the appointment of a Presidential Science Adviser. Apparently the establishment of a President's Science and Technology Advisory Committee (PSTAC) is also being considered by President Carter.

It is doubtful, however, that restoration of this advisory mechanism is, in fact, what the President needs. Today science and technology have so pervaded our society that every department of the government has high level representation of science and technology, and most have staff groups and laboratories which have as their principal functions bringing science and technology to bear on fulfilling the responsibilities of the department involved. At the present time, for example, the Secretary of Defense, the Administrator of NASA, at least three Under Secretaries, and a number of Assistant Secretaries have scientific or engineering backgrounds and President Carter himself has a technical background.

Special Knowledge and Tools

There can be no question of the fundamental importance of science and technology in a contemporary, highly-developed nation such as the United States. But,

Patrick E. Haggerty (b. 1914) was a member of the President's Science Advisory Committee from 1969 to 1973. He is an electrical engineer, served in the U.S. Navy in World War II; is Honorary Chairman and General Director of Texas Instruments, Inc., Chairman of the Board of Trustees of The Rockefeller University, Trustee of the University of Dallas; member of the National Academy of Engineering and of the Business Council; Fellow of the AAAS.

from the viewpoint of the needs and responsibilities of the President, science and technology are not ends in themselves, but are significant for the special knowledge and tools that they can bring to bear on national problems and toward the attainment of national goals. This is not to argue that there is not a need for special pleading for science and technology (often characteristic of the old PSAC), but rather that it belongs elsewhere in the government and not as part of the President's advisory mechanism.

The needed mechanism is probably one which avoids providing science and technology in segregated form, but rather which integrates science and technology into the overall advice that the President needs, seeks, and receives. Integration and balance are requisites if the President is to resolve appropriately the often conflicting objectives and recommendations of the several government departments. (Consider the potential for contradiction between the objectives of the Environmental Protection Agency and the Department of Energy, for example.)

The goal of providing integrated advice to the President could be achieved by unifying—in a Council of National Development Advisers—the Science Adviser, the old PSAC and OST, the Council on Environmental Quality, and the Council of Economic Advisers.

The proposed Council of National Development Advisers would include economists, scientists, engineers, industrialists, environmentalists, sociologists, educators, etc. Its members should, as a result of training, experience, and achievement, be exceptionally qualified to analyze and interpret developments in a variety of pertinent areas and advise the President in a coherent and integrated fashion with respect to national development—with national development understood to be not merely material development, but development in the full sense of an improved quality of life.

A Similar Approach

Among the advisory mechanisms to the President, the Council of Economic Advisers would appear to have most persistently contributed coherent and useful recommendations. What is envisioned is a council with a similar approach but with its objectives widened from the more narrowly economic and material to the overall development of our nation. Thus, the National Development Council would be established by a new Act of Congress to replace the Employment Act of 1946 which created the Annual Economic Report of the President, the Council of Economic Advisers, and the Joint Economic Committee of the Congress. This new Act of the Congress, in addition to establishing the Council of National Development Advisers, would also provide for an annual National Development Report from the President and a joint National Development Committee of the Congress. This enlargement of scope from the narrower economic aims of the original Employment Act of 1946 would offer an improved mechanism for assuring that not only science and technology, but economics and environmental considerations would all be brought to bear in generating coherent, integrated, and balanced reviews and recommendations to assist the President.

The Life, Death, and Potential Future of PSAC

James S. Coleman

My perspective on The President's Science Advisory Committee and the function of the President's Science Adviser is somewhat different from that of most who were involved with that committee, for I was one of only three social scientists who at some time served on PSAC. (The other two were Herbert Simon, who preceded me, though our terms overlapped slightly, and Daniel Moynihan, who was appointed shortly before PSAC's demise.) That difference in perspective will affect the comments I have to make in the last section on the future of PSAC-like activities. It will probably influence less the earlier sections of these comments. Altogether, my comments will be divided into three parts:

a)   How did PSAC function, and what can we learn from that?
b)   What was the crucial issue that killed PSAC, and how is the generic problem raised by this issue to be solved?
c)   What is the future role of PSAC-like activities in the President's Office, and what can be said in particular about the social sciences?

How Did PSAC Function, and What Can We Learn From That?

PSAC had, at the time when I was a member, a split personality. It had an intended function of providing advice to the President, from a somewhat broader sense than that which the President's Science Adviser alone could provide. For example, President Nixon had read reviews of *The Limits of Growth*, which had been recently published, and was concerned about some of the dire predictions made in that book. He asked PSAC, through Edward David, his Science Adviser and Chairman of PSAC, to study this and report to him. PSAC members read the book, and then invited Jay Forrester, on whose ideas and computer models the book was based, to a monthly meeting of the committee. After this meeting and

James S. Coleman (b. 1926) was a member of the President's Science Advisory Committee from 1970 to 1973. He has been Professor of Sociology at the University of Chicago since 1973 and previously was on the faculty of Johns Hopkins University. He is a member of the National Academy of Sciences, the National Academy of Education, American Academy of Arts and Sciences, and the American Philosophical Association. He is the author of many books and has been Visiting Professor at Cambridge University (and Fellow of Churchill College) and at the Institut fur Hohere Studien in Vienna, and a Guggenheim Fellow and a Fellow of the Center for Advanced Study in the Behavioral Sciences.

further discussion, the committee met with the President, and in a general discussion reported the general reactions, on which there was some consensus (i.e., that the problems dealt with by the book were not trivial, but that the book itself was quite misleading concerning the character and urgency of the problems).

This exemplifies one half of PSAC's split personality: to provide advice to the President on policy-relevant issues that required technical expertise in science, ordinarily covering a range greater than that spanned by a single discipline. A more common type of example of this half of PSAC's personality involved providing independent advice to OMB concerning programs proposed by Federal agencies, such as the Defense Department or NASA. For example, PSAC was asked to evaluate alternative future programs of NASA, including the ambitious program of NASA and lesser alternatives (like the Space Shuttle). OMB depended heavily on these evaluations, which were regarded as disinterested independent advice, and they played, I believe, a strong role in the decisions on some NASA and DOD programs during the time I was a member of PSAC.

The other half of PSAC's split personality was that of advocate for government expenditures on science, i.e., a representative of "science" in the government to help insure the health of science. PSAC reviewed allocations for scientific research and development in the government's budget, and generally advocated greater expenditures through its chairman, the Science Adviser.

It may be that such a role is inextricable from the technical advisory role I described as the first half of its split personality. For example, in the problems for which disinterested advice is sought by OMB, such as a NASA program or a DOD program, there is likely to be included a set of research and development activities. These are activities in which it is difficult for a body like PSAC to give advice without being concerned about the impact of the decision on the health of science. It may be that even if the two roles are separable, it is valuable to have them performed by the same committee. Yet PSAC, and the President's Science Adviser, sometimes found itself in the position of confusing what was important for healthy science with what was important for a healthy country. PSAC was sometimes close to the position of saying, "What's good for science is good for the country."

Two Separate Roles

I believe the two roles can and should be separated, and that the advocacy of healthy science can better be carried out by the National Science Foundation in some capacity, leaving PSAC's definition of function to be unalloyed with the self-interest of science. This does not mean, of course, that self-interest of science (or of a particular branch) will never influence the evaluations of a member of PSAC, but rather that it does not legitimately do so. It should be very clear to whom PSAC is responsible; and its responsibility should not be to the science establishment as an establishment.

If we leave aside the self-interest function of PSAC as a representative of science, there are various modes in which PSAC operated, and it is useful to say something about the contrast between two of these. One mode was to deliberate, review, examine, and evaluate at the monthly meeting of the entire committee. The review of Forrester's work and the review of the NASA program proposals were carried out in this fashion. The other mode was to work through panels, of which the chairman and perhaps some other members were from PSAC, but which also included non-PSAC members. These panels had several different types of origin. In some cases, they were originated by a request from OMB or another Federal agency, on a technical problem that required rather intensive investigation. The reports of those panels were often made directly to the interested agency, and sometimes these reports were classified, depending on the nature of the problem posed. In other cases, panels were initiated by PSAC itself, ordinarily acting upon a proposal by a member to examine a topic or an area he felt was of particular importance. PSAC showed some tolerance of members' interests in this, though such proposals were sometimes argued at length, and the member was sometimes dissuaded from his plan. When PSAC approved the proposal and set up the panel, the member or members initiating it were given a staff member from OST, and a rather free hand to pursue the topic.

In some cases, the panel's product was a report; in others, it was less tangible, such as changes in agency policy resulting from informal discussion with the panel. For example, when I joined PSAC, Frank Westheimer, though a chemist, was chairman of a panel on education, which was particularly concerned with early childhood education, and with the question of what kinds of programs were most successful. The products of this panel did not include a panel report; instead, the panel aided in the development of curricula for such programs, and also influenced policies of Federal agencies dealing with early childhood education.

When reports for publication were produced by panels, they were ordinarily reviewed by the committee as a whole, and once they were accepted, were typically published by GPO as reports of PSAC panels. For example, the panel I organized after joining PSAC, a "Panel on Youth," issued a report entitled *Youth: Transition to Adulthood*, which reviewed the organization of American secondary schools, and raised questions about the design of alternative environments for youth.

Some PSAC panel reports were highly influential in the evolution of government policy in areas related to science. For example, a panel headed by John Tukey prepared and published a report on the impact of modern chemicals on the environment. The report was published before the environmental movement had begun, and was quite influential in establishing the level at which the discourse about this subject was initiated.

PSAC's activities can be roughly divided into two: the panels and the activities of the committee as a whole. The *Limits of Growth* question and the NASA question exemplify matters that were addressed by the committee as a whole; the panels described above exemplify the Committee's functioning in its panel mode. In which way did PSAC function better? From my own observation, there is no comparison; PSAC functioned rather poorly as a committee of the whole, and very well through its panels. When sitting as a full committee, PSAC considered issues about which almost none of the members were highly knowledgeable, and about which none felt a strong and continuing responsibility to inform himself. The issues were not ones that were pursued at length, but were issues that occupied the committee only at the meeting during which they arose and one or two subsequent meetings. The result was discussion among a set of intelligent men only semi-informed about the issue at hand, and with little incentive to think about the issue between monthly meetings. The scientific expertise which constituted the reasons for their appointments seldom came into use. Similar outcomes would have resulted from discussions among any other set of persons with similar analytical approaches to questions but with totally different backgrounds.

Makeup of the Panels

The panels differed from this in three ways: they consisted primarily of persons with rich backgrounds in the special areas covered by the panels; each member of the panel felt a direct responsibility to use that background to address the questions at hand, and was thus more likely to devote time to it between meetings; and the panel had a specific product, often a report, which imposed responsibilities and demanded individual efforts. Thus the panels, I believe, were much more likely to show evidence of the special scientific expertise that should constitute the hallmark of a science advisory group.

However, another disturbing question arises concerning the panels. When one looks at the audience for the panel activities, and especially the published panel reports, it is immediately clear that the primary audience was not the President, but professionals, in and out of government, who had some involvement with the area addressed by the report. The question then arises, just how is this related to the presumed function of PSAC? The committee was a *President's* Science Advisory Committee, not a Professionals' Science Advisory Committee. Yet each of these reports was aimed at affecting the general discourse surrounding an issue, not principally among those on the White House staff, and not only among those in government, but also (and perhaps primarily) among the much larger number of professionals outside government who had some interest in, and small influence over, the general direction of policy in the area.

Was this a proper function for the PSAC? It is not a function of advising the President; yet it seemed to engage the best of PSAC members' energies, and to constitute the best of PSAC's fruits. I believe it is a proper function, viewed in the following way: there are some Federal policy questions with a large scientific component in which the decision is largely that of the President, or of executive agencies in conjunction with the President. This class of policies is perhaps best exemplified by major weapons programs in DOD, on which a high-level executive decision is necessary, or a major expenditure for a new particle accelerator at a national laboratory. This was the class of science-laden policies for which PSAC was initially formed; and this class of policies continued to occupy a significant portion of PSAC's attention until its end. (That attention was most productively engaged, as I've indicated earlier, not in meetings of the committee as a whole, but in panels, though panels that did not publish reports that circulated outside government.)

But, increasingly in government policy, another set of science-laden policy questions began to arise. These were questions in which the policy was not made wholly within the executive branch, but included the participation of Congress and thus the large set of interested and knowledgeable parties outside government. In some policy areas, such as those addressed by the Westheimer education panel and my own panel on youth, a portion of the policies were even made not at the Federal level at all, but at state and local levels.

A Change in the Structure

The change was not merely a change in type of policy. It was in part a change in the very structure of government decision-making. Defense policies that would, in the early days of PSAC, when World War II was still in recent memory, have been made wholly within the Executive Branch, were now, in the late 1960s and 1970s, thrown wide open to debate in Congress and throughout the country as a whole. Perhaps the best example was the ABM issue, which, had it arisen in the 1950s, would probably have been decided within the Executive Branch alone.

Given this expansion of the class of science-laden policies that are open to public debate, and the contraction of those that are made by the Executive Branch, then there are, I believe, two possible directions for the PSAC function: one is to narrow its compass to that greatly restricted set of decisions that are made by the President or by agencies in the Executive Branch. The other is to encompass within its scope the expanding class of science-laden policies that will be made in part outside the Executive Branch, that is, by Congress or by other levels of government.

It is clear that the emergent form of PSAC activity, that is, reports in policy areas published openly for an audience of professionals, was a response to that expanding class of policies, with the reports as attempts to provide the "PSAC function of advice" to the new, large, vaguely-defined set of actors who would participate in the open decision-making from which a policy would result.*

Is this modified function, that of informing a broad class of participants in policy making, compatible with that of informing the President and Executive agencies, that is, PSAC's original function? The answer is not a simple one, as I believe is illustrated by the issue which killed PSAC, the matter to which I now turn.

What Was the Crucial Issue That Killed PSAC, and How is the Generic Problem Raised by This Issue to Be Solved?

It has come to be evident, from literature on the Presidency, that the President, whoever he is, has as his strongest need some persons who are close to him, in whom he can place absolute trust. I believe this is a result of the fact that the necessary duties of the President far outstrip the time and attention capacities of a

---

* The formation of the Office of Technology Assessment as an agent of Congress was a response to the same shift in locus of decision-making on issues involving science.

single person, so that a President has a need for persons who will perform some of those functions *as if they were he.* *

PSAC, of course, is not in this capacity. PSAC members had primary affiliations elsewhere, and were only in the service of the government for a couple of days each month. The President's Science Adviser, a full-time position in the Executive Office of the President, is potentially in this capacity, that potentiality depending on the President's degree of trust in him. Few Science Advisers have enjoyed the degree of trust which would lead the President to regard their advice as absolutely and fully in his interests. And probably this is as it should be, for "the President's interests" include not only those associated with the position, but also those associated with his own future in that position, e.g., his future power or his reelection. The President does have a right to expect that his Science Adviser will not be a representative of or advocate for the science establishment, at least where this role conflicts with his responsibilities to the Presidency. ** This means that once a policy position has been established within the Executive Branch on an issue in which science is involved, he should either support the administration's position, keep quiet, or resign; but certainly not oppose it. ***

But what about PSAC as a committee, and PSAC members as individuals? If the function of PSAC is to be expanded to that of informing a broad class of participants in policy making, as I have indicated did occur in PSAC's later years, then what responsibility does PSAC have for supporting, or not opposing, Executive Branch policy in performing this function?

The Death of PSAC

The issue which more than any other precipitated the killing of PSAC by President Nixon was closely related to this question. Some members of the committee were strongly opposed to Administration policy positions on anti-ballistic missiles and on

* Leonard Garment, commenting on those closest to President Nixon, once said to me that one of that set of staff members acted "almost as a second skin to the President," by which he meant that he expanded the President's capacity for contact with others beyond the constraints imposed by the 24 hours available to the President himself. Although Nixon probably felt this need somewhat more than Presidents who had a greater trust of the more distant environment, the need is a function of the position of the Presidency. For example, President Carter, with a far more outgoing and less suspicious personality than that of President Nixon, totally revamped his cabinet and personal staff, using as the primary criterion for continuation the degree of loyalty to Carter that he felt the Cabinet or staff member had.

** I say "Presidency" here rather than "President" to distinguish the interests of the incumbent President from those of the position. The Science Adviser should, in my view, be responsible to the position without being responsible to serving the personal interests of the incumbent. If he finds these two interests in conflict, he should make his case as forcefully as possible, but retain his responsibility to the position.

*** Recently I (presumably along with many others) received a letter from Frank Press, the current President's Science Adviser, urging support for SALT II. I have no idea what Press's personal opinion is on SALT II, but his letter is wholly consistent with the role requirements of the Science Adviser, as I have just described them, quite independently of his personal position. If he personally feels that SALT II is bad, then he should have no obligation to actively support it as in the letter he sent, but he would have an obligation to remain silent on the issue, or to resign if he wished to actively oppose it.

support for development of a commercial SST. This opposition was in one case publicly expressed in writing, in interviews, and in testimony before Congress. President Nixon came to see, I believe, PSAC as no longer responsible to him, and as symbolic of opposition to his policies by the scientific community.* This was the essential reason for PSAC's death, although of course the Freedom of Information Act was beginning to make the deliberations more difficult, and would have required some accommodation if PSAC had continued.

This incident that caused PSAC's death, while it arose because of particular, personalities, reflects the underlying problem alluded to above. Insofar as PSAC attempts to inform an audience outside the Executive Branch, what are its responsibilities vis a vis the science-related policies of the Branch? In most cases, the problem does not arise, that is, in all those cases in which no Administration policy has crystallized, and the discourse is in its early stages, with no positions to be defended or opposed. But in some cases, those when there is an Administration policy position, it does arise. What is to be PSAC's role in such a situation, and what are the individual responsibilities of the PSAC members who spend a small fraction of their time in their PSAC role?

I believe it should be clear that PSAC as a committee has no responsibility to actively support any such Administration position; and if the committee as an entity has no such responsibility, then certainly the individual members as members have no such responsibility. This position was generally understood at PSAC during the period of my tenure, and probably throughout its existence. The answer to the second part of the question, however, has had less general consensus: does PSAC, and do the members, have a responsibility not to actively oppose, outside the Administration, a policy position held by the Administration? There was, I think, general consensus that PSAC as a body should not actively oppose Administration policy, but it is clear, from the differing actions of different PSAC members outside their PSAC roles, that there was not consensus about the members' responsibilities outside PSAC.

A Difficult Question

The question is a difficult one, for there are fine gradations in the degree to which an Administration is committed to a policy, as well as in what constitutes "active opposition" to the policy. I think, however, that at the extreme, when there is a clear commitment and when the PSAC member proposes to engage in active public opposition, he has a responsibility to resign his membership on the committee.

What about the committee itself? Does it have any role to play for those controversial issues in which there is a high scientific component and on which the Administration has taken a position (either with or without PSAC consultation)? In the past, PSAC had no function in this situation; it turned to other issues. But it is possible, I think, to envision a role which would be extraordinarily useful without embarrassing the Administration within which PSAC resides. That role would be

---

* Different Presidents react differently to scientists. Nixon seemed to hold them in some awe, did not feel comfortable with them, and had a certain degree of general distrust of them.

one of establishing and holding hearings on the scientific aspects of the policy issue—not hearings concluded by a decision on the part of PSAC, but hearings in which PSAC members' role was limited to that of posing questions. Such hearings would be conducted in somewhat the same fashion as Congressional hearings or Executive Agency hearings, but with crucial differences from both of these. The differences from the Congressional hearings lies in the level at which the questions would be posed, and testimony given. The difference from the agency hearings lies in the fact that the agency is itself an interested party: it has a set of regulations to make, enforce, and defend, regulations which often are attentive only to one policy goal to the exclusion of others. In addition, agencies sometimes have their own technical staffs and research activities, and they are unlikely to be dispassionate when their researchers' results differ from those of others.

EPA, for example, becomes an interested party on both these counts, as does the Food and Drug Administration, the Department of Energy, and other agencies. Actions taken by these three agencies, and a number of others as well, involve a large amount of technical input, yet there are no arenas where the scientific assumptions which underlie the actions can be independently examined.

There are other potential candidates for such arenas, and questions about the appropriateness of a PSAC-like body carrying out such a function; but all this moves toward the topic of the third section of this paper. I now turn to that section.

What is the Future Role of PSAC-like Activities in the President's Office, and What Can Be Said in Particular About the Social Sciences?

Some indications of the general outlines I envision for a PSAC of the future should begin to emerge from the first two sections. There should be an explicit recognition and acceptance of PSAC's expanded role that developed over time of informing not only the President and the Executive Branch, but also the broad set of parties in and out of government whose interests and pressures go into the shaping of policy. The panel mode of activity, as I have described it, worked well for both functions, that of direct scientific policy advice to the Executive Branch, and that of informing a broad set of parties through publication of PSAC panel reports.

For particularly controversial areas, however, a new mode of activity of the sort I began to describe in the preceding section, seems both feasible and potentially fruitful. The hearings would have some of the character of the "science court" which has been proposed as a way of focusing and resolving scientific debate in areas of policy importance.* They would not, however, result in a decision. Rather PSAC would constitute an independent, disinterested arena within which opposing scientific arguments and evidence would be aired, with the only outcome being publication of the proceedings. There would not be a report, as in most National Research Council reports on such policy areas, setting forth the committee's conclusions. Such conclusions are often not the product of the committee as a whole, but that of a particularly active member who may reflect those prejudices with which he began

---

* See *Science*, 193, August 20, 1976, pp. 653-56; Technology In Society, Vol. I, Issue 3 (Autumn 1979), pp. 229-237.

the investigation; or in the worse case, they are the product of a staff member, not only reflecting such initial prejudice, but also often exhibiting a lower level of technical expertise. The proceedings would, rather, be an outcome of the stimulus imposed on the interested scientific and technical community by PSAC. And they would be designed as a compact and focussed input for those who directly or indirectly may shape policies in these science-laden areas.

For example, there has been strong controversy over the use of additives in food, with the current controversy focused on meat additives such as nitrites for curing of pork. The Food and Drug Administration, which has the responsibility of establishing regulations on additives, would be one party from whom testimony would be taken, but with clear recognition that it was an interested party, interested if for no other reason than the constraints imposed by its regulatory mandate. As another example, there is no arena within which environmental and energy tradeoffs can be examined, since EPA's and DOE's mandates limit them to considering only one of these. Thus hearings on such an issue would be an important input to the policy discussion surrounding environmental and energy questions. Even within the mandate of either of these agencies, their role as regulatory bodies and as initiators of research make them ill-fitted to hold hearings in which they are both judge and one of the contesting parties.

Who Would Be Better Suited?

One can well ask whether another body might be better fitted to such a role as I've outlined above. Two in particular suggest themselves: Committees of the National Research Council (NRC) set up for the purpose of investigating exactly these kinds of policies in which science input is important; and second, the Office of Technology Assessment (OTA) originally designed as a Congressional counterpart to PSAC.

NRC committees, while appropriate as issuers of reports which come to some conclusions and make recommendations (reports which have been of highly variable quality, and have sometimes had more attention from their audience than went into their preparation by committee members), seem less appropriate as bodies before which hearings are held and testimony given. OTA, on the other hand, could come to have such a function, and does constitute a potential alternative locus for hearings of this sort. OTA has not, however, found a wholly satisfactory mode of operating, and it is currently not clear in just what directions it is going.

What of the social sciences in all this? Is there a role, in the kinds of functions I have described as a scenario for the PSAC of the future, for social scientists on PSAC and for social scientific content being addressed in the same way that content involving the natural sciences is addressed? The few examples of social science content in past PSAC deliberations (such as the education and youth panels I mentioned above) were not much more than accidental, depending upon interests of PSAC members.

At the time when PSAC was first formed, and for at least 10 years after that, the answer would have had to be no to the question of introducing social science content. The reason is quite clear: there were few policies of the Federal government (other than fiscal and taxation policies which required technical input from economists) in purely social areas. Then in the mid-sixties, a whole range of such policies came into being, and the quantity has continued to increase since that time. There are education policies, public health policies, housing policies, welfare and income maintenance policies, and many others. These policies sometimes are attended by extensive systematically-collected data which can provide scientific evidence relevant to the policy. And, as in the case of policies related to natural science, the agency which could constitute an independent forum for reviewing results is itself an interested party and thus unable to provide such a forum.

A recent example illustrates all this very well. In the late 1960s, a set of income maintenance experiments were designed and initiated to learn the effects of an alternative to welfare payments. These experiments were sponsored initially by OEO and then taken over by ASPE (Assistant Secretary for Planning and Evaluation) in HEW. The principal question was what would be the effect of such a policy (in effect a guaranteed annual income) on labor supply. Would a high proportion of persons (and the question was asked primarily about men) stop working and live off the guaranteed income from the government? Additional information about other possible effects of the policy were gathered as well, and the experiments were carefully designed with differing levels of payment and control groups receiving no payments.

It was found that there was some effect on labor supply, though not a large effect. There is general consensus among all investigators who have analyzed data from the various sites (an area of New Jersey; an area of rural North Carolina; Gary, Indiana; Seattle; and Denver) about the size of this effect.

However, one set of investigators, sociologists analyzing the Seattle and Denver data, found a surprising result: the income maintenance program tended to lead to dissolution of existing marriages, and reduced rate of formation of new marriages. The effect was a major one, and the projected impact of such a program nationwide on low income families would, if the investigators were correct, be very large. But were they correct? There was an extended controversy between, on the one side, those investigators who had found the result in Seattle and Denver, and on the the other side, investigators who had not found such an effect in New Jersey, together with the officials at ASPE. Why was ASPE on one side rather than attempting to discover which result was correct? In part at least because those in ASPE were in favor of an income maintenance program, and were not eager to see evidence that might impede such a policy. Later, they were involved in the design of such legislation, and one justification of the policy that had been given by President Carter was that it would "help preserve the American family."

The finding was first made in 1974; after some years it was confirmed by the other investigators at the New Jersey site; and only in 1979 did the result emerge in Senate hearings on the proposed legislation. (In this case, the result was brought to light through Senate hearings, which might seem to negate the argument that hearings under other auspices would be desirable; but the result only came out under the questioning of Daniel Moynihan, one of the few social scientists in Congress and a former member of PSAC.)

This is an example in which the technical arguments were at a high level of statistical complexity, and parallels quite closely science-laden policy issues in the natural sciences. It would have been an ideal issue for PSAC hearings designed to hear all sides of scientific controversy. If there had been a PSAC engaged in such activities, the matter could have been resolved as early as 1974 or 1975, rather than only in 1979, after extensive work had been done in designing policy based on incorrect assumptions about its effects.

No Difference in Principle

I have used this extended example to suggest that there is nothing different in principle between the kinds of functions that a PSAC-like entity could carry out for natural science and for social science. After World War II, the Federal government began to initiate many policies requiring extensive scientific input, and PSAC came into being as a useful aid. At that time, there were few social policies requiring comparable technical input from the social sciences, and a social scientific component of PSAC would have been of little use. But in the 1960s, such policies did come into being. The social sciences are, on the whole, less well prepared than the natural sciences to supply the useful technical input; but this is far less true than it was even 12 years ago (when a social scientist was first appointed to PSAC), and for some policy issues there is a strong base of systematic information that can aid in assessing the probable effects of various policies.

Thus altogether, I believe there is an important future role for a PSAC-like entity. Its functions would be expanded in two ways beyond the original conception of the committee: it would encompass a broader range of policy issues, including not only those requiring natural science input, but also those requiring social science input. But more importantly, given the changed structure of decision-making in the Federal government on many of these policies, the committee's role would include not only that of advising the President and the Executive Branch, but also that of informing the broadly dispersed set of parties who play some role in policy-formation, and have pressing need for clarification of the scientific questions relevant to the policy area. Its mode of operation had already evolved in that direction before its demise, with its panels and their published (and often widely circulated) reports. It would necessarily evolve further once that function is explicitly recognized as a proper part of the committee's activities.

Federal Science Policy and Support of Autonomous Universities: A Modest Proposal

Gerard Piel

My concern in this paper is the role of the President and of the Congress in the most fundamental aspects of long-range science policy, namely, the advance of human understanding and the education of the youth of our nation—therefore, with the welfare and independence of our universities. A seemingly radical, but truly conservative, program is advanced.

If the object of our national science policy is a flourishing and advancing scientific enterprise, then we must give highest priority to the restoration and fortification of our universities. That means substantial Federal subsidy of those institutions. It should go without saying that science—rational, verifiable inquiry into nature and the nature of man himself—belongs in the university. This is necessarily so, because no other institution in our society is chartered to promote such an inquiry in the absolute freedom that is its very life. Our actual universities do not always provide the environment of freedom in which science lives. Other institutions, notably certain government agencies and a few industrial laboratories, have sponsored significant work in science. Admittedly, I am speaking of the idea of the university, in the words of Alexander Heard, as the seat "of inquiry and ultimately inquiry about anything."

This recommendation is more than pious; it is practical. It gives us an operational approach to policy for the public support of science. As we are so often assured, the nation cannot enjoy the fruits of applied research without basic research. Other approaches to the definition and distinction of pure and applied science lead into aimless refinement and dispute about the taxonomy of science, in the happy phrase of Victor Rothschild. They provide no handle for policy to allocate the human and material resources to the one or the other kind of research.

To recognize that science will thrive as we make our universities thrive clears the ground for policy in another significant way. It decouples pure science from

Gerard Piel (b. 1915), publisher of Scientific American *since 1947, is a trustee of the American Museum of Natural History, of New York University, and of other educational institutions. He was twice elected a member of the Board of Overseers of Harvard College. He is a member of the Institute of Medicine of the National Academy of Sciences and is the author of* Science in the Cause of Man, The Acceleration of History, *and articles on education, science, and public policy.*

applied and disengages the justification of science from the threadbare and so often corrupting arguments in favor of its utility. Science in the university can be seen more plainly for what it truly is: an enterprise we carry on for its own sake, as an end itself, as the supreme expression of our humanity and of our success in the attempt at civilization.

Pious and practical as this recommendation may be, it is not popular. In the tightly-bound positive feedback loop of the present project-grant system, the professoriate, the university administrations and the Federal granting officers are nearly unanimous in their opposition to institutional grants. The last attempt to leaven the flow of project funds with money for the universities themselves, nearly a decade ago and at the initiative of Congressmen, died for lack of response from the university community. Now, with the universities at the nadir of their fortunes and repute, the presidents of the country's 15 foremost universities—after weekends in retreat at the Yale University Seven Springs conference center, which that university has since had to convert to cash—have declared again the community's commitment to the project grant and its resolute rejection of the institutional grant.

Nostalgia comes along with déjà vu when one reads this brief for what the COSPUP (Committee on Science and Public Policy) reports of the National Academy of Sciences in the 1960s called "the permanent inter-related system." The project grant is extolled again as the arms-length transaction by which the government buys research that the scientist happens, in any case, to be doing. There follows the same argument for the plurality of Federal granting agencies. No embarrassment is found in the fact that 85% of the funds still issue from mission-oriented agencies; that more than half the total originates in the so-called health agency, and that the contribution of the military and paramilitary agencies continues to outweigh the benefaction of the National Science Foundation. As the late Lloyd Berkner said, instead of one National Foundation, the nation has seven or eight.

This preference for the purchase order and penchant for plurality stem from the familiar fear of "government control" and wariness of "politics" found so often among the citizens of our democracy. Such attitudes comport with the national ethos of free enterprise. Americans like to rely upon what they think of as the "natural" and "self-regulating" play of the market as against the deliberate framing of public policy. In Patrick E. Haggerty's well-chosen words ". . .by public policy these days we generally mean some kind of governmental intervention, probably Federal." But the unseen hand of the market has not protected science, as Robert Morrison has observed, from "the unseen hand of the Office of Management and Budget."

A Policy of No Policy

This science policy of no science policy did, of course, work for a time. It worked with enormous success. The achievements of American science under this kind of public sponsorship have been celebrated sufficiently to require no prolongation of the celebration here. As the 15 university presidents argue, the rate of bestowal of Nobel prizes on US scientists correlates well with the rising tide, in the 1950s and 1960s, of Federal funding. The percentage of prizes brought home to this country rose from the average of 10% established in the pre-World II period to 60 or 70% in the most recent decades. The lustre of these statistics is somewhat dulled, however, by an observation of Lord Rothschild; he was able to cheer his countrymen with this rhetorical question: "Have we not got 4.6 Nobel Prize-winners per ten million of our people in comparison with America's 3.3?"

The enormous success of American science still does not suffice to establish the prevailing arrangement for its public support as the ideal or ultimate. Nor can those arrangements be evaluated without due regard for their effect upon the universities engaged in them. Even now with the flow of funds throttled down, the Federal outlay for science must be reckoned with as a Federal subsidy not alone for science but for our great universities. It still amounts to more than 10% of the total national expenditure for all higher education and makes up from one-quarter (Harvard) to three-quarters (MIT and Cal Tech) of the total budgets of our 100 largest and most distinguished universities. Such radical revision in the relations of these institutions to the Federal government deserves the deep and thoughtful study and reflection that has so far been resolutely avoided by the university community.

Despite the success of science, the great universities of America stand lower today in popular esteem than they did at the end of the war and the beginning of their "Federal period." It was not to be thought, in fact, that institutions dedicated to such high purpose could ever find their way into such low standing. A tide of anti-intellectualism and disillusion with science runs strong among the people, especially among the graduates of our universities. Science and the great universities continue to be identified in the public mind with the overhanging, unthinkable catastrophe of World War III. A profound disruption of the faculty community—by the overwhelming external support for the sciences as opposed to the arts and, within the science faculties, by the competition for loyalty as between the invisible college and the university—was exposed in the instant collapse of the grandest institutions a decade ago before the challenge of the student rebellion. The university commitment to the outré technologies sponsored by the military and paramilitary granting agencies has weakened, apparently, the nation's command of domestic industrial technologies. Except for computers and aircraft, US manufacturing industries carry no weight in international trade and lose customers to overseas competition even in their home markets.

In Washington, science and the universities face what is, perhaps, their most humiliating loss of status. It is not only that Federal funding in real (say 1968) dollars has stopped increasing and has been seriously discounted by inflation. It is also a fact that neither science nor higher education have a claim in their own right on the national treasury.

The funding of university science came by overflow from R&D expenditures for other purposes—for weaponry, for space, for health. Science prospered as those expenditures increased through the 1950s into the 1960s. Such funding financed the great work done in that period; it also brought work into university laboratories that did not belong there. But now that the research budgets of the weapons and space agencies have stopped growing or declined, science in our universities—especially in our great universities—is in crisis. Federal patronage, supplied from ulterior motives, has fallen away from equally irrelevant motives. After 30 years and a cumulative $30 billion in Federal funding, science and higher education have no claim on the national treasury because the case for their claim has never been made.

Misgivings About Performance

All of this goes to excite deep misgivings about the performance by our universities, in this period, of their constitutional function in American society. In *The Higher Learning in America,* Thorstein Veblen defined the university as "ideally and in popular apprehension. . .a corporation for the cultivation of the community's highest ideals and aspirations." Speaking to the idea of the university, Alexander Heard said: "A university, to fulfill itself as a university, will seek to make it possible to examine the assumptions, the conventions, the taboos of the time, and the structure of knowledge and belief that rests upon them." In finding their way into service as contractors for the Federal government—as "instruments of national purpose" in the exultant phrase of one university president—our universities have jeopardized their role as centers for the independent criticism of public policy and the formulation of human purpose.

The commitment to the space adventure in the 1960s supplies the most immediate example of what I have in mind. There came from the universities no real consideration of the advisability of this enterprise and no vision of alternative goals to which such extravagant outlay of human and material resources might have been committed. Space, as a paramilitary exercise, helped to offset the decline in direct military outlays for research in the universities.

At this juncture, looking toward a more hopeful future, it may be useful to recall how our universities got into their present insecure and unsatisfactory as well as demeaning relationship to the Federal government. According to the late J. B. Conant, "the close connection between university research and the armed forces was in a sense an accident." He recalled that Vannevar Bush had been appointed by Franklin D. Roosevelt in 1945 to draw the "blueprint for the Federal subsidy of scientific investigation in the postwar world." Bush recommended "the establishment of an entirely civilian agency"; his report, *Science and the Endless Frontier,* "contained no indication that the Navy, the Army, or the Air Force would be involved in the furthering of scientific research by the Federal government." When, in 1947, legislation for a National Science Foundation on the Bush design arrived on the desk of Harry S. Truman, however, it was promptly vetoed for the reason that "the administrative arrangements were unsatisfactory to the most influential of Truman's advisers." Three more years had to pass before a satisfactory bill got through the two houses of Congress. This time, the President signed but, as Conant said, "it was too late. The armed forces had taken over."

A Lack of Faith

The armed forces did indeed take over; military and paramilitary agencies provided as much as 65% of the Federal funds for university science when these funds were at their peak. I think it took more than the Truman veto, however, to bring this situation about. The President found in the Bush design for a National Science Foundation features that reflected, in his opinion, "a distinct lack of faith in democratic processes." The Bush bill sought, in fact, to insulate science and the universities from politics. It called for a National Science Foundation under a board of part-time directors to be appointed by the President, to be sure, but for staggered terms that would place the appointment of a majority beyond the reach of a President's term in office. This board was to appoint the management of the foundation.

I add this footnote to Mr. Conant's concise history because I think it illuminates the strategy of university administrators and scientists since then. Their skittishness about engaging in democratic processes has been expressed in their preference for the closed politics of their relations with the Executive branch and its plurality of mission-oriented granting agencies, including the military; their contentment with an undernourished National Science Foundation, and their satisfaction in the policy of no public policy for science.

The record shows it was politicians, not scientists or university presidents, who first raised the alarm about the impact of Federal patronage of science on the universities. Until Congress invited them to come testify, the entrepreneurs of university science had kept a cautious distance from the Legislative branch, where politics is necessarily open. To this day science has no spokesman on the floor of either house and no committee formally charged with its overseeing except a sub-committee of the House committee that is more interested in the big money that goes into "astronautics."

What we know today about the Federal funding of science we owe to a succession of hearings in the House of Representatives, beginning early in the 1960s, organized by Carl Elliott of Alabama, Henry Reuss of Wisconsin, and Emilio Q. Daddario of Connecticut. The National Science Foundation was supposed to generate such information; the best it had to offer was estimates, predicated upon uncertain taxonomic criteria, of the relative flows to pure and applied research. It was the House hearings that developed the complete matrix of the source and application of funds. They revealed for the first time the entangling alliance of the universities with the military. This exposed, in turn, the correlation between the concentration of university research funds and the geographic distribution of military procurement contracts; the resultant slighting of the important graduate schools of the Big 10 and the private universities of the Midwest; the consequent distortion of emphasis by fields of inquiry, and the de-emphasis on education, even graduate education, in some of the most handsomely subsidized institutions. To these proceedings witnesses from the universities contributed little more than testimony to their satisfaction with the now established arrangements. The suggestion that other arrangements might be considered brought from these witnesses alarmed protest.

The Need for a Policy

The record nevertheless established the need for a policy for support of higher education, if not of science. To the credit of Lyndon B. Johnson and his Science Advisers it can be said that they made the first attempt to adumbrate such a policy. In an Executive order issued in September 1965, the President declared the plain fact that "research and education [are] inseparable" and that "Federal expenditures have a major effect on the development of our higher education system." Department and agency heads were accordingly enjoined "to insure that our programs for Federal support of research in colleges and universities contribute more to the long-run strengthening of the universities and colleges. . .[and] to find excellence and build it up wherever it is found so that creative centers of excellence may grow in every part of the nation."

If the need for public policy in the support of the higher learning in America were still in doubt in 1965 there can be no mistake about that proposition in 1979. While Federal outlays to university science in current dollars exceed the 1968 peak, inflation has reduced effective, constant dollar, support by one third. In state as well as private universities the enrollment of graduate students has been ruthlessly curtailed. The argument that the country is over-supplied with Ph.D.'s must be recognized as a specious and discreditable rationalization for the shrinkage of Federal support. It is not the predicted decline in undergraduate enrollment, which has not yet begun, but the topping-off of public expenditure, that has shut down the academic job market.

Only in America

The United States is the only industrial country that has not committed resources from its national treasury to the support of science and higher education. Our democracy is the only one that expects each university tub to stand on its own bottom and invites its graduate students to pay their tuition from the family bank account. Every other industrial country makes significant annual appropriation to its universities and has devised one sort of institution or another to administer it. As we consider now the making of our commitment, we have many precedents to choose from.

Such precedents may guide us in the task of institution-building that we must face at last. The hazards in this undertaking must not be underestimated. There is sound common sense in the identification of public policy with governmental intervention. Public support of autonomous institutions implies a paradox, an undecidable question in the best Kurt Gödel sense. Yet our recent experience proves that the underlying issues must be faced and dealt with in the open. The longer we attempted to avoid them, the more dangerously unmanageable they became.

A useful lesson in institutional design is offered by the troubled launching of our National Science Foundation. The defect in the Bush bill lay in its attempt to draw the line of autonomy across the link between the Federal government and its granting agency. That surgical incision must be drawn not there but between whatever Federal agency we create and the autonomous universities. The agency must be explicitly chartered for its mission and face full public accountability for its performance. That mission is to see that the universities have enough resources, to promote their excellence, and to secure their autonomy.

This much is easy to declare. To carry the analysis and design closer to any conceivable implementation calls for the wisdom and devotion of every citizen who cares. The present generation of university administrators and professors has a special obligation to join in the task. They are the trustees of the integrity and autonomy of their institutions that are otherwise so ominously weakened. It is necessary to think past the proposal from the Carnegie Commission on Higher Education, for example, that Federal subsidy should take principally the form of scholarships. That is another subterfuge for no policy. The more simpleminded the formula, it is true, the fewer are the strings tied to the money. Yet no formula that involves human behavior can be expected to work automatically. Policy will make itself if it is not made by conscious, rational design.

The operation of the University Grants Committee in the United Kingdom commends itself for study at this juncture. This agency allocates its funds in accordance with a sufficiently simple formula; it makes institutional grants on a straight capitation basis, so much per student. In blandly British style, however, it employs the formula to enlist the universities in the design of the evolving national policy for higher education. Its so-called recurrent (capitation) grants are offered by the quinquennium. This induces each university to come forward with a five year plan. Those plans provide the basis for the non-recurrent grants to capital projects. Toward planning at all institutions the U.G.C. offers cues that embody its own estimate of the demands to be laid upon the universities by the society around them.

The academic year 1971-1972 started one of the quinquennial planning cycles; the "Preliminary Memorandum of General Guidance" then issued by the U.G.C. is instructive. As a "working hypothesis" it set out a "suggested capacity" of 320,000 students for 1976-1977, an expansion of 30% between "arts-based" and "science-based" students. Forecasting "strong pressure to reduce unit costs," it counseled certain economics of scale. The average size of arts and social studies departments "should not be below about 100 full-time equivalent students"; while science and technology departments "should not average less than about 120 to 150." In this connection, it advised that "larger departments have the great academic advantage of making it easier for staff to preserve time for their research."

With cost pressures still in mind, the document proffered the hint that the "enlargement of existing departments in those universities which currently have them is more economical than the creation of new departments and new institutes." By way of indicating the range of opportunity in this direction, the document went on to observe that demand for graduates in "the less common languages" is likely to be small. On the other hand, the committee urged "universities. . .engaged in management education [to] submit plans for expansion"; that priority be given "for increasing the flow of Mathematics, Science (particularly Physics) and French teachers to schools''; that "instruction in mathematics and computing is desirable for as many students as possible," and that high priority be given to developments in educational technology.

Sensible Policy Making

My purpose in this recital is to show that even a simpleminded formula can serve as a vehicle for sensible policy-making, principally by encouraging competition among securely independent institutions. Conversely, I want to show that policy need not trench upon autonomy. I do not deceive myself with the idea that the U.G.C. model, so appropriate to a homogeneous society that has a generous regard for its elite and does not yet send as much as 10% of its young people to college, can be translated to our diverse and often divided American culture, that now attempts to extend the opportunity of higher education to half of the rising generation. The essence of the U.G.C. model deserving serious consideration here is that it does work to secure the autonomy of the British universities against the stresses exerted upon them by public funds from other quarters, especially by scholarships and fellowships and by project grants.

A commitment to the primary support of universities by institutional grants does not, therefore, exclude the public funding of scholarships and fellowships. It is time our country dismantled the economic hurdles that discriminate against talent from low income families and relieved those families of the humiliation of the means test.

Nor does institutional granting exclude project funding. The instruments of Big Science call for capital inputs in big lumps, and sound precedents have been established in the funding and management of our national laboratories. Project funding—even from mission-oriented granting agencies, under invisible-college administration—is equally appropriate for little science and for research in the arts and humanities as well, providing the recipient professor is secure in his own estate in his own university.

The great healing promise of institutional support for American universities lies in the provision of economic incentive for their reconstruction as communities. Such incentive can be amplified by Federal grants that match institution-building funds from other sources, from State, municipal and private funds. It is significant that the University of California—need I say, a State institution—stands regularly among the first three in the list of universities receiving private benefactions.

The leadership in the setting of rational Federal policy for the support of higher education ought to come from the great privately endowed universities of the Northeast. These institutions have provided the model that secures the independence of our State universities. They can no longer pretend that they are not Federal universities. Harvard, with its princely endowment, ought to recall that it was a college of the Commonwealth only a little more than a century ago, supported by annual appropriations from the General Court of Massachusetts. In placing—even burying—the support of science in the larger context of support for higher education, I have sought to reinforce by a clearly visible institutional identity the difference in our motives for sustaining pure as distinguished from applied science. This proposition collides at 90° broadside with the model advanced by the last effort from the White House to articulate a national science policy. Just before Richard Nixon abolished the office of Science Adviser to the President, Edward David declared the country was learning to allocate its support for science in accordance with its commitment to a range of National Objectives. According to David, the end-need for a technology should establish the size of the subsidy for the support of its pure science.

In much the same spirit at about the same time, Sir Harold Himsworth proposed that the experience of his own Medical Research Council be extrapolated to a policy for the support of all the sciences. The funding of basic research should be tied to technology in each autonomous "province of learning"; that such grandly defined end-needs as "health," "energy," "materials," and so on determine the allocation of resources to the learning that answers to each. At home, Sir Harold was contending with Lord Rothschild's proposal that the autonomous Research Councils be dismantled and their applied research enterprises made to answer to Her Majesty's several Ministers; in Rothschild's words: "to democratic society itself and its elected representatives." But I have my own grounds for dissent from the Himsworth as well as the David model. Against both models, I urge disengagement of the support of pure science from the financing of applied research. I would let expenditures for higher education set the general level of primary funding for university science. This implies further that the making of policy for pure science—the choosing of objectives and the setting of priorities—be returned from Washington to the universities.

The Time Has Come

With equal urgency, I argue that the time has come to disengage science and higher education from dependence on the Executive branch. To take on the solemn responsibility for the restoration and fortification of our universities I propose that we create an entirely new kind of agency, a Congressional agency. Let us call it The Congressional Endowment for Promoting Useful Knowledge.

It has been argued to me that this proposal runs counter to the principle of the separation of powers. Congress is already, however, a billion dollar enterprise. It has created, is served by, and manages the Library of Congress, the General Accounting Office, the Office of Technology Assessment and the Congressional Office of the Budget. The budgets of these agencies make up a considerable percentage of the total expenditure of that billion. Put the Congressional billion against the expenditures of the Executive department and we have a cost for thinking, for goal-choosing, for legislation that is less than a quarter of one percent of the total. If my proposal were to be taken seriously, the Congressional budget could be multiplied by five and still give us a bargain costing less than 1% of total Federal expenditure.

In my fantasizing, Congress would delegate the management of The Endowment to an appropriate, representative and competent panel of university administrators, faculty and private citizens. Responsible to Congress, they would arrange to get the decisions as to allocation of funds made by peer review conducted in broad daylight. The funds would flow principally in the form of long-term institutional grants. The granting criteria would be designed to fortify the autonomy of those universities that best defend the autonomy of their scholars and scientists and to restore decision as to research priorities to the university community where those decisions belong. Inside the universities, we may count upon the institutional grant to lend economic incentive to the latent centripetal forces and, at the very least, to unite the community around an apple of discord.

The Executive department could still make its contribution to higher education. The 15 university presidents have set out an ambitious program of scholarship, fellowship and project grants. Fortified by the Congressional Endowment, however, universities and scientists would be placed in a secure bargaining position, especially as to the choice between work the Federal agencies want to push and questions that happen to interest them.

The proposal that universities should be funded by a Congressional agency has not drawn the instant enthusiasm of the entrepreneurs of university science. Even though their access to the White House has become uncertain, they are experienced in what C. P. Snow has called "closed politics," and they know their way down his "corridors of power" in the Executive branch. They argue that the geographic distribution of power in the Congress corresponds not at all to the geographic location of universities, that Congressional funding is subject to pork-barrel and log-rolling politics, that the know-nothingism that blights our public discourse has its endemic home in the Congress.

This response belies the fact that the first people to discover that we had to reckon with science policy were members of Congress. Against the arguments that turn upon the frailties of Congress, it can be said that a closer relation with the universities, if the universities set the example on their side of it, could do much to elevate the quality and integrity of discourse on Capitol Hill. The Congress, no less than the White House, requires informed and disinterested counsel on the momentous decisions, mined with scientific and technological booby traps, that are before it every day.

That brings this discussion to still more compelling considerations in favor of a Congressional agency for the support of university research. The right way to bring the informed counsel of the scientific community into the deliberations of our Federal government is in the place and at the time of the formation of choices, values and policies, that is, in the legislative process. What is more, a close connection to the Congress in this function comports better with the autonomy of the university than engagement with the Executive Department in the execution of policy. To which it may be added that it would be more becoming of the universities to conduct their relations with the taxpayer through the necessarily open politics of the Congress. The Congressional Endowment for Promoting Useful Knowledge would give the electorate, represented by its universities, a continuing presence in the national government. For this mission, the university is the appropriate institution because it is the place where the next generation of citizens re-discovers and re-shapes human identity and purpose.

Politics in the Science Advising Process

David Z. Robinson

Introduction

I am convinced of the importance of science to government—and vice versa. I am also convinced of the need for a broker in this relationship and of the importance of the Science Advisers, past and present, in serving this brokerage role.

I spent the years from 1961 to 1967 working for Jerome Wiesner and Donald Hornig who were the Science Advisers to Presidents Kennedy and Johnson. When I left my industrial job to go to Washington, I had an idealized—even naive—image of the role of a scientist in government and of the Science Adviser in particular. The scientist should use his (no one even imagined "her") expertise to analyze long-range technical problems that faced the government and to give advice to the politicians who would accept or reject it. The scientist was disinterested and rational. The politicians had to deal with messy concerns such as regional politics, short-term "brush fires," patronage, and bureaucratic maneuvering, including allocation politics within the Executive branch.

After I had worked in the office a while, it was clear that the science advisory apparatus, no less than the politicians were affected by these messy concerns and that a successful apparatus was one that could deal with them effectively. Where the Science Adviser failed, he failed often because he did not take into account the politics of the situation.

The Science Adviser's most potent weapon was and is a reputation for clear, disinterested technical knowledge. If he can keep that reputation and at the same time understand and use the political process, he can be effective. If he does not use or understand the political process, he will be bypassed and ineffective, regardless of the quality of his technical advice.

This paper illustrates this thesis by reviewing how three Science Advisers encountered and used "political" concerns such as regionalism, quick-fix solutions, bureaucratic maneuvering and patronage. I also discuss briefly some "failures" of

David Z. Robinson (b. 1927) is vice president of the Carnegie Corporation of New York. He was on the staff ot the President's Science Advisory Office (1961-67), leaving to become Academic Vice President of New York University. His earlier career as a chemical physicist included service as Assistant Director of Research, Baird-Atomic, Inc., Boston, and as science liaison officer, Office of Naval Research, in London. He is a member of the New York City Board of Higher Education.

the science advising process in the sense that the advisers were not successful in persuading their advisee, the President of the United States. The ultimate failure, of course, was the abolition in 1973 of the position of Science Adviser, the staff of the Science Adviser (by then known as the Office of Science and Technology [OST]) and the President's Science Advisory Committee (PSAC). Although President Ford reestablished the position and staff in 1975, PSAC no longer exists. If PSAC is to return, the political problems it raised would have to be dealt with.

Regional Politics

Our government has been based on geographic representation since its beginning—both in the electoral college and in the legislature. Since the beginning, there have been arguments among regional representatives about how Federal dollars should be distributed. Inland states look for dams and river projects; coastal states for harbor improvements. Urban states look for better housing and unemployment benefits; rural states for increased crop supports. Traditionally, the Army Corps of Engineers appropriation became known as the "pork-barrel" bill because Senators and Congressmen could and did arrange to obtain a share of the expenditures for their districts.

When it comes to government investment in science support, of course, decisions are supposed to be made on the basis of technical capacity and research quality, not regional political considerations. However, scientists are as human as politicians. Thus Eastern scientists and their Congressional representatives would like to see facilities in the East even if in their hearts they know that the Western group is a little better. The geographic distribution of support for science has been a perennial problem for the Science Adviser.

"Centers-of-Excellence" Program

The Science Development Program of the National Science Foundation (NSF) was to some extent the product of regional politics. The original legislation that set up NSF posed conflicting goals. It specifically stated that the awarding of grants by the foundation should be done on the basis of scientific merit. On the other hand, it also encouraged widespread geographic distribution of funds. When, in the early 1960s the NSF (and, more importantly, the Congress) examined where grants and fellowships were going, it found that a substantial majority went to the major research universities, particularly in California, New York, and Massachusetts. The conflict in goals was exemplified by an exchange between a Southern Senator and an Ivy League university president at a public hearing. The senator said he had heard that a major Eastern university had hired a professor from the South who had an NSF grant—and then had paid him out of the grant that was moved with him. He accused the university of using federal funds to steal faculty from regional institutions. The university president replied to the effect that it was the American way for a faculty member who wished to better himself by moving to the Northeast to do so. This response did not please the Southern Senator.

Following the hearings, under pressure from the appropriations subcommittee charged with the NSF budget, the science foundation changed its policies on fellowships in a way that would lead to a wider institutional and geographic distribution. But it did little to affect the distribution of research grants. When Lyndon Johnson became President, he expressed concern to his Science Adviser about the narrow distribution of research grants.

In response to this Presidential interest, the National Science Foundation established a Science Development Program designed to improve institutions that were not in the top rank and to make them more competitive for research funds. (One member of the foundation staff described it to me as a program designed to "get more institutions into the top 20"). The development program put $3-5 million over a five-year period into institutions and departments that were deemed to be on the edge of greatness. The expectation was that the impetus of the development program would enable these institutions to compete at the highest level for funds. One of the major premises of the program was that since the regular funds for university research were increasing at a rate of 15% per year, growth would take care of the needs in the developing institutions as well as the established ones. ("A rising tide lifts all the boats.") Unfortunately, funds for research did not grow in real terms between 1967 and 1975, and so the competition with more established institutions became more intense, and the program is not generally regarded as a success.

This program was not terribly popular with the science establishment within the President's Science Advisory Committee which felt that the development program was reducing funds necessary to maintain the first quality departments. Donald Hornig, sensitive to the realities of regional politics and convinced that it did not affect the budget for traditional science research, was a forceful defender.

High Energy Accelerators

One of the major tools for research in fundamental physics is the particle accelerator. The complex forces between the various components of the atomic nucleus may be best studied by causing collisions between accelerated elementary particles and various targets and observing in detail what happens. *

Accelerators are very expensive to build and require a dedicated and competent team for their operation. They are also discrete items that can be put in or knocked out of an executive budget. As the energy frontier of high energy research moved from the million volt (MeV) to the billion or giga-volt (GeV) range in the 1950s, the costs increased to over $100 million per accelerator.

The first major accelerator was built by E. O. Lawrence at the University of California, Berkeley, before World War II. After the war, physicists, particularly nuclear physicists, had unusual prestige because of their work on the Manhattan Project, where basic research in nuclear physics had led to spectacular practical results. They could attract the funds necessary to do research in elementary particle physics. A number of accelerators were started at that time and geographic politics entered early. Influential Eastern scientists moved to develop East Coast

---

* Accelerators are defined by the particles they accelerate (usually electrons or protons), the maximum energy of the particles (in units of electron-volts) and the intensity of the beam, or number of particles per second that are accelerated.

competence by obtaining funds to build a very large laboratory accelerator at Brookhaven on Long Island.*

During the 1950s the cost for accelerators on the scientific frontier was going up so fast that the number of accelerators on the frontier would almost certainly decline. Their location became of great importance to scientists as well as Congressmen. Furthermore, persuading a President in times of budget stringency to add an amount that would increase the reported national budget deficit by a full digit (for example, change it from $19.3 billion to $19.4 billion) was a major problem.

The first accelerator to come under Presidential scrutiny because of its high cost was the one proposed by Stanford University in 1957 (SLAC). This was to be a two-mile long installation which would accelerate electrons to 99.999999975% of the velocity of light—an energy of 20 billion volts (20 GeV). The facility was estimated to cost $100 million to build and $25 million per year to operate.

George Kistiakowsky's memoirs refer often to the Stanford project. He was persuaded that the accelerator was important and had to find a way to get the necessary government support. He began by appointing, together with the Atomic Energy Commission (AEC), a panel of distinguished scientists to make a recommendation. The panel recommended that the Stanford accelerator be built, while pointing out that SLAC should not be allowed to absorb funds needed for other fields of research. Kistiakowsky saw to it that the panel's report was delivered personally to President Eisenhower. After President Eisenhower made the decision to go ahead with the project, Kistiakowsky reinforced the effect in the scientific community by getting him to announce the go-ahead in a banquet speech to the American Association for the Advancement of Science on May 14, 1959.

Thus, by understanding the dynamics of the executive branch and the value of the timely use of outside experts, Kistiakowsky skillfully brought about an unprecedented expenditure on basic research. Unfortunately, the Democratically-controlled Joint Committee of the Congress refused to appropriate the funds, raising questions about the planning of the project. Kistiakowsky and the AEC appointed another panel to respond to those questions, and it remained for the Kennedy Administration to push through the appropriation in 1961. (SLAC, of course, turned out to be a smashing success from a scientific point of view, even though most of the important experiments that have emerged were not foreseen when the justification for building it was put together.)

One year later, in 1962, the Kennedy Administration received a proposal from the Midwest Universities Research Association (MURA) for an imaginative installation that would accelerate protons to about the same energy as other large accelerators but would greatly increase the number of particles produced per second. The proposal called for a 10 GeV proton accelerator to cost $120 million. MURA was a special consortium of midwest universities created to design and eventually operate, an accelerator which would enable those universities to compete with the laboratories at Brookhaven and Stanford. It was probably necessary for its own survival as a laboratory that the accelerator be built. The consortium felt embattled in its effort to secure an accelerator and had allies all over the midwest.

---

* Other accelerators were established at major universities around the country. The costs had not yet become significant in budgetary terms.

When the MURA proposal arrived, Jerome Wiesner, Kennedy's Science Adviser, knew that there were two other design projects almost at the proposal stage. One was a 100 GeV (later changed to 200-400 GeV) proton accelerator being developed by a group at the Lawrence Radiation Lab at the University of California, home of the original cyclotron (and a number of later and larger versions). This new machine would cost $100 to $200 million. The second proposal, being worked on at the Brookhaven National Laboratory, was for a 600-1000 GeV accelerator (costing $600-1000 million). Brookhaven at that time was operating the highly successful 30 GeV accelerator, the largest in the United States.

Wiesner was faced with a difficult problem. The political history of the Stanford project suggested that it would be difficult, if not impossible, to get more than one accelerator through the administration and the Congress within the next three or four years. He decided that the three proposals should be considered together and that a long-range plan should be developed which would be acceptable to the Atomic Energy Commission as the funding agency, as well as the Office of the President.

Following the procedure that Kistiakowsky had used successfully, Wiesner decided to set up a joint panel with the AEC. The new panel—consisting of scientists from high energy physics and other fields and chaired by Professor Norman Ramsey of Harvard—was asked to develop a 10-15 year plan for the field as a whole, and most importantly, to suggest research priorities. Wiesner knew that scientific panels tend not to consider priorities or choices when really good scientific proposals are at stake. Thus, while he told the committee that he wanted their best judgment, he also told them that they could not choose "all of the above"; approval of MURA, for example, might hold up the construction of the other two accelerators.

The "Ramsey Panel" worked very hard for eight months. Its major conclusion was that the MURA accelerator would be very valuable, but that the two other accelerators (which were some years away from being ready for construction) were more important. The carefully worded conclusion recommended that "the Federal Government authorize the construction by MURA of a supercurrent accelerator without permitting this to delay the steps toward higher energy." The implication was that if "the steps toward higher energy"—i.e., the more powerful accelerators—would be delayed, the MURA accelerator should not be built. Incidentally, although the final recommendations were unanimous, the panel experienced its own geographic politics with the midwest members supporting MURA and the western members favoring the California proposal.

On the basis of the lukewarm recommendation, the AEC in 1973 requested permission from President Kennedy to go ahead with construction of the MURA accelerator in Madison, Wisconsin. A decision was needed in December as part of the President's budget proposal for the fiscal year beginning July 1, 1964.

Once the Ramsey Panel report was published, scientists started lobbying. MURA scientists reached midwestern Congressmen who felt very strongly that their area of the country had not been treated fairly in the distribution of money for science and who mounted a very strong campaign in support of the proposal. California Congressmen (using information supplied by Berkeley scientists who feared that construction of MURA would delay their machine) argued against MURA.

When President Johnson assumed office that tragic November of 1963, he had to make a decision on the accelerator proposal as part of the new budget. He found himself facing not simply a technical and fiscal issue but a political one involving legislators from all over the United States. "I devoted more personal time to this problem than to any nondefense question that came up during the budget process," the President wrote later.

Johnson asked Wiesner to write down on two *separate* pages the arguments for and those against construction. Then at a meeting with a special delegation of mid-western Senators and Congressmen, Johnson did not commit himself finally. But he read the delegation only the page with arguments against the accelerator, the major argument being that a better accelerator was in the offing. Since one of the senators glimpsed Wiesner's name on the memo, Wiesner got unfairly blamed for being against the MURA construction. (I suspect that Johnson intended that the senator see the signature.)

Johnson later wrote Senator Humphrey of Minnesota to tell him that the 10 GeV MURA accelerator definitely would not be built. But, by the way he took the step, he essentially committed himself to building a 200-400 GeV accelerator at a later date. The Ramsey Panel had recommended that the 200-400 GeV accelerator be built by Lawrence Laboratory and assumed that it would be built in California. Instead, Johnson asked the AEC to search the country for the best possible site.

With the site question open, 85 detailed proposals arrived from every state except North Dakota and Massachusetts. Visiting teams of scientists inspected all the sites and were greeted by Congressmen. (Many of the Congressmen were willing to say that they supported construction of this great scientific tool even if it did not go to their own district; one site visitor called this disavowal of self-interest "taking the pledge.") The AEC sent proposals from six finalists to the President and Johnson personally picked a site in Batavia, Illinois, outside Chicago. The decision was made in November 1966 just *after* the elections. Despite all of the comparisons and technical considerations, Johnson apparently had in mind that the Stanford accelerator had gone to the West Coast, and that his friend from Illinois, Senator Douglas was retiring. The happy ending is that a 400 GeV accelerator was built in Batavia by a team headed by Robert Wilson of Cornell and has been producing extremely valuable results.

Most high energy physicists believe that the decision to favor construction of a 400 GeV accelerator rather than the MURA proton machine was the best solution to the problem of limited resources for big accelerators, although the factors that went into the decision were not all technical. Wiesner's strategy of picking an outside committee with broad scientific reputation and requiring them to come up with priorities was probably essential to getting the 400 GeV accelerator built when it was. His and Donald Hornig's dealings with the President, the AEC, and the Bureau of the Budget pushed the acceptance of the 200-400 GeV accelerator effectively inside the executive branch after the proposal came in despite the scale of the investment. Despite the political furor over the MURA decision, the end result was to improve the chance for construction of a major research facility. Kistiakowsky, Wiesner, and Hornig, none of whom is a physicist, deserve the thanks of the physics community for the political maneuvering that led to construction of the Stanford and Batavia accelerators.

"Brush Fires"

The President's Science Adviser and PSAC worked hard on long-range issues that had not yet attracted public attention or adequate resources. The work on oceanography in the 1950s, the PSAC panel on the environment in the early 1960s, the energy study in the middle '60s and the world food panel of the late'60s are good examples of this kind of thoughtful work. The studies and reports clarified a number of issues and influenced the thinking and programs of the government either directly or through changes in the public agenda. The office should be judged and was set up to be judged on its long term impact.

As a practical matter, however, the effectiveness of the Science Adviser's office has often been measured not by its careful and objective explorations of longer-term technical issues, but by how well it responded to short-term crises. Three examples of such crises were the cranberry scare, Project Needles and the great blackout.

Cranberries

George Kistiakowsky found himself involved with the problem of contaminated cranberries in the closing days of the Eisenhower administration. Cranberries were being removed from the market because of pesticide contamination and a fearful consumer boycott was developing. The agriculture department and the cranberry industry were on one side and the Food and Drug Administration on the other. President Eisenhower was in the middle and called on Kistiakowsky to help solve the problem. Within two weeks, Kistiakowsky put together a respected and competent group to review the situation, and their objective report helped considerably to calm things down.

Project Needles

Shortly after he took office, Jerome Wiesner received very strong complaints from the astronomy community about a proposed Air Force communications experiment it was alleged would interfere with radio and optical astronomy. The experiment, infelicitously called Project Needles (later changed to Westford), involved setting up a belt of 250,000,000 hair-like one-inch metal filaments 2000 miles out in space. These filaments would preferentially reflect microwave radiation of a particular wavelength. If two antennae focussed on the same part of the belt, it would be possible to communicate securely over long distances. There was obvious military interest in such a secure communications path.

The project had been examined a few years earlier, when it still had a secret classification, by the Space Science Board of the National Academy of Sciences. The Board had concluded that the experiment would not harm astronomical research, but recommended that the experiment be made public internationally to avoid misunderstanding. Despite efforts to explain the experiment internationally (within the constraints of military secrecy), the Air Force was besieged by nervous astronomers. Something had to be done. Again, Wiesner set up a special panel of astronomers. While the panel decided that the experiment itself was harmless, it expressed concern that a later operational system, which would probably use a much denser belt of dipoles, might cause a good deal of trouble.

Despite the panel's report, Wiesner was afraid that the International Astronomical Union meeting in Berkeley would formally condemn the experiment making it difficult to go forward. Accordingly, a letter from Wiesner reviewing the results of the panel study and giving assurances of further consultation with the scientific community before any operational belt was set up was brought personally to the Berkeley meeting by a staff member. The letter was reasonably well received, and the union resolution on "Project Needles" was much milder than had been anticipated. Here was a case where the Science Adviser and the staff had operated quickly to try to stop controversy although they inadequately judged the emotional and political response to large scale environmental experiments.

Ironically, the furor over their experiment was influential in moving the Defense Department to concentrate on active rather than passive satellite systems (i.e., those that radiated signals to the ground rather than those that reflect radiation). The long-run effect of such satellites may well be more detrimental to radio astronomy than an operational needles belt would have been.

The Blackout

Donald Hornig was washing dishes in his kitchen one November evening when President Johnson called to tell him to do someting about the widespread electric blackout that had just occurred in the Northeast including New York City. He dried his hands and called some key staff members. They worked until late that night, and the next morning he was able to convene a group to examine causes and possible preventive measures. He reported quickly to the President about the limitations of the Federal government in dealing with a fragmented electric power industry, and divided responsibilities within various Federal agencies. Although the report analyzed the issues very clearly, it did not make any recommendations for a politically acceptable solution.

The ability to respond quickly to issues of the day carried much more weight with the President and the Executive Office staff than did the ability to carry out thoughtful long-range studies. The Science Adviser and the staff were the ones who had to carry the brunt of the brush fire detail, since the President's Science Advisory Committee met only monthly. Short-range problems in particular required knowledge of Executive office politics and the established bureaucracy. The successful Science Advisers developed that knowledge and used it effectively, being careful not to overstep their authority, but capitalizing on the power which came from working on something of direct and immediate interest to the President.

Bureaucratic Maneuvering

The Science Adviser is a member of the President's staff with no line authority. He must deal with government agencies, all of whom would like to keep the White

House at arm's length. (The recent Cabinet changes are indicative of the ongoing tensions between Executive Office staff and the major agencies.) Under these circumstances, a Science Adviser must be able to find a way to operate effectively in the federal bureaucracy.

One solution is to form alliances. The key organization in the Executive Office for domestic policy is the Office of Management and Budget (OMB). Every agency must bring its requests to OMB and recommendations are made by OMB to the President. OMB has one of the best staffs in Washington, but the members of the staff are not scientists. An alliance with OMB can bring the Science Adviser into issues that are deemed important by the agencies and by the President.

Most of the Advisers have developed a good relationship with the Director of OMB and encouraged continuing informal communications among the staffs.

The role of the Science Adviser and staff in the formal budget review process was very important in the sixties. No other part of the Executive office spent so much time in the special room where agency budget requests were discussed. Whenever a technical issue came up, the chairman of the meeting (usually the Director or Deputy Director) would turn to the Science Adviser's staff, and would always take their recommendations seriously. This type of interaction exists today, and is one of the important sources of influence.

The unusual relation with OMB did not lead to bad relations with the Executive agencies. Some thoughtful members of the agencies realized that they could get a fairer hearing if there was good technical competence on the staff. These people would work with the Science Adviser's staff. Because agencies knew that the Adviser's staff had access to OMB, they were more willing to cooperate.

Science Advisers involved themselves extensively in Defense and Foreign Policy in the 1950s and early sixties. As the staff of the National Security Adviser increased in size, it was necessary to forge alliances. Wiesner and Hornig had an arrangement with McGeorge Bundy and Walt Rostow that put the staff member involved in disarmament issues both on the NSC staff and the Science Adviser's staff. This arrangement worked well until Henry Kissinger abolished it. Here was a situation where the Adviser was outmaneuvered by a master! During the Nixon years, the Science Advisers appeared to outsiders to get less and less involved in Defense and Foreign policy issues.

Patronage

"Patronage"—or the securing of jobs to reward friends or political allies—is usually looked upon as a bad thing. Helping to secure jobs in government for first-rate technical people—and vice versa—may be a good thing. Where a Presidential appointment to major scientific jobs is involved, the Science Adviser who has a desire to be effective should try to influence the decision.

Jerome Wiesner advised on many appointments. He was involved in choosing the Director of the National Science Foundation and the Assistant Secretaries for Science in many agencies. In the former case, he helped to block an appointment desired by a powerful Congressman. This hurt the NSF's budget one year. However, by involving himself in the appointments, he had an opportunity to persuade good people to take the jobs, and he had easy access to the people who were appointed.

In the appointment of technical and scientific personnel, the Science Adviser speaks as an expert and his word, if he has the President's confidence, has major weight. He can have an enormous effect on the conduct of governmental affairs through influencing who is appointed.

Failures

The Science Adviser and the PSAC had failures as well as successes. The Vietnam War is an example of the type of issue that the Adviser could not deal with well. In the early years of the war, scientists and engineers with military research experience, tried to help President Johnson accomplish some of the things he wanted to do—sealing off the border, for example. As the war enlarged, the role of technology appeared limited and the President's Science Advisory Committee began to raise questions about efficacy. This attitude was easy to discern as disloyalty in the Nixon White House and the committee was soon bypassed on many issues dealing with the war.

The supersonic transport was another issue which divided PSAC from the White House. The Science Adviser and the Science Advisory Committee got involved early in the program to develop and build an SST. The committee was cautious partly because of noise. (To be economic, the United States SST would have to be bigger and hence noisier than the Concorde which was then at a more advanced stage of development.) Although the committee was encouraging at first on the possibilities of the development, later information made the committee quite discouraging and negative. The major problem was the apparent impossibility of simultaneously meeting environmental and economic goals with contemporary technology.

A controversy behind the scenes in the administration ended with President Nixon recommending the SST. When PSAC opposition was mentioned at a Congressional hearing by the SST program manager, the committee asked Richard Garwin of IBM, a member of PSAC who headed a panel studying the SST, to testify. Although the panel's report was not public and Garwin did not tell them what it recommended, he described his technical reservations and proved an effective critic of the project. Congress turned down the SST development funds, a judgment that was almost certainly correct in hindsight. Later, the panel's largely negative report was released after a court case under the Freedom of Information Act. Clearly PSAC had failed to be persuasive inside the administration, and Garwin's testimony greatly angered Nixon loyalists.

The ultimate evidence of political failure was the abolition of the President's Science Advisory Committee and the position of the Science Adviser in 1973. The attitude in the embattled White House with regard to the committee was that there was no point in inviting your enemies inside your tent since everyone who was not 100% loyal was an enemy. The Science Adviser was deemed expendable in part because a powerful OMB director did not see the value of independent advice.

Soon after President Ford was inaugurated in 1974, there was pressure from the scientific community to appoint a Science Adviser. Senator Kennedy introduced legislation to that effect. At the invitation of President Ford, Vice President Rockefeller recommended reestablishing the position of Science Adviser (but not the committee), and an appropriate compromise was worked out with the Congress.

Conclusion

The most effective Science Advisers were those with clear understanding of the government, and how to make it work. Technical skill is available outside the government, and a Science Adviser can find technical help. The help needed to accomplish goals is harder to find.

Successful Advisers had long experience with government agencies. They knew how far to push things. They knew the difference between advising and decision-making. They knew how to form alliances, both with the agencies and with key parts of the Executive office, although even the most successful did not have enough direct influence with the Congress.

They also knew that their effectiveness depended on having high-quality people in the executive agencies to deal with. They used their influence on appointments to make sure that appointments—even patronage appointments—were based on merit. They understood the need to have high quality science all over the nation. They knew to balance the short-range with the long-range issues.

In summary, they were first-rate politicians. Their successors will have to continue that tradition if they are to remain effective.

PSAC was a powerful force in its early days when it dealt primarily with secret issues; as it moved to new issues which could generate public controversy, it made enemies. Presidents Johnson and Nixon never looked on it as *their* advisory committee. It disappeared because it was not part of the team.

Reestablishment of PSAC will be difficult to accomplish. The Federal Advisory Committee Act requires that many of the discussions be public. This will inhibit discussion. Until Presidents (and not only scientists) want competent disinterested outside technical advice, there will be no PSAC. It will take an unusual Science Adviser and unusual political circumstances (equivalent to a Sputnik) for a future President to see that need. In recent years Presidents have seen only the political costs.

Executive Leadership for Science and Technology

Willis H. Shapley

In its latest incarnation, advisory machinery on science and technology in the Executive Office of the President has now been in place for more than two years. It is timely to look at what has emerged, compare it to previous expectations and current needs, and give some thought to how the pattern of central leadership in science and technology in the Executive Branch might evolve most usefully over the next several years.

Advice to the President on science and technology is not an end in itself. The effectiveness of the organizational machinery to provide it depends on the context in which it operates. One crucial aspect, of course, is how the Science Adviser fits into the structure of the Executive Office and the operating patterns of the Presidency. Another is how the Science Adviser and his office relate to the rest of the Federal science and technology establishment. Thus, evaluation of the current Office of Science and Technology Policy (OSTP) and consideration of the ways in which it might evolve entail the broader questions of the mode of operations of the Executive Office as a whole and the organization of the science and technology activities of the Executive Branch.

These questions are too big to be dealt with in their entirety here. Some of the considerations which experience and observation suggest may be parts of the answers can be indicated, leaving many blanks to be filled in by others. To anticipate, the basic thesis will be that the current concept of OSTP is generally good and working well, but that there is an additional need for a strong focus of central leadership in Federal science and technology which should not and does not need to be in the Executive Office. This leads to the conclusion that the goal should be a minimal Science Adviser's office in the Executive Office supported by a heftier "Agency for Science and Technology" which, in addition, would perform central planning and leadership functions and bring together a group of federal agencies directly concerned with science and technology.

Willis H. Shapley (b. 1917) is senior consultant on research and development budgets and policies, American Association for the Advancement of Science. For many years as a senior official in the Bureau of the Budget, Executive Office of the President, he was concerned with research and development matters, and from 1965 to 1975 was Associate Deputy Administrator of NASA.

Today's OSTP

OSTP as it now exists can be briefly characterized as follows. A distinguished scientist, Dr. Frank Press, serves as Director of the office and as Science Adviser to the President. He maintains a low profile and evidently has earned and gained the confidence and respect of the President, his immediate associates, and key figures in the Office of Management and Budget (OMB) and other elements of the Executive Office. His staff has been held to about 20 professionals by Presidential edict, in order to limit growth of the Executive Office. Extra help is provided by detailees from other agencies, by drawing on a large stable of consultants and by assigning the National Science Foundation (NSF) and other agencies to do studies and reports. Some standing groups and numerous ad hoc teams on particular problems have been formed, comprised of government or non-government people or both, but there is no standing general advisory committee like the President's Science Advisory Committee (PSAC) of the 1960s.

In its first two years, Dr. Press's OSTP has been active across a wide front of problems in science and technology. Budgets for research and development, especially for basic research, have been a prime concern. Dr. Press and his staff have established working arrangements with OMB whereby OSTP is apparently accepted as a responsible partner in the decision process, rather than regarded with suspicion as a special pleader for science and technology. The range of other activities of OSTP is impressive. Those that are matters of public knowledge include the initiation of a major study of innovation in industry, strengthening basic research policies and organizations in mission agencies (notably in the Departments of Energy and Defense), leading an interagency review of national space policy, establishment of interagency programs on earthquakes and climate, development of policies in important health and environmental areas, promotion of a new program of competitive grants for agriculture research, and establishment of a new Institute for Science and Technical Cooperation with developing countries. OSTP staff participates in the Executive Office system of preparing, reviewing, and endlessly revising the Presidential Review Memoranda (PRM's) and other documents through which policy options are developed and presented to the President. On important issues, Dr. Press also gives his personal advice and the independent views of his advisory groups directly to the President and his associates.

In terms of visible results of general interest, two accomplishments stand out. One is the wholehearted acceptance by the Carter administration, after an initial period of ambivalence, of the importance of Federal support of basic research and of budgetary policies protecting basic research and recommending substantial increases each fiscal year. This was a continuation of policies instituted by President Ford at the urging of his Science Adviser, Dr. H. Guyford Stever, but there is no doubt that the new Science Adviser was largely responsible for the special emphasis given by President Carter to basic research, in many public pronouncements and in specific budget recommendations.[1]

The other outstanding accomplishment is also there for all to see but has gone almost entirely unnoticed. This is the President's special message to Congress on science and technology of March 27, 1979—a 7500-word document that summarizes the rationale, objectives, concerns, and future directions of federal activities related to science and technology.[2] It contained little that was new—no spectacular new proposals or goals. Its significance, both within and outside the government, lies in the fact that it is the first comprehensive and authoritative articulation of Federal policy on science and technology. Previous Presidential statements on science and technology or research and development have been in the nature of catalogues of programs[3] or ad hoc urgings for support of budget requests, usually for basic research, in trouble in Congress.[4] Reports of the National Science Foundation, the National Science Board, and other federal agencies, even those attempting to be comprehensive and carrying a Presidential endorsement, have generally reflected a scientific and technical outlook on national policy rather than national policy on science and technology. It is surely to the credit of the Science Adviser and his office that they were able to produce a wide-ranging and thoughtful statement of Federal policy on science and technology that the President and his administration could accept.

Expectations Fulfilled?

How well is the current OSTP fulfilling the expectations of those who believe in its importance and urged its recreation after the earlier Office of Science and Technology was abolished by President Nixon in 1973? As we have seen, science advice is back in the White House and making its presence felt in budget decisions and other important matters. OSTP is tackling some of the important programmatic and policy questions and has produced a comprehensive statement of Federal policy on science and technology. Have we now reached a viable long-term organizational pattern for advice and leadership on science and technology in the Executive Branch?

The answer to the first question, I believe, is that OSTP is doing a first-rate job in meeting the general expectations to the degree possible within the external constraints. The current OSTP does, however, in some respect fall short of the expectations of the Congressional authors of the legislation mandating its recreation.[5] It is a much slighter organization than seems to have been envisaged. Several functions, spelled out in great detail in the law, have been down-graded or eliminated entirely. The initial reluctance of the Carter administration to accept any science and technology presence in the Executive Office, because of a strong commitment to reducing its total size and suspicion of special pleading for science and technology, was overcome only at the price of restricting the size of OSTP and of removing some of its statutory functions.

By reorganization and other executive actions, responsibility for two major reports was transferred from OSTP to the National Science Foundation: an annual Presidential ''Science and Technology Report'' and an annually updated ''Five-Year Outlook'' of problems and opportunities in science and technology; parts of the latter were subsequently passed on by the NSF to the National Academy of Sciences. The new President's Science Advisory Committee (PSAC), set up in the law with an initial two-year life to study Federal organization for science and technology but with the possibility of being transformed by the President into a born-again 1960s, PSAC was abolished outright on the grounds that the President's overall reorganization project could perform the necessary functions.

It is too early to judge how well the Carter administration's stripped down OSTP and the reassignments of responsibilities will approach the legislated expectations. The NSF produced the first annual Science and Technology Report about six months late.[6] The report consisted of a bland "strategic overview" by OSTP, a presentation of budgetary information on Federal research and development that had already been rendered obsolete by Congressional action, and chapters on a variety of other topics that contributed little that was new. It was a disappointing performance, not Presidential in a meaningful sense—in contrast to the subsequent special message of the President in March 1979 discussed above. A decision was made to skip the second annual report due in February 1979; NSF says that it will try to do better in 1980. As for the Five-Year Outlook, the National Academy of Sciences and NSF are still at work at this writing, and it remains to be seen what is ultimately produced.

With respect to both of these reports, one must ask whether the expectations written into law really were reasonable in the first place. It is hard to see what real need there is or significant use would be made of the Science and Technology Annual Report, even if it can be produced at the beginning rather than the end of each session of Congress. The President's major recommendations on research and development and other aspects of science and technology are spelled out in his State of the Union and Budget messages. A special analysis in the budget identifies the amounts for research and development; further details are provided in the voluminous justifications of the agencies. There is always the option of a Presidential special message, if needed. In any case, it is difficult to visualize the National Science Foundation, with its central focus on research, especially basic research in universities, as the President's anointed spokesman for all Federal science and technology.

The Five-Year Outlook raises some of the same concern. Its scope is intended to include technology and the application of science and technology to problems of national significance, not just the outlook for science as such. There is much merit in the notion of a periodically updated road map of where science and technology are going and how they might contribute to dealing with important problems, but there is also room for doubt that the annual paperwork exercise implied by the legislation is the best approach. There do not seem to have been compelling reasons for loading this function on OSTP in the first place. A stronger case can be made that the type of in-depth study and looking ahead required could best be done by an independent group able to work with experts in many fields; from this standpoint the delegation of the task to the National Academy of Sciences and its associated engineering and medical organizations seems to make sense.

Even if one accepts that OSTP should not be expected to take on the workload of these specific reporting requirements of the OSTP Act, it seems to me doubtful that OSTP as now conceived can fully meet the needs of the Executive Branch for central leadership on science and technology. The job is too big for a staff office in the White House. There is a need for dealing with more than just the most pressing attention-demanding problems. Leadership should be available when needed on small policy problems as well as great. The whole area of the management of Federal R&D, technical operations and services, cries out for more leadership attention. There should be time to think, to anticipate future problems, and to try to get ahead of them. Decisions should be based on a depth of understanding unlikely to be achieved when the pace is so hectic. A small, dedicated, but over-worked staff may be able to put on one virtuoso performance after another, but it cannot be expected to do the full advisory and leadership job that Federal science and technology activities require.

The situation could be mitigated, perhaps, by a modest increase in OSTP staff. This might relieve the pressure and enable OSTP to be more effective in its present mode. But there are objections and disadvantages to consider. Holding down the size of the Executive Office is not just a matter of appearances and campaign promises. The last 20 years have seen the growth—sometimes by fits and spurts, sometimes by slow accretion—of a superimposed bureaucracy within the Executive Office that mirrors or cuts across the main elements of the Executive Branch, a trend that many responsible officials and observers believe should be resisted at every opportunity. For practical reasons, OMB has to be divided into separate cells to deal with groups of agencies; some special staff is undoubtedly needed to coordinate national security and domestic matters. But on top of this, almost every special interest whose concerns span several Federal agencies is able to persuade itself, and sometimes the President or Congress, that it needs a special office in the White House or Executive Office. Telecommunications, drug abuse, aging, environmental quality, space, and consumer affairs, not to mention science and technology, are examples. With the expansion of specialized staffs in the Executive Office, there is an inevitable tendency for decision-making and decision-blocking power to shift from the responsible Cabinet officials and agency heads to anonymous staff people in the Executive Office. Only a few agencies have the political or bureaucratic clout to resist. Problems of coordination of agencies and programs by the Executive Office are multiplied by the difficulties of coordination of so many staff elements within the Executive Office itself. Each further increase aggravates the situation. More staff might help OSTP but, from a broader stand-point, could rightly be viewed as a step in the wrong direction.

The Need for Institutionalization

The most critical shortcoming of the current OSTP concept is that it fails to provide institutional continuity for science and technology advice in the Executive Office and policy leadership in the Executive Branch. The present arrangement is in-herently unstable. It depends on personal understandings among the principals in the Executive Office and on the acceptance of OSTP by the other bureaucracies there entrenched. All this could vanish with a change in administration or even a change in some key individuals. No one will deny a President the right to organize, staff, and operate his own office as he and his closest associates see fit. The same reasons that first pointed the Carter administration in the direction of abolishing OSTP, or at least removing it from the Executive Office, might well be persuasive to another President. Even if the shell of an OSTP survived, nothing would be easier than to ignore it if it is not wanted as a part of the official family. We have been down that road before.

Institutional continuity is important for science and technology policy and leadership. The impact of science and technology is a long-term affair. Lead times are long. Studies of important topics can take many months to set up properly, months or perhaps years to conduct, and as long again to evaluate and implement; the whole process often stretches from one administration to the next. Each administration should not have to start from scratch. An institutional memory can help it to learn from past experiences and avoid unnecessary replowing of old ground.

Institutional support is also essential for central policy leadership and advice in science and technology. This can be obtained in part, as OSTP is now doing, by drawing on other agencies in and outside the government and by other indirect mechanisms. But a directly controlled institutional base with a broad range of scientific and technical resources available on call would provide a much stronger capability, both for performing studies and analyses and for managing and evaluating those performed by others.

A New "Agency for Science and Technology"

With OSTP too slender to do the total job and facing an uncertain long term future in the Executive Office, what should be done to provide for effective advice and leadership in science and technology in the years to come? The answer is to establish, in addition to OSTP, a new, permanent, strong, independent agency with broad central responsibilities for advice and leadership in science and technology in the Executive Branch. The essential features of the new agency would be fourfold.

First, it would be headed by a Presidential appointee of Cabinet rank, although the agency would not need to be a Cabinet department—let us call it the "Agency for Science and Technology" (AST). The head of AST—call him or her the "Administrator"—would be the principal official in the Executive Branch responsible for policy, planning, advice, and general leadership on matters involving science and technology that are not clearly within the scope of responsibilities of other departments or agencies. He or she would assume many of the responsibilities currently performed or expected of OSTP, and be a principal supporting arm to the President's Science Adviser and OSTP for the functions they retain. The relationships of the Administrator and AST to the President, the Science Adviser, and OSTP would correspond to those that have existed for many years between the Secretaries and Departments of State and Defense on the one hand and the President, the Special Assistant for National Security and the National Security Council staff on the other. The Administrator, like the Secretaries, would be the senior official under the President in his areas of responsibility. He or she would have full management responsibility for the resources and activities of the agency and direct access to the President to give advice and receive guidance. The President's Science Adviser, like the Special Assistant for National Security, would be the President's personal adviser at the White House, a second independent voice when wanted. He or she would see that science and technology matters receive proper attention in Executive Office decision-making and that interagency policy and program problems are appropriately dealt with. The Science Adviser and the OSTP staff would draw on the resources of AST for technical and analytic support, just as the Special Assistant for National Security and the NSC staff draw on the resources of State and Defense.

Second, the Administrator of AST would have a central policy and planning staff office as his or her principal arm for carrying out the agency's responsibilities for advice and leadership in science and technology in the Executive Branch. This staff would include a permanent cadre of persons qualified to provide leadership and good judgment in the handling of the policy and planning functions of the agency. The permanent staff would be augmented on an ad hoc basis, as needed, to deal with particular matters by temporary assignment of personnel from other parts of the agency, other Federal agencies, or the university or private sectors. The central policy and planning office would carry out on a broader basis many of the policy functions OSTP is now performing on a shoestring basis. It would take on the general planning and reporting functions in science and technology for the Executive Branch, including the five-year outlook and perhaps the annual report specified in the act reestablishing OSTP. In both policy and planning matters, the role of the office itself would be primarily leadership, guidance, coordination, and evaluation. Detailed studies, reviews, and planning would be assigned by contract or otherwise to qualified outside groups to the maximum extent feasible. The service functions of maintaining and improving the data bank of statistical information and other science and technology indicators would be transferred to AST and attached to the policy and planning office.

Third, the principal science and technology agencies now independent or not integral to the missions of other departments or agencies would be transferred to and become operating elements within the new agency. These would include as a minimum NSF, the National Aeronautics and Space Administration (NASA), the National Oceanic and Atmospheric Agency (NOAA), the National Bureau of Standards (NBS), and perhaps some general research and technology activities now lodged in the Department of Energy. The aim would not be to extract and consolidate all the science and technology activities of the Federal government into a single agency. The principal objective would be to provide Cabinet level leadership and advocacy for some independent agencies now outside the mainstreams of government and for some general science and technology agencies now submerged in departments with other primary interests.

A second important objective would be to provide the Administrator of the new AST a strong operating base of scientific and technological capabilities to support his or her roles in science and technology advice, policy, and planning in the Executive Branch. By being able to draw directly on the broad capabilities of the staffs, laboratories, and research and development centers under the control of NASA, NOAA, NSF, NBS, and perhaps others, the Administrator will possess capabilities for both quick reaction and in-depth responses to policy and planning needs in science and technology—capabilities well beyond those of a small OSTP in the Executive Office. The Administrator would also be in a position to deal authoritatively with management problems of the agencies put under his control, problems that are now left to be dealt with or to fester in the individual agencies or are handled sporadically and remotely by OMB or OSTP.

The fourth essential feature of the new agency is more difficult to define—its relationship to the science and technology activities of other Federal agencies. The Administrator of AST would not control activities or programs of other agencies. Defense R&D would remain the full responsibility of the Secretary of Defense, health R&D that of the Secretary of HEW, and so on. As the Cabinet level official for science and technology, however, the Administrator of AST would have functional leadership responsibilities for science and technology throughout the Executive Branch. His positions on science and technology would be generally similar to those of the Department of Defense with respect to defense-related activities of other agencies, the Department of State with respect to international activities, the Department of Labor with respect to labor-related activities, to mention only a few parallels. As in these cases, practical accommodations would have to be worked out. The guiding principles should be that the Administrator of AST should not be circumscribed on jurisdictional grounds from effectively discharging his general advisory, policy, and planning functions, but, at the same time, should not undertake to duplicate or interfere with the activities and responsibilities of the other agencies.

The Timeliness of Action

These, then, in broad outline, are the essential features of a new Agency for Science and Technology to institutionalize in a permanent and constructive way the functions of advice, policy, and planning for science and technology in the Executive Branch. The ideas it embodies are not new, but have not received much advocacy recently, either separately or in the combination suggested above.

The suggested role of the Administrator of AST and his central policy and planning staff is very similar to some of the original conceptions of the role NSF would play within the Federal government. These conceptions faded when the early NSF turned away from broader responsibilities to focus on basic research, especially in universities, and was eclipsed by the explosive growth of other Federal science and technology programs—in Defense, the National Institutes of Health, and later NASA. The notion of NSF as the "lead agency" for science was kept alive over the years by its friends in OMB. There have been attempts to broaden NSF with greater emphasis on applied research, and staff offices and support programs for general policy planning in science and technology have been established. This trend has been reinforced by numerous policy study assignments to NSF by OSTP and OMB.

The time has long passed, however, for recasting NSF into an effective central leadership agency for Federal interests in science and technology. Its focus on long-term academic research makes NSF too remote from current problems of operational concern that require prompt judgments and action, not just further research. NSF's "bag" is science, not technology; but the most crucial Federal policy concerns run primarily to technology, not to science. It can even be argued that efforts to make NSF something more than a foundation to support basic research may serve to detract from this vitally important function, by blurring the focus and by introducing extraneous criteria for research support. The concept proposed above is designed to provide a fresh start and new home for broad-based policy in science and technology while preserving and providing new Cabinet level leadership for the NSF in its primary role of supporting basic research.

The concept of the proposed new Agency for Science and Technology is also very similar in many respects to that proposed in H.R. 4461, one of several bills introduced in 1975 when Congress was considering science and technology policy legislation to establish advisory machinery in the Executive Office. Title III of this bill, which was very carefully conceived and deserved much more attention than it ever got, would have established a "Department of Research and Technology Operations." The composition and functions envisaged for the department correspond closely to those set forth above for the proposed AST and its administrator.[7]

H.R. 4461 was put forward by its sponsors, Congressmen Teague and Mosher, the ranking leaders of the Committee on Science and Technology, as a stimulus to discussion, not as a firm legislative proposal. Title III drew little or no support in the scientific and technical community, was opposed by the administration, and was assumed to be anathema to the agencies slated to be folded into the new department—although at least some leaders of those agencies are known to have recognized the advantages of merging their common interests to achieve a stronger position in the Executive Branch. In the legislative process that led to Public Law 94-282 reestablishing OSTP, the idea of reorganization along the lines of Title III was quietly dropped. All interest and pressure were focused on getting science and technology back in the White House; answers to other fundamental questions that might have delayed this objective were deferred. As we have seen, study of the organization of Federal science and technology activities was made the mission of the newly recreated President's Science Advisory Committee (PSAC) and later was supposed to become a function of President Carter's reorganization project when he abolished PSAC in 1977. There have been no signs of official interest since that time.

For the reasons suggested above, however, the idea of a major new central agency for science and technology ought to be revived. The time is ripe to move toward a stable, long-term pattern of advice and leadership in science and technology. With an able and effective Science Adviser in place, and having now gained a good appreciation of both the values and the problems of science and technology, the Carter administration is in an unusually favorable position to advance and secure acceptance of such a proposal. The President and his Science Adviser could leave no legacy of more lasting value from their stewardship of United States interests in science and technology.

References and Notes

1. For details see the reports prepared by Willis Shapley, Don Phillips, et al., *Research and Development in the Federal Budget: FY 1978* (1977), *Research and Development: AAAS Report III* (1978), and *Research and Development: AAAS Report IV* (Washington, D.C.: American Association for the Advancement of Science, 1979).

2. "Message to Congress on Science and Technology," *Weekly Compilation of Presidential Documents*, Vol. 15, No. 13 (March 27, 1979), p. 529.

3. Federal Council on Science and Technology. *Annual Report on the Federal Research and Development Program: Fiscal Year 1976.*

4. "The President's Message to the Congress Urging Approval of His 1977 Budget Request and Creation of an Office of Science and Technology Policy": *Weekly Compilation of Presidential Documents*, Vol. 12, No. 13 (Monday, March 29, 1976), p. 474.

5. "National Science and Technology Policy, Organization, and Priorities Act of 1976," *Public Law 94-282*, approved May 11, 1976.

6. National Science Foundation, *Science and Technology: Annual Report to the Congress* (Washington, D.C.: National Science Foundation, August 1978).

7. H.R. 4461, 94th Congress, 1st Session. See also Committee on Science and Technology U.S. House of Representatives, *A Proposed National Science Policy and Organization Act of 1975* (Washington, D.C.: Committee Print, 1975); and Committee on Commerce, Science, and Transportation and Committee on Human Resources, U.S. Senate, *A Legislative History of the National Science and Technology Policy, Organization, and Priorities Act of 1976* (Washington, D.C.: Committee Print, April 1977), pp. 273-320.

A Historian's View of Advice to the
President on Science:
Retrospect and Prescription

A. Hunter Dupree

Science policy and technology policy are separate and distinct. In most periods of American history, the distinction has remained implicit. Only since 1957 has the Federal government had at its apex an explicit institution for science policy, and that structure was, at least temporarily, abolished in 1973. Amid all the definitions of pure science, basic science, applied science, research, development, and technology, the scope of science policy has often been confused with technology policy or swallowed up within it. The only perspective from which one can be certain that a distinction exists between the two is the one which emerges from their separate histories. The National Science Foundation, given responsibility for developing a national science policy by its organic act of 1950, chose the route of history when it sponsored in 1953 the project that became my book, *Science in the Federal Government: A History of Policies and Activities to 1940* (Belknap Press of Harvard University Press, 1957).

The United States government has always concerned itself with the direction of technological development. From Alexander Hamilton's report on manufactures and the whiskey rebellion to President Jimmy Carter's speeches on energy, political and economic measures have been perceived as affecting the technological capacities of the country. Protective tariffs, patents, land policy, taxation, subsidies, and regulation have all had obvious technological consequences, and have had a prominent place in political debate, legislation, administration, and judicial interpretation. The interest groups which seek to influence government policy are aware of technological consequences to some degree, and the voters make up their minds with them in view.

A. Hunter Dupree (b. 1921) is George L. Littlefield Professor of History at Brown University. He is author of Science in the Federal Government *and* Asa Gray. *He is a Fellow of the American Academy of Arts and Sciences and has served as its Secretary. He is a member of the Smithsonian Council and has been a member of the NASA and AEC History Advisory Committees and of other governmental panels. He served in the U.S. Navy in World War II. In 1963-64 he was a consultant to the Committee on Science and Public Policy of the National Academy of Sciences. Before 1968 he was a Fellow at the Center for Advanced Study in the Behavioral Sciences, Professor of History at the University of California (Berkeley) and Assistant to the Chancellor of the Berkeley Campus, 1960-62.*

Science policy has taken place in a very different arena. The government has always faced problems which are not definable in terms of the play of economics and politics and has regularly turned to scientists and scientific institutions not only for answers but for the very formulation of the problems themselves. Politicians have trouble shaping science policy because neither they nor their constituents know how to ask answerable questions, much less proceed to answer them, unless they first consult the scientists as to the state of their stock of information and techniques. An occasional President (notably Thomas Jefferson and John Quincy Adams) was sensitive to the special nature of science policy and defined its sphere with clarity, although in the nation's early history the institutional structures were lacking for acting upon the vision, however clear. For a century after the time of John Quincy Adams, the statement and coordination of science policy took place at lower levels than the Presidency.

Many talented scientist-administrators (Alexander Dallas Bache, Joseph Henry, and John Wesley Powell, for example) shaped implicit policies for science within the government in the 19th century. Only with the Great Depression and World War II did science emerge as a separate area of concern worthy of the attention of a President. From Franklin D. Roosevelt onward, every President has had to take a position on how to introduce science policy into his administration and how to shape a posture toward Congress on relevant legislation. When Sputnik signaled its challenge in 1957, Dwight D. Eisenhower formally took science policy into the White House by appointing James R. Killian, Jr., as Special Assistant to the President for Science and Chairman of the President's Science Advisory Committee (PSAC).

The following personal chronicle of the years 1957 to 1970 was written from a viewpoint of a time when the office of science adviser to the President was in abeyance. Since much of the activity of the Science Adviser had dealt with military subjects, and since I was an independent historian without special access to the national security area, I was able to follow only that fringe of science policy which bore on the larger issue of the place and function of science in American society and specifically in the federal government itself.

The Historical Background in the Aftermath of World War II

A distinction needs to be made between the confidence which Vannevar Bush enjoyed in the wartime period under Franklin Roosevelt and the amount of actual personal contact which he had with the President. Bush often went for months without ever talking with Roosevelt, and he could, for example, bring Dr. Lewis Weed into line with the Committee on Medical Research simply by threatening to talk to the President. However, Bush's connection with FDR was much solider than the relationship which Karl Compton had had earlier as chairman of the Science Advisory Committee (1933-35). By using the Board of Trustees of the Carnegie Institution of Washington, Bush was able to get to FDR at will by way of the President's uncle, Frederic Delano. Hence the paradox is that Bush admired almost autonomous government scientific agencies, preferably under the direction of a committee of scientists, at the same time that he insisted on a position close to the

Chief Executive. The history of both of those conceptions is continuous from Bush's time to the present.

In dealing with the Bush Report, *Science the Endless Frontier* (1945), one must emphasize that the National Research Foundation proposed therein was a comprehensive one, including both a medical research division and a military division. Bush once denied when talking with me that he had ever felt that the comprehensive foundation would take over all military research, as indeed the Office of Scientific Research and Development had not during the war. However, he seems, at least for a brief time, to have entertained the idea that the comprehensive foundation would have enough leverage in all fields where science operated effectively to form some kind of national policy. It does not follow that he conceived of an organization which could make effective policy as being directly subservient to the President. When I talked to him in 1954 he emphasized his admiration for the Royal Society in Great Britain and expressed his opinion that the National Science Foundation had not achieved such a position.

In 1954, after the NSF had had some time to get started, Eisenhower issued Executive Order 10521 to clarify the status of science policy. William Carey, then in the Bureau of the Budget, played a creative role in writing that order, which he described in an interview as having a double meaning. On the one hand, it was meant to bolster the position of NSF and if possible encourage the director, Alan Waterman, to come into the Bureau of the Budget with general advice once in a while. On the other hand, it legitimized a category of research which might be called mission-oriented basic research. On this side of the order the mission agencies were given specific authority to support basic research, with all that it implied in the way of freedom from short-term applications, as long as the work was somehow relevant to their overall missions. A plural system could now emerge, with NSF as a minor peer of six or seven other foundations in the government, many of them in the Department of Defense. One of the effects of this tranformation was to make the National Institutes of Health look just like another mission agency and thereby mask the lack of coordination between medicine and science under the NSF.

The Creation of the Office of the Special Assistant to the President for Science and Technology

The academic year 1956-57 was one in which most of my public statements focused on the problem of central scientific organization. Before Sputnik the reception I got was usually one of boredom. In the fall of 1956, I flew back from Berkeley to Philadelphia to present to the American Philosophical Society a paper, "The Founding of the National Academy of Sciences—A Reinterpretation." Long afterwards, I found out from George Corner that Detlev Bronk, the President of the National Academy, had taken offense at my interpretation. I remember on that occasion meeting Lloyd Berkner for the first time. Also I remember people who spoke approvingly of my paper, including A. Baird Hastings and Oswald Veblen.

In January 1957, in the week in which *Science in the Federal Government* was published, the *Saturday Review* took special notice of it. There were lengthy excerpts from the book and a cover with a picture of the White House and the caption, "The President's New Power." Thus, at least in the mind of John Lear, the science editor, the whole story which I had to tell linked science and the Presidency. In the spring of 1957, in connection with the hazing that I was receiving from the history department at Berkeley, I had to give a public lecture, and I chose as my subject the idea of a department of science. The lecture was considered a dud by the historians who heard it, and the only favorable or even interested comment came from Forest Hill, the economic historian then at Berkeley. Despite the indifference, I had my research done and my views published before the advent of Sputnik.

Immediately after Sputnik went up, a panel discussion was organized on the Berkeley campus. Tom Kuhn was on the panel though I was not. Tom saw the whole thing as a demonstration of America's continuing inferiority in science. Kenneth Pitzer, a member of the panel, expressed himself as in favor of a department of science. It seemed as if most scientists not very familiar with the OSRD tradition saw the department of science idea as a status symbol and something which would one day come into being of its own accord. They would not fight for it, but they would say nothing against the idea.

In order to get some input, I wrote an article, "The Real Challenge to Dr. Killian," and tried unsuccessfully to get it into *The New York Times* and the *Atlantic*. My nearest miss was the *Saturday Review*. Bob Kreidler later told me that it had been much read inside Killian's office.

The Quest for a Department of Science

My first touch with Hubert H. Humphrey's subcommittee of the Senate Committee on Government Operations came in 1958 when I sent in a copy of my article. Senator Humphrey not only had the article printed *in toto* but put me on the list for the hearings on a department of science held in the spring of 1959. Since Senator John McClellan was the chairman of the full committee, the staff work as I remember it was done by a fantastic Arkansawyer named Walter Reynolds. Younger, more sophisticated staff people such as Julius Kahn were not then in evidence as far as I could see.

A Lloyd Berkner article on the department of science had come out just before the Humphrey hearings. I called up Glenn Seaborg, who by that time knew of my existence, and asked him what he thought of the bill regarding a department of science as put out in the committee print. He responded very enthusiastically that he was all for a department of science and specifically cited the Berkner article as saying essentially what he believed. When I pointed out to him that the Atomic Energy Commission was left out of Berkner's proposal but was included in the committee proposal, he backed down.

Since I was naive about the ways of committee hearings, I took the proposed legislation in the committee print at face value and largely shaped my comments as an attack on the idea of a department of science. In addition to Seaborg, I had spoken briefly with Dwight Waldo, a political scientist. He seemed completely oblivious to the importance of the subject or the existence of the field of science policy, an attitude which has been an enduring one among political scientists.

When I got back to Washington I was astonished at the difference in the atmosphere around the Senate committee and that which I had envisaged before I went East. Nobody was taking the legislation proposed in the committee print seriously, and Walter Reynolds was going around saying, "It took 10 years to get a Department of HEW, and it will probably take us 10 years to get a department of science. We recognize that the legislation proposed is imperfect, but we have simply put it up for discussion." The big-time science establishment was conspicuous in its absence, and for the first time I realized that the Pitzer-Seaborg type reaction was only a surface one. The hearings were being used to parade appointees of the Eisenhower administration who were under fire, and they took different sides on the issue. Lewis Straus, who probably spoke for the administration, was up for Secretary of Commerce and opposed a department of science. On the other hand Clare Booth Luce, with whom I could not compete in glamor and who was trying to become ambassador to Brazil, came forward in a stunning dress to testify that back in 1946 she had introduced a bill providing for a department.

The only scientist to show his head in those hearings was Wallace Brode, who was an outcast, both as a career government scientist in the Bureau of Standards and because of his uncomfortable position as Science Adviser to the Department of State. As one on the outside of the establishment, he wanted to redress the balance by creating a strong department of science. He quoted quite a bit from my book in his testimony, but when he realized that the weight of my testimony was against a department, he took all of those quotes out in the revision of his statement that became his Presidential address to the AAAS.

In spite of my astonishment at ending up in bed with both Straus and Waterman and finding myself in alliance with the Eisenhower Administration against the liberal Democratics who had invited me, I think my position against a department of science had a solid historical basis. Some points may be made about my impressions at that time.

1. The political atmosphere around the committee was extremely heavy. The Democratic Congress versus the Republican executive seemed to dominate the mind not only of Walter Reynolds, but of Senator Humphrey as well. The issue of executive privilege was prominent in the light of a letter from Dr. Killian refusing to testify. In my judgment there was much more steam behind the issue of executive privilege than there was behind a department as such.

2. The distinction between the White House and the Executive Office of the President was seen as a general bar to the Congress's adequate use of the President's emerging science advisory complex. I had at some meeting discussed this point personally with Don K. Price, who seemed to feel that, by this time, a Science Adviser was so necessary that a President would not dare to dispense with one and that, therefore, the flexibility gained by being in the White House overweighed any permanent advantage in having a more secure place in the Executive Office of the President. I tended to be on the other side from Don on this issue. In the light of Watergate, I think more than ever that there is something to be said for my point of view. If the structure were set up by legislation and then placed in the Executive Office of the President (in parallel with the Council of Economic Advisers), a White House out of control could not destroy the mechanism even if

the President did not listen. No one at that time could conceive of the Presidency out of control in quite the way that it was under Nixon. The possibility of such an event was more conceivable to a naive historian than to a practical administrator such as Don Price.

3. The absence of any real tie between the science establishment and the liberal Democratic Senators was evident. It was not just a matter of Walter Reynold's "cornpone" approach. That Waterman liked things better under Eisenhower than under the Democrats can be documented. The apolitical stance of the scientists and their assumption that they could work with either party prevented their being considered in that period a tool of the Eisenhower Administration, but the comfortable fit on the Republican side did not exist, in spite of Humphrey's rhetoric, between the Democrats and the scientists.

The idea of a party split on science generally and on the department of science specifically should not be pushed very far. As a test I went over to the House side and called up George Mahon, who took me to lunch in the House restaurant in the capitol. The reputation of my family name in Texas, which I always keep carefully separate from my public career based on my academic position, secured an immediate and cordial welcome from Mr. Mahon. He always seemed a member of the appropriations committee first and a member of the House generally only a distant second. His habitual answer on any general question was, "I have not yet studied that."

In this case I pushed on, describing the idea of a department of science as put forward by Humphrey. He said he was opposed to swelling the bureaucracy any further, but finally he came to life a bit by saying that, if somebody did not make some firm decisions, the Russians would be the first to fly a nuclear airplane. I was getting over the shock of that example when he corraled Speaker John McCormick, who knew about the Humphrey hearings and said that a department is surely coming. Then Mr. Mahon brought over Gerald Ford, who was extremely open and friendly. He began with the formula, "I haven't studied that," but then added, "It sounds to me like a good idea." In summary, the political leadership of the Congress, Democrats and Republicans, had the same surface response to the idea of a department of science as Seaborg and Pitzer among the scientists.

The AAAS Takes a Look at a Department of Science

The Christmas-New Year season of 1959-60 must have been a very busy one for me. It may have been just before Christmas that I prepared my first memo for Seaborg on science in international affairs, which eventually became part of the Berkeley campus position, to the Jackson subcommittee on national security policy. Very shortly after Christmas I went to Washington to the Shoreham Hotel at the invitation of Dael Wolfle to discuss the issue of a department of science. On of his motives for calling this *ad hoc* group was to be ready in the event that the issue of a department of science became a major one in the election of 1960. In the light of the background of the missile mess, I suppose that this was a not completely impossible proposition, although in hindsight the points I have just made above militate against it. The conference was one that was entirely in tune with the science establishment. Among the people that I remember there were Lloyd

Berkner, Emanuel Piore, James Mitchell (previously of the NSF), Don Price, and a man of whom I had never heard before but who made an immediate impression on me, Major General James McCormick.

McCormick blew in from the Pentagon a little late, excusing himself by saying he had had to talk to Secretary Gates. Piore, Chief Scientist at IBM, was very much in on what was going on in PSAC and at the White House. He began the proceedings by stating, "We shall have a department of science in five years." (Some three years later I reminded him of that statement and asked if he thought there would be a department of science within the next two years. His answer was, "Well, I was just trying to attract attention.") The main lead in the conference in the early going was taken by Berkner, who came on strong for a department of science. However, the more he talked, the more his reasoning appeared to be: "I want a decent place in the government for geophysics; I define geophysics so expansively that it includes everything that counts below the earth and above the heavens. Therefore, a department of geophysics should be called a department of science."

These discussions took a usual course. A certain number of orphans in the Department of Commerce and the Department of the Interior are always the candidates for a department of science, but the big boys always opt out. Agriculture, space, AEC, Department of Defense will have none of it, although it is perfectly all right for the Bureau of Standards. I see Berkner's thrust in this period as actually a follow-on of the International Geophysical Year, an attempt to make that *ad hoc* venture into a permanent agency. Looking on down the road, I suppose the main result of this initiative was NOAA, but the more he talked, the less it had to do with a central scientific organization. One other aspect of Berkner's position that emerged clearly was that he was virulently anti-university. He felt that universities were ossified structures in which departments that had no possible function were protecting themselves in perpetuity. One could already see the kind of generalization that he was putting forward at the end of his life down in Dallas.

Symbolically, one of the great problems of the whole discussion was NSF. It already had the name and the function in its organic act, so that any step taken toward providing an effective central scientific organization must imply that the NSF is ineffective and that something must be done with the carcass. Since NSF was by this time firmly established in both research support and in education, the question was what to do with its unused powers to make national science policy. Someone eventually would delicately imply that it would all be solved when Waterman retired, an event which then seemed more imminent than it actually was. In this connection Berkner said that he had been offered the directorship of NSF back in 1950 before it was offered to Waterman and asserted that if he had taken it he would have destroyed NSF in the early going by not being sufficiently submissive and prudent. Everybody else more or less joyfully agreed with this assessment; it says a lot about the personalities of both Berkner and Waterman.

PSAC's So-Called Seaborg Report on Science and the Universities

During the spring of 1960, Glenn Seaborg was chairing a PSAC panel on research and graduate education. The report was supposed to say that research and graduate education go together and that the government has a duty to support them and thus the universities—a somewhat broader proposition than the NSF usually put forward in defense of basic research. It emphasized that, in the minds of many of the leaders of the science establishment, the university was the home address of the scientific community—a belief of those who were mainly concerned with weapons, missiles, and nuclear warheads as well as those who were trying to foster something outside the defense area. Bob Kreidler was the staff man, assigned by George Kistiakowsky as chairman of PSAC to help Seaborg with this panel, and he spent a number of weeks in Berkeley working on it. In addition, the whole panel met in the chancellor's office at Berkeley for a full day, and Seaborg, as a reward for my statement for the Jackson subcommittee, invited me to sit in. The other Berkeley professor who was there as a guest was Curt Stern.

President Coles of Bowdoin made a presentation on the plight of science in the private colleges, and the Dean of the Chicago Medical School practically wept at the piteous condition of medical students. In addition to lots of administrators, a number of big-time scientists were there as well. Roger Revelle and George Beadle put on quite a show, virtually destroying biology as previously understood. Curt Stern finally could not stand it any more and spoke up for classical genetics and animal behavior. Seaborg had some idea that, because I could write, I should help Kreidler write the report. I talked to him a few times, but a collaboration was quite impossible for me and quite impossible for him.

In the end, the report, of which I eventually saw a draft, was, in effect, written by MacGeorge Bundy. Thus the Seaborg report supposed to have been written by Kreidler was in actuality the Bundy report. Since it had been commissioned personally by George Kistiakowsky, he could not understand why it was not called the Kistiakowsky report. Bundy, of course, was still Dean at Harvard at this time. Since Revelle was a rising star in the University of California, Seaborg was Chancellor at Berkeley, and Beadle would soon be president of the University of Chicago, this panel gave a bit of a look at the way in which the President's science advisory complex and the research universities acted as a loose but efficient federation.

The President's Science Advisory Complex and the Election of 1960

Because I joined Seaborg's staff as faculty assistant to the chancellor in August, 1960, I was able to see a certain flow of documents within the chancellor's office. Seaborg with characteristic enthusiasm suggested that my interest in science policy would be of use, and went out of his way to send information to me. At the same time, as a member of the history department I was a humanist, not a scientist. Hence my duties in the chancellor's office focused on non-scientific activities, while Starker Leopold was the science assistant, even though he was just then discovering for the first time that the Federal government was in the business of supporting university science. Seaborg was a member of PSAC and was also involved in a number of *ad hoc* groups to advise Richard Nixon as the Republican candidate.

The general situation bore out my previous impression that the science establishment was somewhat more at home with a Republican administration than with a Democratic one, although many, if not most of them, were individually Democrats. Seaborg was one such, as was Berkner. Joseph Kaplan of UCLA was one of the few scientists who appeared to be politically oriented toward Nixon. The conception of the establishment was that it was apolitical and would advise anyone who might ascend to political power. The Republicans prepared a number of position papers, and at one point an article in a weekly news magazine dwelled on this activity. On the other hand the assumption was made that science advice was available to John Kennedy and his campaign because of the presence in that camp of one man—Jerome Wiesner. I particularly asked Arthur Schlesinger, Jr., about science advice when he was in Berkeley trying to line up the disgruntled intellecuals for Kennedy. Schlesinger answered that it was well taken care of by Jerry Wiesner. Thus the alignment showed an assymmetry between the relations with the two parties.

It never occurred to me at that time, however, that the scientists in the establishment, for example those serving on PSAC, were in any sense political appointees or that they would consider it a party matter as to whom they would give advice. There was a kind of unspoken assumption around the chancellor's office that, if Nixon won, Seaborg would become the Science Adviser.

Yet when Kennedy in fact became the winner, there was no hint of dejection on the part of any scientist within my purview; a very strong feeling existed that the talent search launched by the Kennedy administration would approach scientists without asking party affiliation and without considering whether they had participated in the Nixon task groups of election time. Indeed, Seaborg returned from a PSAC meeting in November or December 1960, and reported having had breakfast with Sargent Shriver to discuss the then momentous appointment of a new chairman for the Atomic Energy Commission. Seaborg was being consulted about others and remarked that he was able to name a large number of people not to get but had a harder time with a positive suggestion. In early January 1961, word reached the chancellor's office that Seaborg himself had been offered the chairmanship and that he would take it, resigning as chancellor almost immediately. Thus a major possibility for Nixon's Science Adviser—if Nixon had won—turned into Kennedy's scientist chairman of the AEC.

An impression I had in the weeks immediately following was that in the great roundup of Cambridge talent for the Kennedy Administration Wiesner actually hoped to get a major post in arms control and disarmament. He was reported not to be enthusiastic about the job of Science Adviser and took it largely in the hope of using it for leverage for his arms control interests. Thus to an extent he was from the beginning an outsider vis-a-vis the inheritors of the OSRD and AEC traditions.

The Formation of the Office of Science and Technology, June 1962

A number of factors played on the decisions made in the Reorganization Plan No. 2 of 1962, in which President Kennedy set up the Office of Science and Technology, making the Special Assistant for science in the White House also the Director of OST and, therefore, accessible to Congress. In my opinion the main author of this plan was Richard Neustadt. I became aware of his role when I was on a program at Gould House in the fall of 1962 at a conference which eventually produced the Gilpin and Wright volume. Whatever I said there was not preserved for posterity and I am not sure whether Neustadt's comments as chairman of the panel were preserved. I noticed as the discussion progressed that he kept explaining "what we had in mind" in drafting such and such a section.

One factor that led to the reorganization stemmed from the Humphrey hearings on a department of science. While logically the incoming Democrats should have favored a department of science, much of the strength of their advocacy went back to the period when they controlled the Congress while the Republicans under Eisenhower controlled the Executive. After the election the main thing that was left of that branch-of-government split was the issue of executive privilege, with the Science Adviser in the White House immune from calls before Congressional committees. Humphrey and his group would be satisfied by any device that would make the Science Adviser available to Congress. The steam went out of the drive for a department of science, although Senator McClellan kept introducing legislation for a commission to study it all through these years.

The OST then emerged as a working out of Neustadt's pluralistic conception of Presidential power. The President had to bargain, not only with the Congress, but with the heads of the great scientific agencies within the executive itself. He needed some way to balance them off between one another. This function, of course, had long been done by the Bureau of the Budget and would continue to be done there. A sense was emerging, however, that R&D was becoming too important a budget category to be left entirely to the generalists in the BOB. They themselves were apprehensive about the magnitude of the decisions that they were making. This realization came not so much from the dollar totals of the research budget as from an appreciation by the knowledgeable insiders that the leverage of R&D was much greater than any other activity in the government.

The reorganization plan also had to contend with that perennial problem of central scientific organization, what to do with the NSF. One line of negotiation was with Waterman and eventuated in the ceding of the NSF's authority to make a national science policy to the OST and the Executive Office. When I talked to Waterman in the fall of 1962, he expressed himself as extremely well satisfied with the settlement of the reorganization plan. His reluctance to exercise the national science policy provisions in the organic act of NSF made it natural for him to wish to get rid of them. The problem was coupled, however, with what to do with the National Science Board, which had a vestige of power independent of the NSF. Although Waterman professed to be satisfied, John T. Wilson, who had recently returned to NSF from the University of Chicago, referred to the change as the sellout of a part of our birthright.

The long line of failures in the NSF, going back to the old program analysis office, meant that OST should be given a try for general science policy. The theorists were never quite agreed on the size and functions of a policy staff. On the one hand, there was a universal abhorrence of a large and overloaded bureaucracy (memories of the old Research and Development Board) while at the same time a group of six or eight people were in no more of a position to get on top of major policy problems than Kreidler had been on the PSAC panel on graduate education. The problem which dogged all social scientists and historians dealing with science policy in these years was the belief that policy had to be made by scientists and that staffs had to be made up of scientists. Such ironies as the tremendous careers of non-scientists Irvin Stewart and Killian were explained away as freaks or as manifestations of the unique genius of Vannevar Bush in choosing men.

The Role of Kistiakowsky as Science Adviser and After

After I had spent the academic year 1962-63 in Washington, having relatively few contacts with the President's science advisory complex, it came into view again in the summer of 1963. Just as I was leaving Washington Kistiakowsky called me and asked me to become a consultant to the Committee on Science and Public Policy to help with a report on the principles which should govern the government's relations with universities. I went to Cambridge for an initial conference with him and Don Price, and throughout the fall and winter of 1963-64 I met on a regular basis with COSPUP (standing for Committee on Science and Public Policy of the National Academy of Sciences).

As I worked with the committee and got to know Kistiakowsky quite well, I came to the conclusion that he was still functioning as Science Adviser. He had simply moved his operation across a few blocks from the White House to Constitution Avenue and the academy's building. He was probably not deeply wrapped up in military affairs anymore, but the very creation of COSPUP and the nature of its operations made him in effect a non-military science adviser. COSPUP could draw on practically as much in the way of staff as could OST and PSAC. Many of the members of COSPUP were also members of PSAC, and I heard a great deal about what was being said on general questions in PSAC at the time. Melvin Calvin, with whom I rode back and forth from California in the first-class compartments, told me a great deal about the problems they were facing there and their relation to those we were discussing in COSPUP. One could almost say that COSPUP represented the science establishment more thoroughly than did Jerry Wiesner.

In the informal discussion which I heard at the Cosmos Club after the formal meetings of the committee, I sensed a kind of three-cornered struggle going on. Kistiakowsky could be seen as the heir of Bush and Conant. Wiesner could be seen as somewhat an outsider to the scientists who was also being progressively displaced from the traditional role of referee in military affairs by McNamara. In casting about for a new role, Wiesner was not exactly fighting the establishment, but he clearly did not make its interest his major concern. Hence, they sometimes got "crosswise," although usually they kept up a pretty good public front. The other person in the triangle was Phil Abelson, just coming into his own as editor of *Science*; Phil had a chip on his shoulder that weighed a ton. It had probably been there ever since wartime days when he had been stuck up at the Philadelphia Navy Yard without any money and had developed thermal diffusion in spite of the Manhattan District. Through the American Academy for the Advancement of Science he appealed to some of the more conservative instincts of the rank-and-file of the scientific community, praising the good old days of string and sealing wax and criticizing the low state to which science had fallen. He wrote editorials about Weisner which were almost personally insulting. Needless to say, he had no use for a historian hanging around and resigned from COSPUP with a low opinion of the report I was writing, of Kistiakowsky, of the space program, and a lot of other things.

The Role of the Science Adviser in Shaping the Government-University
Partnership in the Late 1960s

From this account it is clear that, from the very earliest years, PSAC and the Science Adviser had considered the relation of the government and universities the theme of the report, *Federal Support of Basic Research in Institutions of Higher Learning* which I had written for COSPUP. Within a few months of its publication, by December of 1964, Berkeley had blown wide open. I was giving a paper at the American Historical Association meeting in Washington, and all the factions from the Berkeley faculty were there in force. At that period many of the Free Speech Movement sympathizers were putting themselves on the market, making it a wild convention with rumors flying all around. I found a penciled note to call a certain phone number posted up on the bulletin board. When I called the number, I found myself talking to James E. Webb, the administrator of NASA. He wanted to talk about science policy and, in particular, the role of the National Academy and the Science Adviser vis-a-vis NASA, not wishing anybody between him and the President. He stopped off at my hotel, the Shoreham, and we sat in the lobby and talked for about half an hour.

Webb was suspicious of the Space Science Committee of the Academy, of the President's Science Adviser, and of the National Aeronautics and Space Council, which seemed to him just so many blocks to be skirted in the furtherance of NASA's mission. Finally I said to him, "All of these fine-tuning adjustments inside the government strike me as being minor compared to the fact that the government-university partnership is in danger of going completely to pieces at the university end." I suggested that the government should have some policy to meet that eventuality, that the government essentially depended on about 15 universities to do the bulk of the nation's research, and that to lose even one of them would be a national disaster. I did not take the position that such a disaster had occurred, only that I now saw how it could happen. Webb's response was double. One was to broach the idea of my coming to George Washington University as president, and the other was that we should go to talk to Don Hornig, the science adviser.

We went to his car, and on the car phone he set up appointments as we traveled. At the Executive Office Building we went in together to see Hornig. I tried to sketch out a picture of what was going on inside the University of California and what potential implications these events had for the system of science support. I took the position then which I still take that there was a definite anti-science bias to the Free Speech Movement. Hornig took the position that he was highly interested but that he had everything under control. Finally he asked me, "Well, what could I do about it?" My answer was that one should try to get ready elsewhere. While in retrospect I am not sure how much Hornig could have done, I still believe that the science establishment could have met the crisis of the 1960s in a much stronger way than it actually did and that the kind of traps which successively snared Columbia and Harvard could have been avoided if the scientists of the country had been more responsive to the moral and ethical issues raised by the FSM.

DuBridge as Science Adviser and the Twilight of the Office

The appointment of Lee DuBridge as Science Adviser to President Nixon has all the appearances of a direct succession from Bush and the OSRD. The Radiation Laboratory at MIT under DuBridge during World War II had been the spawning ground of science administrators and an informal network of friendships which had been second in importance only to the Manhattan Project in creating the infrastructure of the science establishment for the postwar period. DuBridge as president of CalTech also stood for research universities in the government-university partnership, and his long role, first on the President's Science Advisory Committee in the Office of Defense Mobilization and later on the National Science Board, gave him all of the credentials for being the logical leader. The appointment also gave Nixon the plus of appearing to favor the science establishment at the same time he favored Southern California.

Another side to DuBridge should be considered. He sometimes showed a distinct impatience toward disciplines other than his own and toward institutions of a different character from CalTech.

The period of the early first Nixon administration was one of unmitigated disaster for the government-university partnership. Also it was a trying time for the science establishment within the government. The secret operations of the Ashe Commission in moving science agencies around in the government as if they were so many poker chips completely broke the flow of evolutionary development which had characterized the relation of science and government from 1787 onward. Many agencies, among them the Geological Survey and the National Bureau of Standards, had the feeling that they did not know where they would be in the government in a few months or even whether they would be in existence.

So far as I could tell, DuBridge did not interest himself in the whole problem of the organization of science in the government. In this period the Daddario subcommittee of the House Committee on Science and Astronautics was coming up with a kind of revised central scientific organization in the form of an institute of research and education and had factored out many of the old naive problems in the department of science idea. The proposal sought a modicum of shelter from the pressure for immediate results which had markedly increased in the Johnson years and became a thunderous chorus in the Nixon Administration.

The whole fate of the government-university partnership was up for grabs in the week following the Cambodia-Kent State weekend. On the Monday following that weekend, at a special faculty meeting, a motion was made and then tabled that Brown University should cut all ties with the Federal government.

On the next day (Tuesday) I had to give a talk at the Smithsonian Institution in the afternoon. In the morning, when I arrived at National Airport, I called Phil Yeager, the counsel of the House Committee on Science and Astronautics, who told me to come on over there. As I walked into the Rayburn Building and past one of the committee rooms, who should be testifying there but DuBridge. I just slipped in the door and sat down. The subject was the siting of electric power plants. He was obviously not ready with any program, and in my opinion he was not informed on the subject by all the relevant agencies and groups.

For my own part I considered it absolutely appalling that, when the essential interests of science and research were being uniquely threatened, the Science Adviser to the President was off on the periphery. I talked that morning to several members of the staff of the House Committee on Science and Astronautics and found there was a real concern for what was going on in the universities and also a real hostility toward them on the part of some people, especially a considerable portion of the House. Their desire was to punish the universities for their intransigence and misbehavior by cutting funds and by putting punitive restrictions on research. This sentiment came uncomfortably close to coinciding with that of those elements of university faculties who wished to pull out of government-supported research completely. I came away from the Rayburn Building that morning with a feeling that DuBridge did not even see the possibilities of using his position in the interest of science or the universities at a very critical juncture. In comparison Walter Hickle was a first-class hero.

Against such a background I made my statement at the opening day of the hearings on national science policy before the Daddario subcommittee, late in June 1970. I said that the science advisory complex had fallen off of the organization charts of the Nixon Administration. What I had in mind was the front page of Section Four of *The New York Times* from a recent Sunday, which had a picture of the new organization of the White House staff, including such worthies as Egil Krogh. The science advisory complex, not shown, in effect ended up under the joint command of George Shultz and John Erlichman. At my hotel, before going down to testify, I heard on the *Today* program an interview with Shultz and Erlichman, who told what wonderful things they were going to do with the reorganized White House. If they knew that the President's science advisory complex existed, they did not show it.

After I made my statement, Charles Mosher of Ohio, who is from Oberlin and I had thought a kind of personal friend, felt bound to say that I had been pretty rough on the Nixon Administration and tried to defend it. Also Don Price, who as usual came in at the same time I did, said to me afterward that I had made the same points he had, except that I had been more political. Indeed I felt that he had hidden behind a smokescreen of public administration pseudo-objectivity and Bureau of the Budget jargon to mask the fact that anything at all had gone wrong.

I suppose that most scientists and their sympathizers (it must be remembered that Don Price was by this time in the course of becoming president of the AAAS) realized that they still had sixteen billion dollars in the kitty, and that they had better not rock the boat. I was in my old position of being too naive either to have the word on the party line or to appreciate that DuBridge was already on the way out. Therefore, I went ahead and said what I had to say from the point of view of historical perspective and without regard to the signals of a science establishment to which I did not belong but which I studied.

When a story appeared on the front page of the *Washington Post* the next morning, DuBridge was asked about my testimony. He said that my position was very amusing because Nixon had recently had the President's Science Advisory

Committee to a garden party. I still believe that, if DuBridge had really been a successor of Bush and a person who was using the office of Science Adviser as it should be used, he would have welcomed my remark and reinforced it and put into the record of the Daddario subcommittee a very strong plea for the refurbishment and re-elevation of the science adviser to a position of power. When he did not do so, as far as I was concerned the science advisory complex was already dead, and I was the least surprised person in the country when, in 1973, Nixon formally killed it off.

Prescriptions of a Historian—1957, 1970, 1979

The prescription of a medicine that would revive the President's science advisory complex must be based on a recognition of the tradition of science policy separate both from military policy and from technology policy. The period 1957-73 saw the creation of a science advisory complex which on the surface was mainly concerned with emergency advice on military matters, especially in the wake of Sputnik. My personal account shows the continuous existence in that period of a tradition which made the advisory complex a two-way link between the highest levels of the government and the total scientific community. The problems of policy which should concern this advisory complex have to do with the marshaling of intellectual resources to provide unprecedented answers for the unprecedented technical, social, and biological questions generated by contemporary society. In every period from that of Jefferson to the present, science policy has had to resist being swallowed up in economic, political, and military spheres. The business of science policy is to generate options for the society which are not available in the normal course of economic and technological development and which can be uncovered for general consideration only by imaginative and creative research, often in fields which the previous generation has slighted.

In 1957 my prescription for the President's science advisory complex under Dr. Killian comprised three points:
1. a formal place in the Executive Office of the President, clearly defined by an act of Congress and supported by funds earmarked for it in the budget;
2. a partnership with Congress; and
3. a secure line of communication to the nation's scientists.

In 1970 those original prescriptions were still valid, and I added a number of new ones:
1. a reordering of the relation of the scientific community to the Department of Defense;
2. increased attention to environmental problems;
3. definition of a role for the space program with predominantly scientific objectives;
4. a meshing of the social sciences with sensitive social programs and also with projects heretofore considered as a preserve of the natural sciences without destroying integrity;
5. a policy for the universities emphasizing healty institutions, including the humanities and those parts of the social and natural sciences which do not

conform to the narrow model of mathematics, physics, and engineering which dominated the scene since World War II;

6. a policy of educational support for a research program with a radically different mix of disciplines from that recently prevailing.

In 1979 those 1970 prescriptions are all still good and, thanks to the confusions of the decade, still unfulfilled. What is now needed is not so much organizational tinkering but the development of a science of human systems which can operate at the level of the Presidency—that is to say at the level of multi-national culture and society. This systems science must comprehend but far transcend the systems analysis of the engineers and go on to include an ecological, an anthropological, and a historical perspective. A systems science which has not yet coalesced is the one which is now needed. Such inadequate past attempts as technology assessment have not recognized the distinction between science and technology and have fallen into the hands of small groups of disciplinary imperialists. The creative energy and vision which Bush and Conant applied to the choices of 1940 are the elements which are now needed for a reformulation of the President's science advisory complex, even while the solutions of the leaders of the World War II generation are severely modified.

Science Advice and the Presidency

An Overview from Roosevelt to Ford

William G. Wells, Jr.

Introduction

One of the most important aspects of the broad subject of science and public policy has been the convergence of science and technology and a unique political institution: the Presidency of the United States. Unfortunately, this convergence and its implications largely have been ignored by most historians and other scholars in their studies of the Presidency; moreover, with few exceptions and until very recently, analysts of the broad problem of science and public policy have done little better.

The modern Presidency has evolved into a powerful institution as part of the changing context of our national life resulting from economic, scientific and technological progress. This is not to argue that science and technology in themselves account for the emergence of the Presidency as "our one truly national political institution";[1] however, nuclear and electronic technologies alone have given mid-20th century Presidents power undreamed of by those of the 19th century. Furthermore, all of the Presidents of recent decades have been faced with major problems either directly or indirectly related to science and technology.

The convergence of the Presidency with science and technology has led to the evolution of new functions and organizational forms at the Presidential level which fall under the rubric of "Science Advice and the Presidency." Intertwined with the evolution of these new functions and forms has been the emergence of the central role of science and technology in the modern world. Clearly we must become more "concerned with the relationship between expert knowledge and political power as revealed in the advisory function."[2]

William G. Wells, Jr., (b. 1923) is Staff Director, House Subcommittee on Science, Research and Technology. He has been a member of Congressional staffs for 14 years, with responsibilities including legislative oversight and budget review of NASA, the NSF, and other agencies, along with a wide range of science policy matters. Previously, as an officer in the U.S. Air Force, he was associated with the evolution of the ballistic missile program and with the central planning, direction, and management of Air Force research and development activities. This paper draws on the unpublished dissertation of the author, "Science Advice and the Presidency: 1933-1976." Occasional comments on the Carter Presidency appear, but in general the analysis goes only through the Ford Presidency.

Certain Policy Implications of the Central Role of Science and Technology

Many of the most significant issues of contemporary life have arisen from the pervasive influence of science and technology. These issues include not only major clusters of difficult or intractable problems—such as technological change and its impact on society—but also many topics of organization, programs, budgets, and other science policy matters over which, with varying success, institutions such as the Presidency and the Congress have acquired certain measures of knowledgeability and control.

Most of the issues arise from this crucial point: In a very brief time science has proved itself an incredibly powerful revolutionary force which has swiftly and dramatically affected society's beliefs and values, created and destroyed industries, revolutionized war, transformed and overturned political and social organizations, and modified man's conception of his place in the universe. In short, tremendous changes have taken place with bewildering speed in political, social, intellectual, economic, international and military institutions as science and technology have come to occupy a central place in the life of the entire planet. And there is little reason not to accept Bertrand Russell's view of science "that we are only at the very beginning of its work in transforming human life."[3]

Regrettably, an examination of seven Presidencies from FDR onward reveals less than wholehearted acceptance at the political level either of the above ideas or of a concept which, it is argued, is central to coping successfully with the problems of a rapidly changing world: science now appears as one of the great social institutions coordinate with the other major institutions of society—the economy, education, religion, the family, and the polity.[4] More regrettably, an evolving series of Presidential science advisory organizations has shown little interest or capability in even considering broad philosophical issues comparable to those raised by Russell.

Yet, there has been consistent support of the concept of science as one of the great social institutions and the argument that its voice should be heard in the highest political circles. Moreover, there are large numbers of subjects and issues which have been understood and acted upon, in varying degrees, by Presidents and their staffs alike. Examples abound in each Presidency and include the establishment of science-based organizations, the evolution of nuclear energy and its many implications, weapons development and major changes in the organization for defense and war, oceanography, support of federal research and development, health research and organization, and many others. Thus, it seems fair to say that Presidents have, each in his own way, demonstrated a grasp of complicated public policy issues arising from or connected to science and technology.

As representative of the kind of thinking which needs to take place at the Presidential level more frequently one can point to Truman musing at midnight with John Steelman on what the world would be like as a result of the spectacular scientific and technological advances of World War II and to the far-reaching studies initiated under Roosevelt and Truman. For a different, but equally important, statement on the Presidency and the implications of the growing importance of science and technology, it is possible to turn to Eisenhower. He—as did Roosevelt—somewhat belatedly, it seems, developed a keen interest in science and technology. Even so, most of his long discussions with his Science Advisers were devoted mainly to specific problems in the space and military areas. Only at the very end of his Presidency, and perhaps because he had found the time to think about such ideas, did Eisenhower express publicly some of his uneasiness about science and technology in connection with political and social issues.

After observing that research had come to play an increasingly crucial role in society and acknowledging that scientific research and discovery should be held in respect, Eisenhower warned about the danger that public policy could itself become the captive of a scientific-technological elite. The thought is strikingly similar to the pessimistic concerns of Victor Ferkiss and Jacques Ellul.[5] In Eisenhower's view, "it is the task of statesmanship to mold, to balance, and to integrate these and other forces, new and old, within the principles of our democratic system—ever aiming toward the supreme goals of our free society."[6] This was no small contribution to the concept of the Presidency as the logical focus in the government where the broad problems of science and society should be considered. But occasional Presidential statements, however important they may be, are not enough.

Coming to Terms with the Future

It is imperative that the White House extend part of its attention beyond the immediate array of problems confronting it at any given time. In doing so, it must come to terms not only with what has happened in the past but also with Russell's view of the future. Restating a point made earlier, for better or worse, science and technology are altering man's life and his world; indeed, it has been observed that to speak of technological change on contemporary life has become almost a cliché.[7] Unfortunately, this is correct and herein lies much reason for concern; clichés tend to be accepted at a superficial level with no effort to examine what lies behind them.

Even more unfortunate, Presidents, the Congress, and American citizens alike often have been disinclined to act until a developing situation forces decisions and the establishment of a policy. Paradoxically, important though may be the general concept expressed by Russell, it appears most often in disguise—as nuclear energy or space—which results in limited debates, decisions, and policies. In the context of this analysis, aside from academic studies, only three major endeavors related to technological change in a broad sense were undertaken during the nearly five decades covered by this paper: two originated in the Executive Branch, the other in the Congress.

The first, and only major study dominated by social scientists, was conducted in 1936 by the Science Advisory Committee of the National Resources Committee. One of the advisory committee's first comprehensive studies was concerned with the social consequences of invention; but there were more urgent problems in the 1930s and the report vanished into time with barely a trace that it had ever existed. But even if the report had reached the Presidential level, it is not likely that the Roosevelt White House could have done much with it: the Executive Office of the President had not yet been created and Roosevelt had only a handful of close aides.

Later, as one of the responses to a series of labor-management disputes involving issues of adjustment to technological change in the late 1950s and 1960s, President

Kennedy promised to establish a Presidential Commission to determine the impact of automation and technological change on the economy. On a less than high priority basis, Congress took a year to enact the legislation and Johnson six months to appoint the members to the National Commission on Technology, Automation and Economic Progress.

By the time its report was submitted in 1966 the original reason for the commission's establishment—unemployment—had gone away. President Johnson said, "thank you," and filed away a report which contained 20 major conclusions and recommendations ranging across the entire landscape of American life. They involved a complex interweaving of scientific and technological factors with political, social, legal and economic factors. The Johnson response, while depressing, should not be a surprise. At the Executive Office level there has been little decision-making machinery which consistently and effectively could bring together these diverse factors. Thus, with very good reason, the final third of the commission's report was devoted to the subject of improvements in decision-making mechanisms in American society. The absence of strategic ways of coping with overarching issues—such as technological change—at the Presidential level is evident; the requirement to go beyond the *ad hoc* and the tactical approach is apparent. Indeed, how to do so is considered to be one of the paramount problems of government today—here and throughout the rest of the world.

The third endeavor—the technology assessment movement—originated in the Congress in the late 1960s. How this came to be is beyond the scope of this paper but Congress took action independent of the Executive Branch by establishing an Office of Technology Assessment in 1972. Moreoever, there has been little interest in technology assessment at the Presidential level in the broad conceptual and organizational sense; only in terms of specific issues such as environmental pollution and energy environment trade-offs have Presidents and the Executive Office become involved.

As a practical matter, science and technology in general have been seen for the most part by Presidents as the means to achieve specific political objectives. And there can be no quarrel with this view per se; it is a logical, pragmatic approach to the decision-making and conflict resolution which comprise much of the substance of the Presidency. However, it is argued that such a view in itself is less than adequate for the long-run. A better understanding must be achieved of the implications of this earlier stated crucial point: in a very brief time, science has proved itself an incredibly powerful revolutionary force.

A More Sustained Effort

While it is possible to take some encouragement from occasional Presidential attention to the broad, philosophical aspects of science and society, it is argued that a more sustained effort must be mounted. A major implication for public policy is that continued growth and development of the United States and the rest of the world requires the adoption of a strategy of maturity for concentration on the vital issues that in the long-run will have first-order impact on society's genuine well-being.[8]

Widely divergent views have been held, and continue to be held, about the effects of science and technology on society. These views range from enthusiastic optimism to despairing pessimism about the future and it seems clear that the effects of the contending views will influence public attitudes and government decision-making about science and technology. Presidents, their staffs, and especially those involved with a Presidential science advisory apparatus must be fully conversant with this new dimension of public involvement and its potential impact.

In recent years, former Presidential Science Advisers Edward E. David, Jr. and Lee A. DuBridge, among others, have clearly articulated the implications of the growing public involvement in decisions related in some way to science and technology, although outside the government the most persistent voices have been those of the public interest groups. There is full agreement with William D. Carey, who contends that one of the outstanding features of American science and technology in recent decades is that "they are being *secularized* as lay publics participate in negotiating their right uses."[9]

This growing public involvement brings one to another significant public policy aspect of science and technology. A terrible fear in the scientific community of the 1930s was that becoming involved with the Federal Government and accepting financial support would lead to political domination of science. This issue is far too complex to be examined in detail and goes far beyond the scope of this paper, but it is clear that large-scale Federal involvement in and support of science and technology results in major public policy implications which are at the center of science advice and the Presidency.

In a broad, general policy sense, science and technology have come under political domination. It must be of concern at the Presidential level—from a public policy point of view—that the Federal government is the largest provider of research and development funds in the world and is engaged in roles as a performer, a manager, as a stimulus and as a policy and direction shaper. And it is true that Presidents have intervened directly from time to time in making large decisions affecting the use of science and technology for what they perceive to be in the national interest. However, there seems to be no sound basis for the fears of the 1930s that politics would take over and decide entirely what science would do. In practice, the multitude of decisions made every year on what specific research and development tasks and projects should be undertaken and how they should be conducted are decided by or strongly influenced by the scientific and technological communities—within the broad context of general guidelines and budgets provided by various Federal agencies. Indeed, Presidential intervention more often than not has been to make decisions to provide more support for basic research. The type of political intervention at the laboratory level which exists in the Soviet Union does not occur in the United States. Yet, a different kind of intervention may be emerging as a result of various kinds of public and political interest in areas where the potential effects of work in the laboratory are perceived as dangerous (e.g., the recent controversy over recombinant DNA research).

Shifting Patterns

Further, from a public policy perspective, large-scale Federal involvement has led to broad political domination—as noted above—in the sense that an examination of Federal support of R&D reveals a number of shifting patterns which have resulted from decisions by government. And it is argued that, at the Presidential level, these patterns must be analyzed, understood and adjusted when necessary. For example, from the early 1960s to the mid-1970s the aerospace-defense-military-atomic energy sectors dropped relatively while various civil sectors increased relatively and absolutely. On the other hand, an important trend from the mid-1970s has been that defense and energy R&D have accounted for the largest increases since 1974. These various trends reflect the durability and high priority of defense R&D as well as the emergence of new concerns about energy, world food production, the quality of the environment, the cost and quality of health care delivery, inadequate transportation networks, and an entire array of urban and rural problems—and they often reflect the intervention of presidential decisions beginning with the time of Roosevelt. This is not to say that presidential decisions alone have determined the patterns—far from it. Many factors and forces are active in influencing not only *policy for science* but *science for policy*.[10] Congress has increasingly asserted a greater role in every facet of science and technology; the organizations of the Executive Branch, often in alliance with the Congress or outside interests, exert powerful influence; and in certain areas, the scientific community and industry groups each have played significant roles in decisions.

As a result of analysis similar to that presented here, the Congress made a strenuous effort to place a "strategic tone" in key provisions of the National Science and Technology Policy, Organization and Priorities Act of 1976 (or "The Act"). The underlying purposes were to exert pressure on the Executive Office and the Executive Branch to devote more attention to thinking ahead and to urge that decisions involving the use of science and technology would be considered in the broadest possible context. Unfortunately, the Carter Presidency's performance in this respect has been little better, if any, than that of its predecessors.

A Presidential science advisory office generally will do what the President wants it to do, whatever the Congress may have written into law. And this leads to a concern related to the lack of attention to strategic thinking and "horizon scanning." In the high pressure environment of the White House, great value is placed on rapid responses and "fire fighting" and one fact is quite clear: the high pressure environment is a function of the Presidency, not only of the President. It is just that the pressure has been higher under some Presidents—for example, Lyndon Johnson—than under others.

In concluding this section, it is suggested that a continuous thread can be seen throughout the decades of writing and public debate and the operations of seven Presidencies: science both affects public policy and is affected by it. New administrative devices and organizational approaches have been required with each substantial increase in the size and nature (e.g., atomic energy, space exploration, energy research and development) of the Federal research and development effort.

Moreover, since the 1940s, there have been various new organizational mechanisms within the President's Office and attempts to bring about an improved central perspective involving science and technology—not only for the development of scientific and technological capabilities but also in terms of relating the R&D efforts of the various agencies to particular social or other national problems that transcend the mission of a single agency. Such thinking has not led to the development of a single, comprehensive American science policy; rather, the literature and actual practice reflect a "many policies" approach. And arrival of "The Act" in 1976 did not change this situation in any significant way—landmark legislation though it may be.

Evolution of Presidential Science Advisory Machinery: Highlights[11, 12]

Despite varying attitudes about science and technology, about concepts of the Presidency, about political philosophies, about Science Advisers, each President from Franklin D. Roosevelt through Jimmy Carter has made important decisions affecting science and technology and important decisions affected by science and technology. In doing so, the Presidents have drawn upon a variety of sources for their advice in making decisions. And while Science Advisers have had at various times and for certain issues much influence, they have had no monopoly on providing science advice.

Roosevelt

The halting steps of Bowman and Compton in the mid-1930s to establish some kind of science advisory capability—while not a total failure because of the experience gained for later use—failed because it was not possible to link the activities of the Science Advisory Board to the President's perceptions of his major political problems in the mid-1930s. In sharp contrast, Roosevelt swiftly saw the connection between his objective of military victory during World War II and Bush's proposals for harnessing science and technology for the war effort.

Moreover, Bush, along with Conant, designed an organization of extraordinary competence which had the flexibility to shift rapidly not only in meeting the needs of the President but also in exercising independent judgment on military requirements if the occasion arose. But, on the whole, Bush used this latter power sparingly and relied on persuasion in working with the military forces. Bush and Conant had finely tuned political sensitivities and chose to work with the President primarily through Hopkins; in the tough, but delicate, task of working with the Congress, Bush decided to handle this himself. Perhaps his only major mistake in the Roosevelt years was to disregard Harold Smith, the formidable Director of the Bureau of the Budget, who eventually got the upper hand in science policy matters during the early part of the Truman Presidency. Indeed, it may be argued that the bureau made an institutional decision of far reaching consequence: it would be at the center of science advice in the future, no matter who else was also providing it.

What could have been a problem with different people turned out well because Bush and Conant worked well as a team; still, Conant had his own independent relationship with Roosevelt which in one sense was closer than Bush's. Conant was very much the "political" Science Adviser, who was called upon in later years for far more than science advice. Indeed, he was more politically attuned to both Roosevelt and Truman than was Bush, and eventually became High Commissioner of the German Occupation for Truman.

In one sense, it may not be fair to compare Bush with some of those who were to follow him in the task of providing science advice for Presidents; the looming war created special circumstances and the establishment of OSRD was a special case in that Bush was both an adviser and a manager. Further, Bush made no pretense of advising Roosevelt on any aspect of science not related to the war. On the other hand, Bush provided the first significant model of a scientist being able to work in a highly successful manner at the right hand of a President and he was instrumental in starting a process that has continued with varying success to the present time: Presidential science advising.

Truman

The institutionalization of science advice which had begun to evolve under Roosevelt changed rapidly under Truman with the ending of the war. This was not because Truman was opposed to the use of science advice or did not appreciate the powerful role that science and technology had played in achieving victory. Indeed, this was far from being the case; Truman understood far better than did some of his successors the potential of science and technology for remaking the world. But at a time when Truman was in great need of science advice of the very highest calibre, a break appeared in the closeness of the relationship which Bush had maintained with Hopkins and Roosevelt.

Slowly at first, then more rapidly, communications began to break down between Truman and Bush—especially after the ending of the war. Political differences and disagreements on the organization and administration of science in the post-war world came to the fore, and Bush was eventually shoved aside by Truman's staff. Relations did not break in the Nixon sense and Truman was always willing to see Bush whenever the latter asked, but Truman did not ask to see him; they were on opposite sides of the major post-war debates on the National Science Foundation (NSF) and on the Atomic Energy Commission (AEC). However, despite their differences, there appears to have remained a great deal of mutual respect between the President and his displaced Science Adviser.

Bush never was fired by Truman; in fact, he stayed on as part of the official White House staff for nearly two years after Truman became President. But Harold Smith and Don Price of the Bureau of the Budget took on for a time the major task of advising the President on post-war science policy and organization. Another key figure—particularly in the atomic energy area—was a young lawyer, James R. Newman. The principles of public administration, espoused by Smith, Price, and Newman, under which the science organizations of the post-war world would be responsible to the President and not to the scientific community, clearly appealed to Truman's sense of his constitutional responsibilities as President. Bush on the other hand, was seeking to protect, as had the scientists of the 1930s, science and the traditions of scientific freedom from political interference. Bush always thought he had been "poisoned" by Truman's staff—and there was some of this—but the fundamental incompatibilities in outlook were far more important in the dissolution of the Bush-Truman relationship.

Truman paid a price for not having a strong Science Adviser more compatible with his political views. Compounding Truman's difficulties, there was little continuity; Smith and Price left not too long after the war was over and were unable to carry through on their intention to develop a comprehensive plan for the organization of science in the postwar world. A very busy John Steelman, who became Truman's senior aide, also became a *de facto* Science Adviser—aided by a young attorney, Byron S. Miller—and took over from Smith and Price as the architect of science policy planning for the White House. In this same time period, two young Bureau of the Budget officials were beginning to become involved in science policy matters—William D. Carey and Elmer B. Staats—and they would play active roles until the mid-1960s.

One result of the acutely fragmented approach to the use of science advice by Truman was that the single most important policy decision affecting postwar science, other than the Atomic Energy Act of 1946, was decided virtually by default: as the foundation debate dragged on year after year, the military departments and the National Institutes of Health—in league with the Congress—established the organizations, the funding patterns, and the management policies which set the major outlines of postwar science organization and Federal funding support for decades into the future. The irony of this situation is that preoccupations of the Executive Office and the Congress during these years were the fragmentation of research effort and the necessity to achieve an effective system of coordination for Federal research and development. And, well into the 1950s, the Executive Office —under Truman and then Eisenhower—kept trying to get the infant National Science Foundation to take on this task. Fortunately, the foundation chose to survive and resisted the mandate given to it by two Presidents and the Congress.

Despite the lack of a broad-gauged Presidential Science Adviser and the absence of focus which could have been provided by a science advisory apparatus, several important steps were taken during the Truman period which would have positive results in later years. First, the President's Scientific Research Board—headed by Steelman—was established in 1946 and eventually produced the "Steelman Report." This was a wide-ranging political document on national priorities, resource allocations, policies, organization and directions for science and technology. All of its recommendations were eventually adopted—including the establishment of the NSF, large increases in support of Federal R&D, the creation of an Interdepartmental Committee for Scientific Research, the establishment of a unit in the Bureau of the Budget for reviewing federal research and development, and the designation of a member of the White House staff for scientific liaison. Along with Bush's report, *Science: The Endless Frontier*, the seeds were planted for much of the institutionalization of science advice and the growth of federal support of science which would develop in the years to follow.

Another institutional contribution by Truman came in 1951 as one response to the outbreak of the Korean War. There had been demands for the creation of a new OSRD; but those who made such demands did not understand that the national capability and organization for science was far different in 1950 than it had been in 1940. William T. Golden was the author of a report to Truman which recommended the establishment of the position of a Science Adviser to the President and a Presidentially appointed Science Advisory Committee, and the preparation for standby plans to establish an OSRD-type organization. But Golden's recommendations, though approved by the President, were modified radically and the end result was the establishment of a Science Advisory Committee placed within the Office of Defense Mobilization (ODM).[12]

Under an ill, ineffective chairman (Oliver Buckley), the committee languished unused during the remainder of the Truman Presidency. Communication channels to the President, although technically available, were never used by Buckley; he appears not to have taken a very active role and certainly did not wish to take on the forceful General Lucius Clay who headed ODM—and who had insisted that the committee report to him rather than to the President. Apparently, for more reasons than one, Truman chose not to enter what had become a major jurisdictional battle. The National Science Board was just getting started at about this time in 1951 and strongly opposed the idea of a Science Adviser to the President or a committee reporting directly to the President. The board saw such moves as downgrading its role and status. Truman was busy with the war and a host of domestic problems and, aside from the jurisdictional squabbling, there seemed little that the committee could do for him; thus, once again, the situation of the 1930s appeared. There was no apparent identity of interest between the committee and the President, so he did not use it. Furthermore, Buckley's approach was passive—he saw his role as waiting until the President asked for something to be done; this was a far cry from the forceful Bush approaching Roosevelt saying this is what the scientific community can do for you and the country.

And the Interdepartmental Committee for Scientific Research—launched with high hopes in 1947—remained nearly invisible instead of becoming an effective tool in assisting the Executive Office to manage the far-flung research and development establishment. Not until 1959, in the post-Sputnik crisis, would the committee be pulled from obscurity and be re-established as the Federal Council for Science and Technology. Despite the repeated urgings of individuals such as Carey for greater attention to the problem of managing and coordinating the Federal R&D program at the Executive Office level, most of the administrative and organizational measures of the 1940s and 1950s were dismal failures or met with very limited success. And the end of the Korean War removed the urgency from such matters and there was little inclination in the early Eisenhower years to undertake any major steps in the direction of further institutionalization of science advice at the Presidential level.

Eisenhower

The President did, however, express interest in better management and coordination of Federal R&D and, at the urging of the Bureau of the Budget, attempted to give the task to the National Science Foundation. But this did not work out.

More specifically, Waterman (the first NSF Director) properly resisted any ideas that the NSF should undertake such major assignments as "developing policy" and "evaluating research programs" for all of Federal R&D. It is easy to conclude that these tasks were important in the early 1950s and are still important today; but it is also easy to conclude that Eisenhower and Truman before him—both urged by the Bureau of the Budget—were trying to get the wrong organization to do them. As an infant organization in the early 1950s, it very likely would have suffered fatal injuries; and today, NSF still would be the wrong choice, even though it is now a large and relatively powerful agency.

Despite the lack of organizational initiatives by Eisenhower with respect to science advice during his early years, it seems clear enough that he was willing to use a revived Science Advisory Committee under Lee A. DuBridge who replaced the ailing Buckley. DuBridge and the other members determined to make it a more active body. After an abortive attempt to escape from ODM, the committee engaged in a number of important investigations which gradually led to a convergence of their work and the interests of the President in certain urgent defense problems.

And the Science Advisory Committee—with its panel on Technological Capabilities headed by James Killian—eventually came together with two outside "gadflys," Trevor Gardner and General Bernard Schriever, to convince Eisenhower to undertake an acceleration of the ballistic missile program in the mid-1950s. However, this process had required several years to force the ballistic missile decision to the Presidential level. And as a result of a combination of Executive Office timidity and a lack of comprehension of the potential of space, Eisenhower was persuaded—over the objections of Nelson Rockefeller, at the time a White House aide, and a number of leading scientists—to approve only a very modest US civilian space effort for the 1957 IGY. It is argued that the presence of a strong Presidential Science Adviser might have had some effect, first, in bringing the ballistic missile decision to the White House earlier and, second, in convincing Eisenhower to place a higher priority on the IGY satellite.

At the time of Sputnik, Eisenhower reacted swiftly—but not in panic. Among his first steps were to appoint Killian as his Special Assistant for Science and Technology and to elevate the Science Advisory Committee from the ODM to be the President's Science Advisory Committee (PSAC).[14] As noted above, in 1959 the Federal Council for Science and Technology was established. Direct communication channels were established between Eisenhower and his Science Advisers as the President took personal charge of the national response to Sputnik—and a close rapport developed between them. Thus, with the exception of one major change by Kennedy—the establishment of OST—Eisenhower finished the work started by Roosevelt and Truman by putting into the Executive Office the basic components of a science advisory structure which would last for nearly 15 more years. Despite the achievement, one is left with this general criticism: more often than not, the nation reacts in an *ad hoc* tactical way rather than in a strategic manner. Eisenhower himself later reflected that he underestimated the impact of Sputnik and its aftermath. It is argued that the shock of Sputnik and the charge of a "missile gap" may well have cost the Republicans the 1960 election.

Killian and Kistiakowsky were similar to Bush in that each of them came to the White House when there was a strongly perceived need for their services by President Eisenhower. The emergency was not as acute as that which had faced Roosevelt, but Eisenhower was experiencing severe political problems over space and various military matters following the national state of shock after Sputnik. Although Eisenhower had placed high value on science and technology in the earlier years of his Presidency, it was not until highly visible political problems crashed about him that he fully embraced science and technology as instruments of national policy. And Killian and Kistiakowsky, each in his own way, responded to and assisted the President in ways that he and his senior staff considered important.

Killian and Kistiakowsky were similar to Bush in another way; they had a great deal of influence in the White House and throughout the government because of their close connection with the President, although they consciously stayed away from any attempt to "manage" any of the space and military programs under their purview. Furthermore, they—and especially Killian—went out of their way to work with and through other individuals in the White House even though each of them had easy access to the President. Another factor which provided underpinning for the high degree of influence and power held by Killian and Kistiakowsky—as for Bush and Conant earlier and Wiesner later—was that all five of the individuals were "old Washington hands" and understood the political aspects of science advising. By virtue of their long Washington service and outstanding records, each of the five had an independent standing in the White House which put them on a peer basis with their respective Presidents' principal assistants.

Again putting Bush, Conant, Killian, Kistiakowsky and Wiesner together, it would seem that each had, in his own special way, a close relationship with his President. This is not mean to convey "closeness" in the sense of friendship. It is intended to mean that they were part of the inner circle of Presidential advisers who have that commodity so precious in the White House—the ability to influence Presidential decision making. Indeed, in no instance was a Presidential Science Adviser part of the President's social circle of friends—except perhaps Conant in his relationship with Roosevelt.

The strong roles performed by Bush and Conant for Roosevelt were entirely in keeping with Roosevelt's view of the Presidency and the role of the White House staff in becoming involved in the affairs of the departments and agencies. Not until the time of Kennedy would such an approach be followed again. In contrast, Eisenhower was very disinclined to intervene except on very special occasions when he had a deep interest such as the "Atoms for Peace" proposal. Therefore, Killian and Kistiakowsky, as noted above, while exercising strong influence on agency programs and budgets, did so by working behind the scenes with the Bureau of the Budget and department and agency heads. Working with a small staff and PSAC, neither of the two nor the President was interested in pushing the institutionalization too far. Eisenhower did worry, however, that the science advisory role might vanish at the end of his Presidency.

Kennedy

The arrival of Kennedy in the White House brought many changes—but not the one that Eisenhower had been concerned about (i.e., that the position of Special Assistant for Science and Technology would not be continued by his successor). Indeed, Kennedy embraced the entire apparatus which had been constructed over the years by Truman and Eisenhower and, working with the Congress, gave it an even more powerful mandate than it had held under Eisenhower.

From the very beginning of his Presidency, Kennedy saw that science and technology were important instruments of national policy which could contribute to the solution of critical national problems. And he was acutely aware of the political power available to him and the international prestige potentially available in the space program. While science advice was an element in Kennedy's thinking, the Apollo decision was made on political grounds. And while he worried about the cost of the program, it did not deter him as it had Eisenhower and Kistiakowsky.

The Apollo decision and his deep involvement in negotiating the Limited Nuclear Test Ban Treaty underscore Kennedy's style in reaching out in all directions for advice. While most Presidents have been disinclined to tie themselves to a single source of advice, Kennedy was more like Roosevelt than, say, Truman, Eisenhower, or Nixon in his insatiable need for information and advice from a variety of sources. Indeed, Wiesner saw this as an important part of his responsibility to Kennedy and played an active role in insuring that Kennedy was exposed to all sides of a problem—and did not appear to be threatened if the President heard advice which differed from his own.

Clearly a mature and experienced Science Adviser must accept that Presidential advisers ply their trade in the highly competitive arena of palace politics—a competition which centers on the giving of advice. One of DuBridge's failures was that, notwithstanding his Washington experience and long years as president of a major university, he seemed not to understand this important characteristic of White House life—or chose not to play the game. And the record of the Johnson Presidency suggests that Hornig philosophically accepted a secondary role in the White House power structure. Finally, David understood very clearly the power realities in the Nixon White House and attempted to reverse the slide downward in influence of the science advisory apparatus—but it was already too late by the time he arrived on the scene.

In the giving-of-advice competition even such towering figures as Bush and Conant had to accept Roosevelt's disagreement with them on the matter of sharing atomic secrets with Great Britain after they had vigorously pursued their point of view. The occasional disagreement, especially over fundamental issues such as the atomic argument, is in the nature of things at the White House level. It is only when the disagreements become and remain persistent and deep—such as with Bush and Truman and between Nixon and PSAC—that the advisory arrangement becomes untenable. Also, a measure of maturity and responsibility on the part of a President and his senior aides is required in dealing with differing points of view from Science Advisers. And the record and actions of the Johnson and Nixon Presidencies are less than admirable on this score. It is one thing to suggest that general compatibility of views is important but it is quite another to demand and expect absolute agreement with the political views of the President on every issue.

Not only were Kennedy and Roosevelt able to handle dissent, their operating styles inevitably created it. And Kennedy was much more activist-oriented than Eisenhower before him or Johnson and Nixon afterward; as a result his White House staff—including Science Adviser Jerome Wiesner—plunged into the operations and management of the Executive Branch. Kennedy was determined to gain control of the bureaucracy and developed a powerful White House staff with the intent to do so. Wiesner's close relationship to Kennedy, based on their prior connection, put him in the inner circle of Kennedy's advisers. And in Washington it soon became clear that in many areas related to science and technology, "Wiesner spoke for the President." The nature of Washington is such that this simple fact invokes no inconsiderable amount of power, and Wiesner was both admired and criticized for exercising it. Bush had held a similar power under Roosevelt and he, too, provoked criticism as well as admiration. In stark contrast, Hornig did not "speak for Johnson," nor did DuBridge and David "speak for Nixon;" and the diminished power and influence of the science advisory machinery in the years after Kennedy reflected these simple facts.

The Kennedy style of making a vigorous effort to take charge of the bureaucracy was consistent with a major advance in the institutionalization of science advice at the White House level. Responding to Congressional initiative, Kennedy and Wiesner were very receptive to a proposal by Senator Henry Jackson to provide the Presidential science advisory function with a statutory underpinning. By agreeing to the establishment of the Office of Science and Technology, Kennedy achieved several objectives: first, the science advisory function was further institutionalized —although future events would show that this action did not guarantee either its future use or its continuance and that unforeseen problems were created; second, pressures from some in the Congress and elsewhere for a Department of Science and Technology were greatly reduced; and third, and very important, one of the very real and very legitimate grievances of Congress—lack of access to a top Executive Office official for matters concerning science and technology and national science policy—was removed.

Thus, with the establishment of the Office of Science and Technology, a new dimension of institution building for Presidential science advice was added by Kennedy to the structure and process built incrementally by Roosevelt, Truman, and Eisenhower. In terms of structure, the institutionalization was completed and no further changes would be made until the Nixon banishment of 1973; however, in terms of process, the institutionalization continued to develop during the Kennedy, Johnson and even the Nixon years. There is more to be said about this later.

The concerns of Truman and Eisenhower about foreign affairs and peace were carried forward into the Kennedy Presidency as major, if not dominant, concerns; but added were new dimensions related to energy, the environment, natural resources, and a plethora of social issues—as Kennedy picked up the domestic mantle worn by Roosevelt and Truman and constructed the programs of the New Frontier on the building blocks of the New Deal and the Fair Deal. With respect to science and technology—perhaps more important than the organizational changes —the Kennedy Executive Office—including Wiesner, as earlier noted—was very much inclined to pick up the reins of control for a sprawling scientific and technological establishment which no other agency could safely grasp—and to look beyond the horizon to new areas such as energy and natural resources. The presence of a "central scientific organization" at the Presidential level became more broadly apparent than during any period since the days of Bush in World War II; and to the present time no science advisory apparatus has had such visibility and status.

Kennedy has been called a truly modern man and his vision and outlook made it easy to understand the products and uses of science and technology; and important, too, was that he understood the social significance of science and technology. It was Kennedy who initiated what would become the Commission on Technology, Automation and Economic Progress as one of the very few efforts at the Presidential level to cope with such broad issues as technological change. It was Kennedy's conviction that science provided vast powers for good, and this led to much of his hopefulness about the future. He was forever pressing to put technology to work: in foreign affairs; for helping other nations (an idea that he had inherited directly from Truman); for insuring national security (in which he followed directly in Truman's and Eisenhower's footsteps); and, in seeking solutions to major domestic problems, he looked to science for clues.

While new dimensions were added to the science advisory function during the Kennedy years for dealing with problems of energy and the environment, civilian technology, natural resources, water resources, and food and nutrition, the Kennedy science advisory apparatus under Wiesner sustained the high level of involvement in arms control, space and military problems that had been the case under Eisenhower. Since the major contributions were in the latter categories, it must be recognized that the new dimensions and new initiatives were only beginnings. However, they represented the beginnings of significant change in the agenda facing the apparatus; also, they represented the beginning of change in the nature of the advisory function as the new dimensions brought new complexity which, it turned out eventually, the science advisory machineries of later Presidencies were not able to handle very effectively.

When Johnson inherited the Presidency, a watershed was crossed in terms of science advice and the Presidency and the use and influence of the science advisory apparatus. A decline began regarding both use and influence of the apparatus and the individuals who served as Science Advisers. Whereas Bush, Conant, Killian, Kistiakowsky, and Wiesner were in the inner circles of their respective Presidents and were essentially on a peer basis with senior White House aides, the situation changed dramatically under Donald Hornig and his successors Lee DuBridge, Edward David, and H. Guyford Stever.

Johnson

As a political outsider both to Johnson and to Washington, Hornig definitely was not in the inner circle of Johnson's advisers, and was not able to play in the same league as had his predecessors. Without doubt, Hornig was not a peer of such top aides as Bundy, Rostow and Califano; moreover, Hornig did not have an effective alliance with any of Johnson's top staff over the five years he served as Johnson's Science Adviser. Hornig's relatively low status in the White House could not have been a positive influence on the important process of Presidential decision making.

The Presidential agenda, while inherited in large part from Kennedy, began to change under Johnson as he attempted to carry out both the programs of his Great Society and to conduct an increasingly major war in Vietnam. While the record of the Johnson years suggests a President who expressed support and a sophisticated understanding of science and technology, it was also a period when R&D growth came to a virtual halt. It was inevitable that the growth rates of earlier years could not be sustained, but in what was probably an over-reaction, Congress, Johnson— and Nixon after him—held essentially to a no-growth policy for the rest of the 1960s. Johnson placed a very high priority on his Great Society programs and, by 1966, Vietnam began to be a major drain on national resources.

Yet, in the face of such obstacles, while the total Federal R&D allocation increased only 10% during the Johnson years, basic research increased by 50%. This kind of anomaly suggests some policy selectivity on the part of Johnson and indicates that Hornig had a degree of influence. But exerting influence at the Presidential level on behalf of basic research—which was nothing new from the 1940s onward—would increasingly be seen in the future as performing an advocacy role in the White House for science; during the Nixon years it would translate into the "kiss of death." Despite the halt in budget increases, Johnson (as had Kennedy) was continually pushing to put science and technology to work: in international cooperation, in health, in environmental quality, in natural resources, and in other domestic areas. In most of his statements related to science and technology, there are echoes of the Great Society, but, despite the social sciences' function as the basis of most of the Great Society programs, there is little indication of effective science advice being provided at the Presidential level in connection with the social sciences, or any aspect of the Great Society.

While Kennedy had sought science advice widely, he was inclined to have Wiesner help him seek out a variety of sources; in contrast, Johnson seems to have simply ignored Hornig on a number of major issues which one might reasonably have expected a science adviser to perform at least some role: the SST, health research, and the automation-unemployment issue. On the other hand, Hornig and PSAC played prominent roles in the years that Johnson sought to delay development and deployment of an anti-ballistic missile system. Yet, as an overview comparison, Hornig definitely did not have the prominent role held by Bush under Roosevelt in large strategic matters—and for the most part Hornig was ignored on Vietman—nor did he have the influence of Killian, Kistiakowsky, and Wiesner in other national security matters.

An important casualty of the Vietnam War was the relationship between Johnson and PSAC and very likely this had no small effect on Hornig. Despite PSAC's role in the anti-ballistic missile debate (referred to above) and major contributions in two areas high on the President's agenda (environmental quality and the world food problem), the relationship between PSAC and the President deteriorated as the war continued and PSAC's opposition increased—notwithstanding the fact that it was privately, not publicly, expressed. But PSAC came to be viewed as the embodiment of the academic world's violent and vocal opposition to Johnson's (and later Nixon's) war policies. Other factors were at work, too, such as the variety of PSAC's work in diverse fields, and there was much less direct contact with the President. But the war was the crucial factor in the breakdown of a long-standing relationship. Inevitably, in combination with the continuing deterioration of the PSAC relationship with Nixon, this raises the large issue of how science advice is to be provided if basic political differences are present, and not solely those related to scientific and technological issues. At the end of the Johnson Presidency, indeed, even to the present time, no good answer has been found to this fundamental issue.

An apparently unanticipated, but important, result of the establishment of OST by the Congress and Kennedy was that the Presidential science advisory machinery increasingly and inescapably became more institutionalized. In certain respects, the apparatus took on more and more of the characteristics of a central scientific organization—although not as visibly as under Wiesner. The range of problems expanded dramatically, staff was increased and new coordination and evaluation functions were performed that had nothing to do with providing science advice to the President. While an "institutionalized memory and capability" are important to assist the President, dangers and problems accompanied the increasing institutionalization.

In solving one problem—being accessible to the Congress—another one was created: requirements to appear before the Congress were extensive and there was less time to spend on providing advice to the President and working on the problems which were important to Johnson, and later to Nixon. Also, the requirements for coordination and evaluation placed upon OST by the Congress and accepted by Kennedy in 1962 led to the establishment of an active bureaucracy in the Executive Office which Hornig admitted was difficult to control and manage, along with his many other responsibilities. By the time David arrived on the scene in mid-1970, the all-important relationship with the Office of Management and Budget had virtually disintegrated, and one of David's first priorities was to restructure OST and re-establish working arrangements with OMB.

Another implication of the growing institutionalization of the science advisory machine under Johnson and Nixon was that more and more of its work took place out of sight of the President and his top aides; in short, beginning in the Johnson years and continuing into the Nixon years, the apparatus was not seen as being responsive to the President's needs. Thus, one encounters the ironic situation in which the science advisory apparatus had become engaged extensively in activities related to management and coordination of the Federal R&D establishment which had been so highly desired by Truman, Eisenhower, Kennedy, and the Congress, only to find itself perceived as not being helpful first to Johnson, and later to Nixon.

Even so, the Congress was not entirely satisfied with the attention given by Johnson's Executive Office and his science advisory apparatus to a variety of areas. Beginning in the mid-1960s, Congress became quite involved with various science policy issues and moved to organize itself more effectively for dealing with science and technology. Hence, Congress became increasingly confident about initiating actions when it did not think the administration was performing effectively. A case in point is the marine sciences: Congress established two separate "science advisory" organizations to carry out its mandate because it did not believe that the Federal Council for Science and Technology was giving the field proper attention. In effect, the science advisory apparatus was being pressed to do more by the Congress along the lines of managing the Federal R&D enterprise at the same time that the President and his senior aides saw it as less helpful in working on *their* problems. Hornig has seen the dilemma more clearly in retrospect but already had begun to think out loud about what should be done about it not long before he left office. For example, he called tentatively for a new look at a Department of Science and Technology. This analysis of the problems of institutionalization is connected directly to a general analysis of the problems of mixing advisory and management functions within the Presidential science advisory apparatus at the Presidential level (to be discussed later).

Nixon

At the beginning of his Presidency, it appeared that Nixon would raise the science advisory apparatus to the former influential status that it enjoyed under Eisenhower and Kennedy. Among his very first appointments was DuBridge and among his first press conferences was one devoted to extolling science and DuBridge. DuBridge had outstanding credentials: a brilliant record of public service extending back to World War II, long service as president of one of the country's leading universities, and high standing in the scientific community.

But things did not work out. First, as pointed out earlier, DuBridge seemed not to appreciate the nature of the White House political arena; moreover, he simply was "too nice" to be a member of the Nixon White House. In this respect, he was similar to Hornig, who was a "Mr. Nice Guy" in the Johnson White House. One is forced to the conclusion that both lacked a certain inner toughness which is required of anyone who plays on the "first team" in the White House. While both are good, decent men, and performed creditably under trying circumstances, they were not very effective in the ferocious competition in the White House which centers on the giving of advice. Inevitably, DuBridge, as had been the case with Hornig, was increasingly seen—along with his staff—as being outside the main stream and, therefore, as somewhat irrelevant to the principal business of the White House and the President.

DuBridge was the essence of integrity but nevertheless became increasingly ineffective as time passed. The alienation which took place was due in part to his not understanding how the White House worked, to decreasing Presidential access, to not developing alliances; however, probably as important as all of these factors combined was that he came to be seen by Nixon and his senior aides as the "representative of science." And, as was noted earlier, in the "them-against-us" psychology of the Nixon White House, this was a fatal perception for DuBridge.

This perception developed on the part of the President and his aides, despite the fact that DuBridge was actively involved from the early days of the Nixon Presidency in problems important to the President: environmental quality (including a major oil spill in California for which Nixon very publicly made a show of giving DuBridge responsibility for reviewing the situation and recommending corrective actions), the earlier noted Nixon decision on chemical and biological warfare, and the SST.

Unfortunately, initially DuBridge was on the "wrong" side of the SST issue and was opposed to proceeding; subsequently he went to Congress to defend the President's decision to go ahead, citing, in response to criticism about his earlier views, that the President has the total responsibility for a decision while an adviser, as in his case, provides only one perspective. But the Nixon White House was not a place for dissent—even when it was private, and, when the initial opposition of DuBridge and PSAC became public knowledge (long after the Congress had voted to cancel the program), this was the beginning of the end.

More than one individual has voiced the opinion that the DuBridge period actually marked the beginning of the end for a science advisory apparatus in the Nixon White House. There is good reason to believe that the position of Science Adviser and the associated apparatus would have been abolished upon DuBridge's departure—if it had been judged politically feasible. But, in 1970—before the overwhelming 1972 victory—Nixon was still concerned politically about the academic community and the uproar which such an action would provoke.

Before the end came, the President made what was an apparent attempt to change the nature of the advisory apparatus and, more specifically, the nature of the role of Science Adviser to the President. The appointment of David, an engineer from industry, caused no small surprise. Despite his outstanding credentials as a manager, scientist, and engineer, he was not well known in Washington and represented a break in the links with the past; he was not a member or protege of the fairly small group that had dominated the upper levels of government science advice since 1940. At David's swearing-in ceremony, Nixon gave an indication of just what kind of change he had in mind for the science advisory function: he expressed strong emphasis on the applications of science and technology and referred to David as "a very practical man." Indirectly, Nixon was making it plain that he had had more than enough of the antagonistic scientific community and its advocacy of more and more support for basic research. Ironically, David turned out to be a strong supporter of basic research and was successful in arguing for limited increases.

David had no personal relationship with the President and dealt primarily with Ehrlichman and Kissinger—although occasionally he had access to Nixon when he thought it sufficiently important, as on such issues as nuclear proliferation, or when the President was particularly interested, as on health policy matters and environmental health. The lack of close communications with the President did not deter the energetic David from trying to make the science advisory apparatus work. For example, he took swift action—as noted earlier—in repairing relationships with the Office of Management and Budget which *was* working on the President's problems. Also, he set out to establish better working arrangements with the Ehrlichman and Kissinger on domestic and national security matters, respectively.

There seems to be little doubt that David had the inner toughness lacked by

DuBridge and Hornig, and, while he did not achieve peer status with Ehrlichman and Kissinger, he was able to work with them without being intimidated. Moreover, as suggested above, to the extent that any Science Adviser would have been able to communicate with Nixon, David apparently was able to get through when necessary. But the tragedy of David's service was that as hard as he tried to turn things around, in all likelihood by the time he arrived (as suggested earlier) it had already been decided by Nixon and some of his top aides that the science advisory apparatus would go after the 1972 elections.

By the early 1970s, Nixon and his top advisers were not really interested in whether or not the science advisory apparatus was performing effectively—and the record shows that David, as well as DuBridge before him, turned in performances that by any objective evaluation would be considered effective in an institutional sense. However, objective evaluations were not what the President had in mind and the studies performed by the Domestic Council and OMB on the value and effectiveness of OST and the rest of the science advisory apparatus must be taken with some skepticism—given the barely concealed hostility of the President and his senior aides to PSAC and OST. And apparently they were so unaware of the Federal Council in Science and Technology that it escaped the Nixon "axe" quite by accident.

The scientific community, with its strong opposition to the war policies in Vietnam, and its personification in the form of PSAC, came to be seen, as had been the situation with Johnson to a lesser degree, as the "enemy within." On too many issues, as seen by the President and his senior aides, PSAC opposed the President. But beyond the opposition, PSAC had become anathema to the Nixon White House because its opposition on at least two major issues considered vital to the President—the SST and the ABM—had become public information. This exacerbated an already deteriorating relationship and permanently damaged the relationship between Nixon and his science advisory apparatus.

The record of the Nixon Presidency is not one that shows the President hostile to science and technology *per se*, even though he came to hold a hostile attitude toward the scientific community and, hence, his own science advisory apparatus. The evidence suggests he simply was not interested and did not have a very deep comprehension of the role of science and technology in modern life; his attention could be captured only by the occasional technological spectacular such as the SST, the cancer crusade, and Magruder's proposed New Technological Opportunities Program. Even in the vital field of energy, he was not much interested in the important preliminary work accomplished for him by David before the energy crisis struck; and then the President leaped unknowledgeably, based on idiotic political advice, to such nonsense as "Project Independence" by 1980.

It is somewhat ironic that Nixon abolished his science advisory apparatus at just the time when it perhaps could have helped him and the White House staff cope more capably with energy problems which were crashing all about them in early 1973 like boulders from a mountain slide. It is argued, as it was argued in the case of Truman in the science organization debates of the 1940s, that Nixon impaired his own effectiveness and that of the administration by not having a strong, stable, full-time science advisory capability in the Executive office during a critical period when decisions were being made that could affect the nation for decades to come. Roy Ash's theories of management and organization turned out to be of little use —perhaps because Ash and his council never really understood an important characteristic of Presidential decision making: in the public sector the problems are fundamentally political and not managerial.

It is possible, only possible, despite the increasing paralysis brought on by Watergate, that the effects of the ludicrous "revolving door operation of energy czars" and the associated "crisis policy making" could have been moderated by the presence of a science advisory capability which had a sense of continuity, history and the politics of the energy area. The science advisory apparatus had been active in the energy area since the days of Wiesner, but had not been successful in translating their projections of coming energy problems into action at the presidential level. At least one result of the energy "shambles" in the Nixon White House was that Congress virtually took away the overall problem of energy organization from the President—including the establishment of the Energy Research and Development Administration (ERDA).

This brings one to an important point about Presidential advising, including advising about science: in large part, Presidents determine the range and quality of the advice they get. A President cannot be forced to take advice—on science or any other matter—unless he wants it. On the other hand, in a number of ways, the science advisory apparatus of the Nixon White House was not successful; for example, it seems not to have had a finely-tuned political sensitivity to its environment and there were some very real deficiencies in its operations and performance —above and beyond the misperceptions of the President and his top aides.

After more than a decade of grappling with civilian-oriented problems in the domestic field, the performance of the apparatus was very mixed and, except in a few areas such as energy, there was relatively little quality thinking to show for the effort. Also, it is important that not only the Science Adviser have good working relationships with senior aides but also the staff of the science advisory machinery have effective working contacts at the mid- and lower-levels of the Executive Office and White House staffs; unfortunately, this was not the case—until David arrived and attempted to change the situation.

After the abolishment of OST and PSAC and the transfer of certain residual functions to Stever as Director of the National Science Foundation, the science advisory apparatus and the science advisory function at the Presidential level was placed in virtually a caretaker status. In turning to Stever, it is concluded that he did what was possible; he sized up the political realities and performed in a greatly truncated science advisory role by making the best of a difficult situation. He was under no illusions that he was to perform in the conventional Science Adviser's role of the preceding 15 years. Therefore, he did his job and publicly defended the President's decision as does any official who wishes to stay. Finally, he took the necessary steps to preserve a Presidential science advisory capability for such future use as it might be called upon to perform. In the meantime, he focused on reestablishing relationships with OMB, in developing an expanded capability for the Federal Council for Science and Technology, and in providing a center for energy R&D programs and policy.

Ford

After Nixon's departure, Gerald R. Ford came to the Presidency with a significantly different outlook on a large number of issues—including science and technology and science advice. As President, Ford displayed a well-developed sense of using science and technology in achieving national goals. He may not have had Kennedy's intellectual fascination with science, but he was equal to Kennedy in every way that really counted with respect to science and technology: accessibility to a wide variety of views, strong support of R&D budgets (especially basic research), a feeling of ease in dealing with scientists and engineers, and receptivity to science advice.

In another way he was very similar to Kennedy: he was highly receptive to working cooperatively with the Congress to modify the Presidential science advisory apparatus which had been placed in the NSF by Nixon. The initiative for change may have originated in the Congress—as it did in the Kennedy Presidency—but Ford soon took over the initiative as "his proposal." One major difference in Ford's Presidency from Kennedy's was the receptivity of their respective staffs to working with a science advisory apparatus. Ford, and Vice President Rockefeller, had to cope with strong opposition—and even some hostility—on the matter of reestablishing a science advisory apparatus in the White House. And after it was in place, there was no Hopkins, no Steelman, no Cutler, no Bundy who could serve either as an intellectual or administrative coupling for science advice to the President. In the absence of such couplings, except that through OMB, it was fortunate for Stever that Ford's operating style and staff system permitted a high degree of accessibility.

A very positive aspect of the Ford Presidency is that Ford did not think about providing support of science and technology in terms of mollifying the scientific community—in the sense that Johnson and Nixon had considered the absence or presence of political support or opposition. And whereas Nixon had expressed himself on occasion in strategic policy terms under David's influence, he acted strictly in an *ad hoc* manner or not at all. In contrast, by the end of his Presidency, Ford was beginning to think "strategic" and make plans for using science and technology in terms of linking investment in R&D with the future achievement of great long-term national goals. This was no small achievement on the part of the President and he was assisted by an effective working alliance between Stever and OMB.

From time to time, Ford spent a considerable amount of time personally on the matter of science advice and the Presidency. His work with the Congress in bringing about enactment of "The National Science and Technology Policy, Organization, and Priorities Act of 1976" represents a detailed example of how joint Presidential-Congressional lawmaking—particularly in the area of science and technology—has become institutionalized during the past several decades. Ford's approach was directly descended from the work of Truman, Eisenhower, and Kennedy in building the great postwar science and technology organizational establishment in the United States in the form of new agencies and developing new capabilities in existing departments and agencies. Through the decades of institution building, including that at the Executive Office level, the joint, but not always harmonious, efforts of the President and the Congress were highly important.

Even as the Ford Presidency was ending, there was taking place a new beginning for the support of science and technology and for the re-establishment of a presidential science advisory apparatus. During the short and troubled Ford years, one could well have expected the affairs of science to take a back seat. Notwithstanding the fact that Ford's main problems were restoring confidence in the Presidency, coping with inflation and a stumbling economy, and searching for an appropriate energy policy for the nation, Ford turned out to be not only a good friend of science but also developed a long view of the future and of the role that science and technology would play in shaping the world.[13]

Observations and Conclusions

A major conclusion is that there is a close and intimate connection between expert knowledge and political power as revealed in the science advisory function. Notwithstanding periods when Science Advisers and their advice were ignored or were assigned secondary status in the processes of Presidential decision making, this broad sweep of seven Presidencies and nearly five decades demonstrates that science advice at the Presidential level has been an extremely important feature of the modern Presidency. From Roosevelt's decision to develop the atomic bomb to Ford's far more complicated problem of proposing an appropriate mix of energy R&D strategies, Presidential Science Advisers have performed in significant roles. In varying degrees, Presidential science advice has affected or influenced a great variety of Presidential decisions—in war and in peace, on budgets and organizations, on domestic and international problems.

However, it also is suggested that the far greater number of the problems and issues which confronted Presidents Roosevelt through Ford were either not directly involved with science and technology or the degree to which science and technology were involved was seldom clear. As one moves from problems which involve the physical sciences—for example, the decisions to develop the atomic bomb and the H-bomb—to programs arising from the social sciences—the New Deal, the Fair Deal, the New Frontier, and the Great Society—the scientific and technological content in Presidential decisions has not been perceived at all or has been assigned minimal weight. This is consistent with observations that the great social and political changes underway in the world, largely in response to the impact of science and technology, are not really understood. Moreover, the presidential science advisory apparatus—from Roosevelt to Ford—was not well equipped to deal with or understand problems of this kind, despite their great importance.

As newly-emerging societal problems moved onto the Presidential agenda and that of the science advisory apparatus in the late Eisenhower and early Kennedy years, it was not as clear—as it was in national security and space matters—just how science advice could perform an effective role. This difficulty persisted throughout the remaining years covered by this paper. Another important point is the absence of a crisis in problems such as energy and the environment in the early 1960s, say, in comparison with the military crisis of the 1940s, and the space and military crises of the late 1950s. Clearly the agenda of the Presidential science advisory machinery has changed over the years and former Science Advisers are even today troubled by the difficulties in mobilizing the government to cope with the new societal problems.

Some are pessimistic enough to believe that even with a crisis it may not be possible to do anything about some of these problems. While an entirely pessimistic view is not endorsed, there is agreement that there is substantial cause for concern; it arises from the fact that a high degree of political uncertainty and deep divisions of attitude pervade socio-technology. In short, much of the growing complexity encountered by the Presidency and a series of science advisory machineries arises from an increasingly intricate involvement of science and technology with political and social processes. Major implications for the future are that Presidential science advisory machinery will be required to engage in far more sophisticated analysis than was required in the past. Implicit in this growing involvement with social and political processes is the necessity for the Executive Office to develop better capabilities for interweaving scientific, technical, legal, social, economic, and political factors. How to perform this interweaving process effectively is a major problem facing the Presidency.

The social sciences virtually have been ignored at the Presidential level since the time of Roosevelt although a few limited endeavors were undertaken over the years. While PSAC published four or five reports related to the social sciences, and many New Frontier and Great Society programs were based on the social sciences, there is little evidence of science advice regarding the social sciences being provided on any regular, sustained basis by the Presidential science advisory machinery. In fact, it was not until 1968 that a social scientist (Herbert Simon) was appointed to PSAC; a general feeling about the social sciences and social scientists was expressed by one former science adviser: ''They haven't discovered their Newton's Laws yet.'' Only after the science advisory apparatus had been moved to the National Science Foundation in 1973 did a systematic study of what was happening in the social sciences get underway during the Ford Presidency. However, the effort came to naught as the office began to phase down in anticipation of returning to the White House.

This paper underscores an important aspect of Presidential science advising: there should be at least a general political rapport between a President and his Science Adviser. This is not to argue that the post of Science Adviser should be strictly a political appointment in the sense that many administration posts must be. Yet, there is no escaping the reality that the White House is a political place, that the problems of the public sector are primarily political, and that a science advisory apparatus—especially the President's Science Adviser—must be able to function in an intensely political environment. And while a specific research project in a laboratory may be far removed from the political arena and the scientists performing the research may be entirely apolitical, in a large sense science is affected by political outlooks and attitudes. Examples include the NSF and AEC debates of the 1940s, the Apollo decision by Kennedy, the debates on the content and directions of health research of the Johnson and Nixon years, and the energy-environment debates of the 1970s. The closer a President comes to perceiving some facet of science and technology as being related to the national interest, the more likely political considerations will become important and make the task of a science advisory apparatus more difficult.

The formal Presidential science advisory apparatus has not had a monopoly on providing science advice. In many respects the Office of Management and Budget (and formerly the Bureau of the Budget) has been more influential than the actual science advisory apparatus. Other important sources of science advice have been the departments and agencies, with the degree of influence varying according to the nature of the issue and the style of a particular President. Still other sources include individuals and organizations outside of the government. This suggests that a Presidential science advisory machinery must be capable of both cooperating and competing with these other sources of science advice; in doing so, it needs to be concerned with establishing its credibility and reputation by being responsive to the needs of the President and his principal aides.

The lack of a monopoly on providing science advice by the Presidential science advisory machinery is directly related to a general conclusion about the establishment of national policy with respect to science and technology both in the *policy for science* and the *science for policy* senses. A science advisory apparatus potentially can have an important influence as policy is established through Presidential statements, messages and budgets. Numerous examples are available. However, a more general responsibility for a science advisory apparatus is to comprehend and be prepared to react to the emerging of policies from the clash of ideas in the congressional arena in which various factions have allies—in and out of government; and it must be understood that there are complex interactions between various parts of the government with different elements of the academic, scientific, and technical communities which over the years have resulted in an elaborate network of policies.

This leads to a conclusion that one must consider the respective powers and roles of the Presidency and the Congress in considering and establishing national science and the constraints of each of the institutions in relation to the other actors in the and the constraints of each of the institutions in relation to the other factors in the science policy arena—the bureaucracy and their allies and enemies in the private sector. One must understand that the system requires eventually a general convergence of views, if the system is to work at all, and this calls for an ability to tolerate the long perspective and a diversity of views.

A more specific conclusion, arising from the above, is that the years of inquiry and debate, first in the Congress, beginning in the early 1970s, and later including the Ford Presidency, which led to "The Act" should not be considered merely, as did many, an examination of the Presidential science advisory system. The evolution of "The Act" is more properly regarded as another, albeit important, episode in the long debate over policy, organization and goals related to science and technology and their uses which has been underway in this country from its very beginning, but particularly since the 1880s. Thus, it becomes clear that the evolution of Presidential science advisory machinery has taken place within the context of this larger debate. There is as yet no final answer, and there may never be one, to the question of how best to devise an organization arrangement in the Executive Branch which encompasses providing science advice to the President, assisting the various units of the Executive Office, and providing some kind of central managerial perspective for the enormous Federal R&D enterprise.

It is believed that A. Hunter Dupree's concept of a long-term search for a central scientific organization is a useful analytical tool for examining the above issue and the evolution of science-based organizations in the United States.[14] In the history of American science and technology and their interactions with government and other institutions, the demands of pluralism have had their counterpart in the quest for some form of central scientific organization. Dupree suggests that two broad types of effort have been involved in attempts to achieve what many perceive to be a desirable unifying focus: (1) the building of a central scientific organization, and (2) the adaptation of a predominant agency to take on the more general business of serving as a central scientific organization. Most parties involved call for coherence, coordination and appropriate relationships between government and science; however, disagreement arises because the centralists believe in the need for decisive action at the center to attain their ends, while the pluralists see these same ends accomplished through both cooperation and competition between the individual institutions themselves.

It is argued that this disagreement is closely associated with and has had a marked impact on the evolution of Presidential science advisory machinery. The centralists have argued for and have seen the placement of a science advisory apparatus in the Executive Office as being essential for decisive action at the center. Moreover, most centralists have endeavored to provide the Presidential machinery with appropriate capabilities to provide for "coherence, coordination and appropriate relationships between government and science." As often as not, the centralists have pressed upon Presidents more capabilities and responsibilities than Presidents have preferred.

Since the early 1940s the power base of the centralists has been the Congress; however, beginning in the early 1950s, important elements of the scientific and technological communities increasingly have allied themselves with the Congress on the general desirability of establishing a strong science advisory function at the presidential level. The pluralists have had their power base in the Executive Branch —with some assistance from a minority of the scientific and technological communities; there has never been a clamor by the departments and agencies for the establishment of such a function. Indeed, there has been hostility and resistance both specifically to the science advisory apparatus and generally to the Presidency over the decades. In contrast, and perhaps as part of their efforts to control the bureaucracy, the Presidents themselves seem to have been more centralist-oriented; of seven Presidents, only Nixon displayed an outright hostile attitude and used pluralist doctrine as his underlying rationale for abolishing the science advisory apparatus from the White House.

Another manifestation of the centralist-pluralist disagreement has been a recurring debate over a Department of Science and Technology which began in the 1880s and continues to the present. Indeed, the initial debate resulting from the Allison Commission of 1884 contained the department idea and the high level science advisory function as integral elements, and the two concepts have been commingled ever since. As one examines the evolution of science advisory organization at the Executive Office level, the idea of a Department of Science and Technology emerges from time to time—particularly when Congress is in the midst of one of its periodic forays into overall science policy matters and the organization of science.

Support of Dupree's concept of the quest for the central scientific organization must be qualified to this extent: the idea of adapting a predominant agency to take on the more general business of serving as a central scientific organization does not appear to be useful in explaining what has actually happened in the post-1940 period. Only the National Science Foundation was singled out unsuccessfully by several Presidents (Truman, Eisenhower, and Nixon) to take on the "general business of serving as a central scientific organization," and by no stretch of the imagination has the NSF ever been a predominant agency. It might be possible to argue that for a number of years the Department of Defense was a *de facto* central scientific organization. This is because it was the largest supporter of Federal R&D and established patterns of support and management procedures which have endured to the present time—with major modifications taking place over the years as other science-and-technology-based agencies evolved as part of the postwar creation of a national R&D structure. Finally, there have been some attempts by the Congress to give the NSF and NASA respectively some of the attributes of a central scientific organization but these efforts have not been successful—either in getting out of the Congress in recent years or in being implemented by the NSF in earlier years. In any event, the predominant agency approach does not appear to be useful as an analytical tool in examining the recent historical record in terms of the quest for a central scientific organization.

A major observation on the "quest" concept is that the commingling of the Presidential science advisory function and the Department of Science and Technology in debates over the years has led to confusion and disagreement about the appropriate responsibilities of the Presidential machinery in terms of how extensive they should be. And confusion has occurred as a result of mixing the advisory function with major responsibilities for the management—with frequent special emphasis on coordination—of the entire Federal R&D enterprise as part of the growing institutionalization process of the science advisory apparatus over the years.

From the early 1940s, when Senator Harley Kilgore pressed for greater coordination of the growing Federal R&D establishment, to the 1970s views of Edward E. David, Jr., that an effective science and technology presence in the Executive Office required statutory management authority, some have argued that, in effect, an organization with certain attributes of a department should be located in the Executive Office. Intermediate views earlier led to the establishment of PSAC, the Federal Council for Science and Technology, OST and, finally, the evolution of the current Office of Science and Technology Policy. In contrast to these two somewhat oversimplified sets of views, Presidents and most former Science Advisers have held to a more limited view of what functions a Presidential science advisory apparatus should perform. Emphasis from this quarter has been on "advice for the

President" even as the institutionalization of the apparatus took place over the years—partly from internally perceived needs but in larger measure as a result of pressure from the Congress.

In the end, one is left with a cruel dilemma which is implicit in the disagreement between the centralists and the pluralists. Virtually all parties can agree that good management and effective coordination of the Federal R&D enterprise are desirable, and many can agree that providing science advice to the President effectively is important. However, on the one hand, a department has never gained wide acceptance; and, on the other hand, it would seem that the mixing of "advisory" attributes with "management" attributes has been inimical to the effective performance of Presidential science advisory machinery in carrying out the central role of advising the President. "The Act" did not satisfactorily resolve the dilemma; as Congress often does, when it reaches the limits of the workable agreement, it passes the unresolved issues to the future. Thus, a dilemma which has been near the center of the pluralist-centralist debate for nearly a century was passed along to the future virtually intact.

Contributing to the difficulties arising from this dilemma is that, on the one hand, the President has been and should be involved in establishing broad policy and must acquire appropriate information for effective decision making. On the other hand, it is not easy to distinguish between the types of activities which must be performed and the kinds of information which must be acquired for the development of Presidential policy and the broad evaluation of programs and similar, but more detailed, activities and information gathering which are more properly in the realm of managing the Federal R&D enterprise, directly or indirectly.

As observed earlier, "The Act" side-stepped this issue and settled for continuing the pre-1973 arrangement with these major exceptions: the Office of Science and Technology Policy was given more "managerial" and "horizon scanning" responsibilities than had been held by the defunct Office of Science and Technology and a specific mandate was given to the administration to examine once more the division of responsibilities by studying the feasibility of a Department of Science and Technology, as well as other organizational and policy issues. President Carter has, however, for the most part, ignored this latter provision of "The Act"; in fact, he abolished the advisory committee which the Congress and President Ford had agreed on as being essential to performing the legislative mandate.

Notwithstanding the Congressional mandate, it is concluded that Presidents will devise arrangements which are compatible with their style and concept of government organization; moreover, it is not believed likely that a Department of Science and Technology will be established within the next few years so that, as a practical matter, the real focus will be on what kind of organizational arrangement can be established within the Executive Office to perform the various functions which have been discussed in this analysis.

DuBridge actively studied an arrangement which generally follows the provisions of "The Act" and the pre-1973 structure but which would allow for a better division of responsibilities to offset some of the negative features of increasing institutionalization. The custom of the Science Adviser to the President being the

Director of OST, Chairman of the Federal Council, and the Chairman of PSAC just happened that way; it turned out to be convenient in the early days and also was effective under an activist such as Wiesner. The "four-hatted" arrangement was not dictated by law or by Executive Order. Although DuBridge and Nixon ended up not making any changes in the structure inherited from Kennedy, it would seem that this subject might be investigated again with the purpose of determining the feasibility of different individuals handling the advisory function and the "management" functions, but within the purview of the Executive Office. It is not considered feasible to place major policy and management functions related to attaining a central perspective in one of the departments or agencies.

The construction of "The Act," while overly cumbersome and loaded with too many responsibilities for the Office of Science and Technology Policy, did set the stage for a continuation of the comprehensive debate on science policy and organization which began anew in the early 1970s. It called for the type of concentrated analysis by the Executive Branch which would have complemented the earlier congressional examination. Moreover, it is believed that the studies called for in "The Act" would be of crucial value to both the administration and the Congress. No comparable study of the various complex facets of R&D has been attempted since the Bush and Steelman Reports of the 1940s—and such a study is long overdue.

Having said this, it is clear that the Presidential science advisory apparatus cannot be considered without reference to the rest of the Executive office. Yet, one is faced with the dilemma that this paper was not directed at a comprehensive examination of the entire domain of the Presidency. Nevertheless, to avoid a complete impasse, it is a general conclusion that the short-term perspective of the Executive Office must be modified to permit the attention of some individuals or units to be directed beyond the immediate set of problems before the President at any given time.

Without specifying in detail what form such a capability should take, it seems reasonably clear that a Presidential science advisory apparatus must be closely connected with this kind of institutional capability. The role of the science apparatus would be to assist in performing what many have called the "horizon scanning" function. The revolutionary impact of science on society—as discussed earlier—demands that more attention be devoted to thinking ahead and to insuring that decisions involving the use of science and technology are considered in the broadest possible context.

This conclusion leads directly to another major point: it is essential that the Presidency, assisted by the science advisory apparatus, be organized in such a way as to permit a more strategic outlook on the problems of the nation and the world. The *ad hoc* "fire-fighting" approach which has characterized the operation of the Presidency in the past will not be sufficient to cope with the problems of the latter part of the 20th century and the beginning of the 21st century—which is now closer than 1950. Yet, the Presidency—and the Congress—still act as though it were 1950. A vital perspective on the future must be attained, and it is argued that a science advisory apparatus is one of the best organizational forms to help achieve this perspective.

Finally, it is possible to state the following overall conclusion. At one time, Bush was able to say that the value of the position of Presidential adviser turns on who the individual is, who is the President, and how they get along together. It is agreed that this remains a fundamental aspect of science advice and the Presidency. But it is also argued that science advising is a highly complex process which is affected by a rich variety of influences, some resulting from the nature of the times, some resulting from the nature of the institutions involved, and others arising from the types of problems facing a President and how he perceives them in terms of the national interest. It is believed that the function of Presidential science advising has proved its worth in the past; how valuable it will be in the future depends upon the extent to which the problems identified in this paper are solved in the context of preparing for the 21st century.

References

1. Clinton Rossiter, "The Presidency—Focus of Leadership," in *The Dynamics of the American Presidency*, Donald Bruce Johnson and Jack L. Walker, eds. (New York: John Wiley & Sons, Inc., 1964), p. 12.

2. Edward S. Flash, Jr., *Economic Advice and Presidential Leadership: The Council of Economic Advisers* (New York: Columbia University Press, 1965), p. 1.

3. Bertrand Russell, *The Impact of Science on Society* (New York: Columbia University Press, 1951), p. 3.

4. Robert K. Merton, "The Ambivalence of Scientists", in *Science and Society*, Norman Kaplan, ed. (Chicago: Rand McNally & Company, 1965), p. 112.

5. Ellul, *The Technological Society* and Ferkiss, *The Future of Technological Civilization*.

6. "Farewell Address," US President, *Public Papers of the Presidents of the United States* (Washington: Office of the *Federal Register*, National Archives and Records Service, 1952-1961), Dwight D. Eisenhower, 1960-1961, p. 1039.

7. John M. Logsdon, *The Decision to Go to the Moon* (Chicago: The University of Chicago Press, 1970), p. 5.

8. Lloyd V. Berkner, *The Scientific Age: The Impact of Science on Society* (New Haven, Conn.: Yale University Press, 1964), p. 24.

9. William D. Carey, "Thoughts on Reorganization," *Science*, October 14, 1977, p. 129.

10. Harvey Brooks, "The Scientific Adviser," in *Scientists and National Policy Making*, Robert Gilpin and Christopher Wright, eds. (New York: Columbia University Press, 1964), pp. 76-77.

11. Acknowledgment is due David Z. Beckler for a number of discussions and "The Precarious Life of Science in the White House," *Daedalus*, Summer 1974.

12. Detlev W. Bronk. "Science Advice in the White House: the Genesis of the President's Science Advisers and the National Science Foundation," *Science*, 11 October 1974, pp. 116-161.

13. William D. Carey, "Science in the Ford Years: Last Things," *Science*, January 21, 1977, p. 251.

14. A Hunter Dupree, "Central Scientific Organization in the United States Government," *Minerva*, Summer 1963.

A Scientist in The White House:
A Sociological View

Emanuel R. Piore

The concept of a Science Adviser to the President has been accepted in Washington for over three decades. Seven Presidents appointed individuals, a scientist or engineer, to such positions, with ill-defined functions. Ten distinguished individuals served the Presidents in that capacity. Seven or eight of those served on a full time basis. Five of the seven Presidents found the President's Science Advisory Committee (PSAC) useful. Nixon during his tenure abolished it. Carter never put it in place. The Science Adviser traditionally was Chairman of the Committee.

The functions and tasks that the adviser and the committee performed changed during the decade. The changes were gradual. The initial pattern was established during the Eisenhower and Kennedy Presidencies. The interaction between the two Presidents and their Science Advisers was unique. The Presidents changed, the Science Advisers changed. There were changes in the White House and the other parts of Washington.

The White House staff grew, the Science Adviser's staff grew, the National Security Council became operative, the Domestic Council was created, the Bureau of the Budget became the Office of Management and Budget. The White House staff became more structured, more formal. The concept of completed staff work took hold. Nixon and Carter felt the need to designate a Chief of Staff. The Science Adviser was given additional functions, becoming head of the Office of Science and Technology. This office and new function legislated by Congress reflected the needs of Congress rather than the operational needs of the White House. On second thought, the legislation modulated the work of the Science Adviser and the President's Science Advisory Committee. The staff of the Science Adviser was increased to comply with Congressional mandate. Administrative duties increased, deflecting the Science Adviser's attention from concerns that may

Emanuel R. Piore (b. 1908) was a member of the President's Science Advisory Committee (1959 to 1961) and was appointed consultant-at-large in 1963. He was formerly Vice President and Chief Scientist and a Director of International Business Machines Corporation; he has also been a member of the National Science Board, Chairman of the Naval Research Advisory Committee, and Deputy Chief and Chief Scientist of the Office of Naval Research; he is a member of the Council of the National Academy of Sciences, National Academy of Engineering, American Philosophical Society, and American Academy of Arts and Sciences.

have more importance to Presidential needs. The composition of the membership of the President's Science Advisory Committee changed.

Outside the White House in Washington new cabinet departments came into being—Energy, Housing and Urban Development, Transportation; independent agencies—National Air and Space Administration, Environmental Pollution Agency, Arms Control and Disarmament Agency. Expansion of the National Institutes of Health continued, and internal reorganization within departments and agencies took place. These additions and changes within the executive function of government had a large component of science and technology in the operating aspects and policy functions.

Finally, a subtle change occurred in the country—a lack of consensus on policy direction. This lack prompted one or two Presidents to raise, through indirection, the question of "political loyalty" of the members of the President's Science Advisory Committee. It resulted in the abolition by Nixon of the President's Science Advisory Committee. Carter did not reconstitute this.

An aside: there was limited sympathy for the man-in-space program among the President's Science Advisory Committee members. The reasons were that all to be learned can be obtained with instruments, without the costly mechanisms to keep man's physiology functioning properly in outer space. Again the political realities precluded the articulation of such sentiments.

The Golden Age

The "Golden Age" began in Eisenhower's administration. One event that was pivotal was Sputnik, the other was a report. The launching of Sputnik required the President's reaction, and, thus, his action. The part-time Science Adviser and the President's Science Advisory Committee were moved directly into the White House from the humble home of the Office of Defense Mobilization in the Executive Offices of the President. A full time Science Adviser was selected and appointed. The report known as the Gaither Report dealt with defense policy in the context of global consideration. Due to Korea, the readiness of our forces was increased, new bases were acquired, et cetera. Nuclear weapons were part of the background. The Director of the Office of Defense Mobilization shepherded the Report to the President. The then Director of the National Security Council was impressed. The President was impressed. Immediately he appreciated the interplay of science, technology and national policy in defense and foreign affairs, and concluded it was a valuable asset in the White House. This was the beginning of an era of a Presidential Science Adviser and the President's Science Advisory Committee.

The adviser and members of the committee had a unique background. Their common background stemmed from the war years. Obviously they were highly qualified, scientifically and technologically. They had made important contributions during the war as group leaders. They came from laboratories. Ten to 15 years had elapsed. They had acquired wisdom; they had acquired a knowledge of how policies were made, and they felt at home in such environment. They had worked together before, and they knew each other's strength and weakness. They had open doors to various Cabinet secretaries.

There were groups in government which were also ten to 15 years older, and who also had risen in responsibility. During World War II, the membership of the President's Science Advisory Committee and groups in government learned to work with each other, acquiring mutual respect. Let me cite one such example. The flag officers in the Department of Defense, the Admirals, the Generals, were also ten to 15 years older. They had known and worked together with members of the President's Science Advisory Committee, when the Admirals and Generals were Lieutenant Commanders, Commanders, Majors, and Lieutenant Colonels. In addition (not a slight matter) they understood the Presidential involvement.

There were fire drills; the first was to determine the vehicle that could place a few US pounds in orbit. The Navy and Army each had a candidate, Vanguard and Jupiter. The question was which vehicle had the greater probability for success.

An aside is worth mentioning. Eisenhower had an opportunity, six months or a year before Sputnik, to augment the Vanguard budget and thus make the missile more reliable and available earlier. He turned the request down. He saw no point in competing with the Russians in this area. He now was forced to consider the public reaction.

Ultimately the National Aeronautics and Space Administration was created. The Science Adviser and the President's Science Advisory Committee were involved. This assured that scientific programs were not neglected, and that the civil and military space would remain separated. Such involvement includes technology, science and national policy.

Similar observations can be recorded in the establishment of the Arms Control and Disarmament Agency. Its Assistant Director for Research and Development was a Science Adviser/President's Science Advisory Committee nominee.

Quite apart from space, it is worthwhile to enumerate some of the chores that the President's Science Advisory Committee performed in the "Golden Age." It presented to the President the possibility of a test ban of nuclear weapons. The President accepted the suggestion and the initial implementation occurred in Geneva, the meeting of two groups of scientific and technical people, one group representing the USSR and the other representing jointly America and England. This initial exploration emphasized what we knew technically, and what we didn't know in detecting large explosions in the air and underground. The nominees for the American group were made by the Science Adviser in consultation with the President's Science Advisory Committee.

There was general agreement that it was feasible to detect explosions in the atmosphere, and there was great uncertainty on lower level explosions underground. It was thought that earthquakes would mask those detections. That was the problem. These conclusions stimulated increased emphasis on understanding the character and the signals generated by rumblings underground, and gave a new impetus throughout the world in geophysics, both theoretical and experimental.

The Science Adviser was very instrumental with the President and the then Bureau of the Budget in bringing about an increase in the budget of the National

Science Foundation (NSF). This increase stimulated the continued fiscal growth of the NSF. There was a joint group created between the President's Science Advisory Committee and the Atomic Energy Commission to understand the needs of high energy physics. The proposed instruments and equipment were very costly. Without consultation with the Chairman of the Atomic Energy Commission, and his opposition, the President determined that we would go forward with the Stanford Linear Accelerator, and the Fermi Lab moved forward with less drama. It is interesting to note that the cost of both constructions were estimated in each case over $100 million, and Panofsky and Wilson finished the construction within the budget. Constructions that cost $100 million or more seldom are achieved at the initial estimated cost.

Another item of lesser importance: the Navy proposed the construction of the 300-foot radio telescope. The telescope had a dual purpose: electronic intelligence and research in radio astronomy. It was obvious that radio astronomy should not be a captive of the Department of Defense, and that the intelligence function was questionable. This was killed. The Committee was uncomfortable with the continued expansion of the National Institutes of Health, but decided not to interfere.

Structure for S&T

The President's Science Advisory Committee was concerned with the structure in the Cabinet departments in regard to science and technology, and it re-activated the concept that each department with large commitments to science and technology have a special assistant secretary for those areas. The President liked that idea, and all departments complied except Agriculture and Interior.

Associated with that philosophy, a director of science and engineering with the protocol title of Undersecretary of the Department of Defense was put in place. This is important in terms of the pecking order in Washington. He outranked the military secretaries. Responsively, each military department then put in place an assistant secretary with comparable functions.

At that time the President's Science Advisory Committee in its deliberations spent a great deal of effort in trying to identify problems that existed in the nation and were of concern to the President—problems and policies having a science and technology component. The identification of a problem usually took a number of President's Science Advisory Committee meetings. These deliberations normally ended with a meeting with the President. This happened once a year, twice a year, three times a year. The conclusions of the President's Science Advisory Committee were presented to the President and most of the time he would select one or two problems that were of concern to him. This assured the Science Adviser and the President's Science Advisory Committee that, if their work was proper and had an element of excellence, they had an audience of at least one, an audience that could take action if it satisfied the needs. Often the Science Adviser reported his conversations with the President to the President's Science Advisory Committee. This was an important stimulant to the members.

With time, not necessarily under Eisenhower and Kennedy, the Science Adviser played a greater and greater role in representing the United States on bilateral agreements that included science and technology. If my memory serves me, the first was an attempt to get an agreement with the Russians right after Khrushchev's removal. With time, every President meeting with his opposite number, lacking any real coupling with the foreign government, would produce an agreement relating to an exchange of information and cooperation in science and technology. Thus the Science Adviser had a continuing responsibility to attempt to implement such agreements.

The difficulty in the implementation was that the White House was not a source of money. The departments and other agencies had fiscal resources. Obtaining such resources depended on the persuasiveness of the Science Adviser. This presented many problems—another burden on the Science Adviser. A career foreign service officer was assigned full time to the Science Adviser and the President's Science Advisory Committee.

In describing the "Golden Age" and citing a number of examples, the selection was to display a spectrum of activity. The operation and function of the President's Science Advisory Committee and the Science Adviser fall into five categories:

- national policies, whether internal or external, that have a large component of science and technology;
- the execution of the function of the administrative structure within government in the departments and agencies depends on science and technology—that proper administrative structure was in place for thoughtful deliberations;
- the content of science and technology, the distribution of effort among the departments and agencies;
- consultation on appointments of officials whose primary responsibility would deal with science and technology;
- rounding out the involvement of the Science Adviser and the President's Science Advisory Committee, it is well to note also that the Science Adviser sat in on the Advisory Committee on Foreign Intelligence.

From time to time immediate advice was required by the President. This in turn resulted in the "fire drill".

Continuing Growth

As one moves away from the "Golden Age" and observes the operation with the passage of time, it is well to appreciate the continuing growth of the staff within the White House. It is important to note the continuing proliferation of departments and agencies having heavy involvement in science and technology; the legislation that makes the Science Adviser available to Congress as a witness at committee hearings; the modification of membership of the President's Science Advisory Committee; and the change in staff of the Science Adviser due to legislation and the need in the White House to understand and make judgments.

Finally, one observes the changes that have occurred in the departments and agencies themselves.

As the influence of the President's Science Advisory Committee and the Science Adviser grew, a concern in some quarters appeared. This concern is best stated that the membership of the President's Science Advisory Committee consisted of the "boys club", and that representation was needed from a broader set of disciplines, with an additional element of geographic distribution. Great men with reputations were asked to serve. Many of them lacked past contacts with government departments and agencies—lacked an intuitive understanding of how Washington operates. This produced a peculiar lack of psychological coherence, but it was important to do this to counteract subtle criticism in the scientific and engineering communities that a cabal existed in the President's Science Advisory Committee.

The contact with the President gradually eroded, although the workload increased, and thus the problems tackled were those that members of the President's Science Advisory Committee, whose motivations were noble in character, thought were important to the nation.

The underlying mood was that as the President's Science Advisory Committee thought and speculated on these important problems, an audience would be forthcoming. And, for one reason or another, there was a lack of appreciation that in the final analysis the audience was one man, the President. Many significant papers were written and presented. The environment certainly was treated, but there was no audience. Biological warfare was studied, and this had a profound effect on United States policy on biological warfare and its containment. An American thrust to design a commercial supersonic transport was killed by Congressional action. In part this action may have been due to a President's Science Advisory Committee report. The report became public. This in some measure added to the erosion of the status of the President's Science Advisory Committee in the White House. The increasing staff was pulled in to do more and more of the report writing, and deal with the increased workload. Staff members played an increasing role in identification of problem areas and articulation of the recommendations. A staff member made his reputation as a great expert on energy. As a result of that he became the staff director of the Ford energy study, and ended up as chairman of TVA. The staff had many outstanding people, and one can observe their careers after tenure on the President's Science Advisory Committee. With time, the staff had greater influence with the Science Adviser than did the President's Science Advisory Committee. In addition, the Science Adviser had additional responsibilities to the Congress, additional responsibility to manage his increasing staff.

All these developments occurred through a period of 20 years. In the process they eroded the usefulness of the President's Science Advisory Committee, and the use that the Science Adviser made of the President's Science Advisory Committee. This does not mean that the President's Science Advisory Committee and the Science Adviser did not make very important contributions to the nation. But it does indicate that prior functioning of the Science Adviser and the President's Science Advisory Committee were profoundly modified. More and more they became a nuts and bolts operation, more and more responding to specific requests from the Security Council. They no longer were close to the deliberations of the

National Security Council. More and more, time was spent with the Bureau of the Budget (Office of Management and Budget) on specific budgetary items, and more and more time was given as the White House increased its concern with programs in its departments and agencies, and was spent resolving conflicting programs between departments and agencies. The President's Science Advisory Committee developed many aspects of a job shop.

During one period when the President recognized the need for greater emphasis on technology, the Science Adviser was consulted after the fact, after the program was approved in principle. That program produced trivial impact on the nation. In the recent deliberations on energy and the meetings at Camp David, the Science Adviser was not present. His present responsibility is to pick up the pieces that the President announced to the nation. Somehow the last four Presidents lacked the input that science and technology can make in policy considerations, and the structure of government to produce more thoughtful and more relevant activities in the areas of science and technology in the individual departments. There is a lack of appreciation that traditionally the technical agencies are best run by technical people. The continued lack of understanding is that, if a problem is identified by non-technical people and technical people are asked to deal with it, the problem may be irrelevant to the discussions that end with establishment of policy positions for the nation, in both domestic and foreign affairs. Nixon, toward the end of his tenure, designated the Director of the National Science Foundation as his Science Adviser. It is well to point out that with time, the Science Adviser and the President's Science Advisory Committee no longer were involved in the overview of the Defense Department, the largest appropriation for research and development in the Federal government.

The Limitations of the Office

The limitations that the Science Adviser faced were due to a number of actions. The abolition of the President's Science Advisory Committee was one of the casualties of Vietnam. The country became divided during Johnson's administration, and this division intensified under Nixon. Thus problems that had great importance to the President did not receive undivided attention in the President's Science Advisory Committee meetings. In fact it was often difficult to obtain the pertinent information. In addition, many scientists and engineers joined the ranks of single issue groups. They went public, at times with a great deal of emotional articulation. The Presidential advisers on national security did not find it necessary or profitable to include the Science Adviser. Specific technical issues were assigned. The definitions of the issues were those of the National Security Adviser or his staff. The Office of Management and Budget was interested in getting reaction to very specific questions that they defined. The White House no longer had a collegial atmosphere, but a highly structured one, with emphasis on completed staff work. Such procedures dampen the conflict between competing ideas.

Even within such atmosphere and administrative structure, the existence of a Science Adviser is necessary, for a number of reasons. There are technical problems that require understanding and illumination. Often departments and agencies recommend actions and policies that contain in some measure a parochial departmental point of view. There are conflicting recommendations coming from one or more departments and agencies. The Science Adviser, as the resident scientist in the White House, will find such situations falling in his lap. The Office of Management and Budget seeks advice. The service staff in the White House needs assistance in scientific and technological areas, and, finally, the Science Adviser is the traditional spokesman for research, both in protecting its budget and in the administration of policy. The Presidents have used the Science Adviser as the President's representative when dealing with foreign powers involving agreements between the two nations in research, science and technology.

Finally, within the White House a broad view of the role of science and technology may evolve. In reconstituting the President's Science Advisory Committee such a broader view will be accelerated. In selecting members for a reconstituted President's Science Advisory Committee, proper weight should be given to wisdom and experience, in addition to respect for the individual's scientific accomplishments.

Organizational Structure and Advisory Effectiveness

The Office of Science and Technology Policy

James Everett Katz

The increasingly complex, technological and interrelated nature of our society, as well as of the problems confronting it, has resulted in an ever-growing emphasis on science and technology as instruments for problem analysis and resolution.[1] This is reflected by the proliferating scientific advisory staffs at the local, state, national, and international levels.

Because of this important role assigned to scientific and technological advice at most levels of government, one important aspect to consider is the structural aspect of advisory mechanisms with an eye to the conditions which enhance the effectiveness of scientific advice. This paper considers the structural aspects of one major advisory group, the United States' highest science policy office, the Office of Science and Technology Policy (OSTP), a part of the Executive Office of the President (EOP). The major objectives of the OSTP are (1) to advise the President, the National Security Council, the Domestic Policy Staff (DPS), and other EOP units on matters concerning science and technology; (2) to assist the Office of Management and Budget (OMB) with reviews of proposed budgets for Federal R&D programs; (3) to provide general leadership and coordination of the Federal R&D programs; (4) to promote a stronger partnership between Federal research funders, State and local governments and the scientific community; and (5) to provide Executive branch perspectives regarding science and technology policy to the Congress.[2]

The Presidential offices for science and technology had been reestablished and strengthened under the Ford Administration, which also lobbied for the passage of the National Science and Technology Policy, Organization and Priorities Act of

James Everett Katz (b. 1948), author of Presidential Politics and Science Policy *(1978), is Research Associate Professor in the Department of Social Sciences, Clarkson College, Potsdam, New York. He previously held research fellowships at the Center for Science and International Affairs, Kennedy School of Government, Harvard, and at the Center for Policy Alternatives, Massachusetts Institute of Technology; and has taught at Indiana University and William Paterson College. The author thanks W. O. Baker, Harvey Brooks, Donald Hornig, Jürgen Schmandt, Eugene Skolnikoff, and Richard Garwin for their helpful comments.*

1976 (P.L. 94-282). Presidential Science and Technology Adviser H. Guyford Stever made organizational and substantive contributions to the effectiveness of the Office of Science and Technology Policy (OSTP). Thus far in the Carter administration, Presidential Science Adviser Frank Press has acquired the confidence of the President and restored to the position some of the prestige lost during earlier administrations.

The OSTP itself has fared less well under the Carter Administration. The OSTP "has almost been completely destroyed," said one former science adviser. The Carter re-organization team had recommended the abolition of the science advisory mechanism in early 1977, but Carter and his staff were prevailed upon to save at least the OSTP's organizational essence, provided the OSTP's role was rigidly circumscribed. The President's Committee on Science and Technology was disbanded, however. While the OSTP did survive, it was in a greatly diminished role and subject to several limitations. The staff had to be very small (which would, it was anticipated, hamper its effectiveness) and was proscribed from undertaking policy initiatives without the express permission of the political staff. The OSTP was blocked from a pro-active role in defense, natural resources, and especially energy. The vision of a vigorous, politically significant science policy office was snuffed-out, largely because the President's top advisers recognized that many areas of science and technology were politically sensitive and hence should be handled at the political level. They wished to see no competing centers of influence arise in the White House that would dilute their control over these issues.

Congress and some leading elements of the scientific community have tried to have OSTP's role expanded in line with the original authorizing legislation; much criticism has been directed at the White House's handling of the OSTP and coordination of science policy.[3] However, few challenge directly the concept that since the OSTP is part of the President's staff, he has the right to arrange his staff however he wishes. Press has aligned himself solidly behind the President's position. Press said recently that the Science Adviser "must recognize that he is not in the White House as a lobbyist, or as a representative of a constituency. He is there as an assistant to the President, serving the President."[4]

Experience indicates that the science adviser must be one of the "President's men." Yet, having limited authority, resources and power to confront the vast Federal science and technology enterprise has created a series of operational and organizational quandaries for the Presidential Science Adviser and his staff.

A Major Dilemma

A major ongoing dilemma concerns the way to best use slender staff and resources to work on the highly-complex, diffuse and intractable problems the staff is called upon to analyze. The size of the White House staff was an early campaign target of Carter and he pledged to pare the White House staffs by 30%. (He only accomplished a 12% over-all reduction from the highest number during the Ford Administration.) This meant that the Congress's original plan for a large OSTP staff of about 40 did not materialize. Under administration pressure to demonstrate good faith, Press reduced his staff by 30%.

Many maintain that the OSTP full-time staff of 24 (which includes support and secretarial assistance) and the 16 people detailed from other government agencies is too small. The OSTP must prepare reports, participate in committee work and proffer science advice, and oversee the breadth and depth of Federal science and technology. Press himself has said, "If we had a larger staff, we probably could do things better," and that Carter "prefers to have small staffs. . .He's my boss. I have to do things his way."[5] While not questioning the ability of the specific OSTP detailed staff, it is a truism in government that when staff are detailed, whether to another agency, or for committee work, the people selected are not necessarily the best but are instead those available.

To conveniently cover the amorphous subject of science and technology, the OSTP is broken into three divisions. These areas are (with representative illustrations of work): (1) National Security, International and Space Affairs, dealing with such topics as the test ban treaty, East-West technology transfer, the MX and cruise missiles, UNCSTD, Antarctica policy, and space arms control; (2) Natural Resources and Commercial Services involved in evaluating earthquake and dam hazards, radioactive waste management, climate research as well as oceans, energy, mineral and mining policies; and (3) Human Resources, Social and Economic Services concerned with research policies for nutrition, bio-medicines, social science, and agriculture. It has also studied drug legislation and radiation standards. Combinations of OSTP staff members have also prepared special reports on issues in response to presidential requests (such as the construction of a sea-level canal in Central America) or in anticipation of future problems that will be confronting the President.

In addition to these activities, the OSTP director chairs the Federal Coordinating Council for Science, Engineering, and Technology (FCCSET) and the Intergovernmental Science, Engineering and Technology Advisory Panel (ISETAP). FCCSET operates as a sub-cabinet group addressing the Executive Branch in R&D-related affairs. ISETAP, while originally given wider functions, now, after 1978 Reorganization Plan No. 1, has the narrower role of advising the OMB on ways to have Federal technology better meet State and local government needs. Some flexibility is maintained since the OSTP director has a free hand in selecting the members of these groups. As worthwhile as these groups' activities are, they do take Press's time and absorb OSTP staff resources and effort.

All these Herculean labors could not possibly be accomplished by the small OSTP staff itself. On many issues the OSTP serves as the lead agency, taking responsibility for pulling together and chairing interagency panels to address the issue. In addition, the OSTP has attempted to develop a network of panels, advisers, and support staff to supplement its own meager staff. The OSTP has tapped outside resources to do much of the work instead of trying to develop staff resources in-house. Often the OSTP relies on the NSF as well as other governmental agencies for manpower, money, and administrative support. Frequently organizations completely outside the government are utilized to do the actual inquiries. The National Academy of Sciences (NAS), the National Research Council (NRC), the Committee for Scholarly Education, the American Physical Society, and several for-profit contractors are among the organizations tapped by the OSTP. In the future the OSTP hopes to expand the list of those undertaking OSTP work to include organizations such as the American Association for the Advancement of Science and the American Chemical Society.

Reliance on Outside Support

Relying on outside support to do the OSTP's work creates problems both for the contractor and the OSTP. The conflict over assigning public responsibilities to private organizations has periodically plagued the government. At times Congress and the Executive have expressed disapproval of this practice; at other times its advantages have been applauded. Thus, in 1962 the Bell report[6] pointed to a pernicious cycle through which the Federal government became increasingly dependent on non-governmental laboratories to conduct its own research. This was because the government itself was underwriting the raiding of manpower and ennervation of work in its own labs. At the same time it is generally recognized that Federal research organizations tend to become debilitated through bureaucratization and because the Federal researchers become isolated from the cutting edge of the research community.

An illustration of the farming-out process is the preparation of the five-year forecast and plan for science and technology as mandated by the law re-establishing the White House science offices. The preparation of this plan was detailed to the NSF, who, in turn, used the NAS to provide major inputs into the study. However, the acceptance of this assignment caused some problems for the NAS, both in terms of its standard operating procedure and its position of autonomy from governmental politics. (The NAS is a private organization established by an act of Congress during the Civil War to provide advice to the government.)

The first major problem revolved around the fact that the NAS was being requested to produce a confidential report for the government's own use. It was feared that the White House could be highly selective about the parts of the report it chose for inclusion in its own plan and that regardless of what was omitted the resulting work would have the quasi-NAS legitimacy. Thus the NAS might be lending its name to something it actually opposed. The NAS sought to avoid this dilemma by agreeing to publish and publicly release its own report. From this report the President and his staff assistants could accept or reject whatever they wanted in preparing the government position. This attempt to safeguard NAS autonomy has drawbacks and dangers of its own. The second problem is that the NAS could arrive at a plan different from that which the administration wanted. Contradictions between the two approaches could be seized upon by the media, Congress, or other watchdog groups and in turn cause political furor.

Of all the non-governmental organizations the NAS and the NRC have taken up the largest share of the work. Their assignments from OSTP have included an evaluation of the scientific exchange program with the Soviets and the drafting of an issue paper for the development of the national position paper to be submitted to UNCSTD. Professional societies representing specific disciplines have also been drawn upon. For example, the American Physical Society has received a contract to explore the future costs and likelihood of success of solar photovoltaic energy.

One OSTP strategy to defuse the problem of relying so heavily on private organizations is to diversify the sources of advice. By expanding the inputs to include the ACS, AAAS, and other organizations, the NAS and OSTP should be able to avoid much of the potential for criticism. To some extent, however, this conflict seems irresolvable given the present strictures on OSTP. Yet OSTP is outstanding among Presidential-level offices because it is the only one of the few offices with a regular outreach program to draw in experts from the private community.

The resources of OSTP have been spread even thinner as a result of its involvement with a growing user group—state (as part of its outreach and public involvement program, mandated by Congress) and local governments. The OSTP supports the Intergovernmental Science, Engineering and Technology Advisory Panel (ISETAP) composed of 19 state and local government officials, and the National Science Foundation and the OSTP directors. Meeting at least 10 times a year, ISETAP works with Federal, state and local governments to identify high priority problems at the state and local levels which science and technology can help resolve and to determine research issues associated with these problems that the Federal government needs to address. ISETAP is also developing recommendations for enhancing the utilization of research funding by state and local governments. There is some criticism that the OSTP is not wholehearted in its ISETAP role—that it is simply acting in response to Congressional pressure. The massive workloads combined with this high rate of dispersal of the tasks to various agencies and organizations has led to problems of coordination for the OSTP. It is difficult for the director and the second-rung assistant directors to know what is going on in each division and the information problem is magnified for those lower down in the hierarchy. In the past the lack of communication has led to inefficiency and overlapping responsibilities within OSTP.

Problems of Coordination

There are also problems with coordination between the OSTP and the other Executive Office of the President (EOP) staff, such as the Office of Management and Budget (OMB), the National Security Council (NSC), and the Domestic Policy Staff (DPS).

Coordination, obviously, is not a problem that can be solved once and then forgotten, but rather is a continuing process that needs constant adaptation to the developing situation. The need for flexibility in policy supervision is even more pronounced in a rapidly changing field such as R&D. There are a host of ways R&D can be divided among agencies—for example, basic research is concentrated in the NSF while applied research is largely conducted through the line agencies such as NASA, DOD, and HEW. Programs and disciplines, such as oceanography, energy, social science, biochemistry, and atmospherics, cross cut numerous agencies as well.

As part of their centralized responsibility for science policy, the OSTP and FCCSET must coordinate four different types of R&D management activities:

- Budgeting for R&D, including planning and analyses. Here there is a special need to review resource allocations in relationship to national goals and priorities. The advisers should be able to recognize opportunities and anticipate future needs in a timely manner.
- Comprehensiveness of R&D programs. This is necessary to reduce duplication and assure that there are no gaps between programs. It should be noted that in crucial areas, duplication tends to be encouraged in order to have several paths to the same problem. This is the case for fusion research.
- Cross functional policy-setting. These are issues which affect several agencies but are not within the bailiwick of any particular agency. Here, for example, are issues of patent policy and laboratory utilization.
- Implementation. Here OSTP coordination involves assuring that R&D performers are coupled to the ultimate users, and facilitating the delivery and utilization of federally created technology. Both the pull of users in the private or public sector and the push of the technology need to be considered.[7]

These four areas of coordination are characterized by two approaches to coordinating science and technology—horizontal and vertical.[8] Horizontal approaches use coordination among individuals of equal standing within their own agencies. This type of research coordination, usually not very dramatic in its activity or results constitutes the bulk of Federal science and technology coordination. A forum is provided to develop personal linkages and exchange information. Minor issues are resolved and low-level questions discussed, often on an informal basis. However, once larger issues are introduced, the games of bureaucratic politics and strategic maneuvering take place. Horizontal coordination is relatively easy where the stakes do not involve vital organizational interests, but when they do, this type of coordination is especially difficult to accomplish, since there is no hierarchical structure immediately available to which committee leaders can appeal against their peers. Horizontal coordination presents a sharp challenge to both committee leadership and the larger managerial organization in numerous ways. First, since the committees operate by consensus, any decisions tend to be at the lowest common denominator. The committee members may agree to particular outcomes or recommendations, but these are not necessarily meaningful results. Each member represents an agency which has its own vested interests and goals and hence avoids confronting issues or making decisions that will adversely affect the prerogatives of the represented agencies. One result is that only general or vague recommendations or decisions are arrived at. As a consequence, national programs involving interagency cooperation fail to develop in one lead agency through interagency coordination. Instead, agencies use the forum to legitimate what they are already doing.

A necessary condition for the effectiveness of these interagency committees is that policy-level personnel participate in them. These people tend to be the most important members, hence, the busiest, and consequently tend to progressively deputize attendance. These high-level policymakers delegate attendance to their subordinates who in turn send their subordinates until (regardless of the individual's specialization in the area in question) the representative will not have the authority to speak on policy changes or compromises his or her agency would be willing to make.

A horizontal coordinating body has no authority to implement its decisions. Only the operating agencies have implementation resources and powers. As a result interagency committees can decide, but not execute. If any agency or section of that agency disagrees with a policy decision, the committee cannot coerce the agency into the agreed-upon action. Similarly, when studies are performed there may be no recipient or specifically intended audience for the committee recommendations. Studies are produced, decoupled from anyone who would or could exercise authority to implement actions recommended.

A vicious cycle takes place in which the usefulness of the interagency committees is downgraded, which in turn leads to frequent absenteeism and delegation of alternates to attend meetings. This reduces the importance of the committee and so on. In sum, horizontal coordination is used most often, but tends to be effective only on issues that matter least. For the most vital issues, vertical coordination is necessary.

The Vertical Approach

The vertical approach to coordination means that the power and prestige of the President is directly or indirectly involved to secure the desired results. Here the most significant vertical coordinators are the OMB and the OSTP. The major vertical activities take place in the EOP, and often involve Presidential budget-making. Less frequent, but also influential, is intervention on the part of Presidential staff offices in bureaucratic operations. Effectiveness here is largely a function of association of the science staffs with either the budget process or with the President and his top operatives. The perceived lack of proximity with the President weakened the OST (Office of Science and Technology, OSTP's predecessor) during the Nixon administration; this reduced OST's ability to vertically coordinate agency operations.

This general situation is in marked contrast to agency reaction to the New Technological Opportunities Programs (NTOP). Unlike the usual lackadaisical reaction to OST initiated proposals, the direct request from the President to move on NTOP conveyed through the OST (prior to its replacement by a subsequent organized group) spurred the agencies to great action.[9] The main source of power for these staffs is not so much persuasiveness, or rationality as much as association with the President. This association can be either direct, as through personal contact with the President and Presidential directives, or derived from association with the budget process.

The American system relies heavily on the budget process to review the activities of the departments and this budget analysis enables issues of effectiveness in actual performance to be raised periodically for Presidential decision. While the creation of a Federal budget in no way assures that science and technology will be coordinated, it does give an opportunity to discuss and integrate R&D issues which crosscut the Federal agencies. The budget has the additional advantage of being an institutionalized tool to coordinate activity annually and routinely.

In this regard the OSTP's relationship with the OMB is extremely important, for the statutory authority of the OSTP in budget matters is sharply delimited. The OMB is the powerful budget agency of the President, and Science Adviser Press and his staff have gone to great lengths to keep an open and helpful relationship between the two agencies. Thus far the relationship seems to be mutually beneficial; laudatory remarks have been made by each group about the other. An example of this fruitful cooperation was a joint study in which the President commissioned Press and the Deputy Director of the OMB to do an in-depth cross-functional study of Federal research. The study panel included Cabinet officers and the Vice President. The end result of this study was a Presidential budget increase for basic research of about 11% across the government.

In addition to substantive budget review and analysis, the OSTP can serve other roles in conjunction with the vertical coordination with the OMB. For example, the OSTP can assume a brokerage or expediting role, as when it facilitated the transfer of unneeded labs from one agency to another that could use them. The OSTP can also play an advocate's role, encouraging agencies to assume programs that fall between agencies or have been overlooked. This is a role that needs to be handled very gingerly since in the past this activity has caused problems for the OST when it became identified as a lobby for science. Great lengths must be taken in order to insure that the advocated programs are truly programs desired by the President and not just something the OSTP itself would like to see accomplished.

Vertical coordination, especially in terms of intervention into operational programs of the agencies, must be pursued to avoid ennervating over-commitment. There are an infinite number of issues the OSTP could address, so very sharp limitations must be levied in order to prevent too much effort being devoted to fire-fighting and managerial intervention, in practical terms trying to fill a bottomless pit. Effort should be given to setting and periodically reviewing priorities in this effort and assessing the potential benefits and the likelihood of their being achieved.

Centralized Coordination

Centralized coordination, both horizontal and vertical, tends to be resisted by the Federal agencies. Interagency groups tend to be seen as having a practical use; agencies have little to gain through participation in interagency coordination bodies and something to lose, especially if their representative is not an astute negotiator. Centralized coordination can yield increased efficiency in governmental operations if handled properly. But the entire concept of coordination, both horizontal and vertical, implies accountability—that the Federal agencies must hold up their practices and policies for review and examination by higher authorities, and that the agencies are also subject to change or discipline as a result of this examination.

In effect a more vigorous OSTP and FCCSET means a greater degree of centralization of policy oversight responsibility and accountability for R&D in the Federal system. One of the most notable characteristics of the US R&D science policy is its pluralism, a marked contrast with other systems such as Japan, France, and to a lesser extent, West Germany. There are dangers in overplanning and inflexibility, yet there is no reason why centralization of some responsibility and the maintenance of flexibility are mutually exclusive. There are costs to centralization but there are likewise costs to decentralization. In an era of scarce resources and sharp foreign competition, R&D must be marshalled as efficiently and effectively as possible.

Here some lessons might be learned from the centralized, but flexible, Japanese science policy. The Japanese have succeeded not in becoming pre-eminent in the production of new scientific knowledge, but in the production of manufactured items. This is in part due to the fact that their science policies are designed to facilitate harnessing the research of other countries rather than the creation of new findings in and of themselves. While certainly the US would not want to forego its leadership in national security related research, a re-direction of R&D towards industrial, social, and commercial applications could be of great benefit. A more closely integrated program of interaction between researchers and industry could be of benefit in ameliorating trade, productivity, regulatory, safety, environmental, and a host of other problems. Some criticism has been directed at the OSTP for by-passing industrial innovation and engineering aspects of science and technology policy in favor of pure research. For example, a GAO report faulted a Federal Council's Coordinating Committee for Materials R&D for its "strong basic science orientation with little or no engineering or other input."[10] A minor point, perhaps, but indicative of the mind-set that was operating was the resistance by some of the scientists associated with the re-structuring of the White House science offices to the addition of "Engineering" in FCCSET and ISETAP titles. Hyperbolizing, science was to remain unsullied by "profane" applied engineering. This situation is gradually becoming rectified, as is exemplified by recent efforts to establish programs to stimulate private sector innovation, but much remains to be accomplished.

Centralized coordination bodies require skillful leaders, careful supervision, and clear objectives in order to succeed. Implementation of recommendations, for example, has been a key issue in the operational success of science offices, both past and present. In the past especially, the OST and Federal Council would issue reports, but would address no particular recipient who would or could exercise authority to implement the actions recommended. This weak coupling between policy recommendations and the R&D management was a source of frustration for both the Science Advisers and policy-makers and of course reduced a report's impact and the adviser's influence.

Likewise, inadequate leadership permits the potential drawbacks of committees to come to the surface. David Beckler, the long-time former executive officer of PSAC and later acting OST director, argues that the science coordinating mechanism operated "on the basis of self-interest, consensus and compromise, and its proposals have been implemented only to the extent the agencies wished to take advantage of its conclusions."[11] In many cases log-rolling took place when representatives would agree to support favored projects of others in return for a similar endorsement.

While personalities are important, some of the organizationally-based problems could be reduced through the limited centralization of authority over R&D programs in the hands of the Science Adviser and through delegation to the OSTP. This centralization would mean greater power for the Science Adviser and his staff. Such enhanced authority would improve the Science Adviser's ability to have recommended policies implemented. In the past, even though the appropriate official in an agency might agree that a particular program should be implemented, no action would be forthcoming because of agency resistance. Money can even be put in the budget, but still, because of bureaucratic resistance, the program may not be properly implemented. While being far less than a "science czar," increased authority for the Science Adviser could permit him to be more effective in identifying areas where more work or a changed emphasis would be helpful and to have the wherewithall to see that his recommendations are carried out.

Centralization also means that the OSTP could more effectively aim at harmonizing the various aspects of "high policy" and assuring that the Federal R&D management environment is structured in a way that will motivate the most efficacious allocation of resources and programs. A primary goal of OSTP centralized coordination should be oriented towards strengthening science and technology programs of mission agencies in the civilian and domestic spheres. The management of these programs has lagged in contrast to weapons and space research. Unlike national security and related areas, civilian issues such as urban transportation, welfare and health tend to be "moving targets," rather than "stationary" ones. The result is that flexibility needs to be a hallmark of the centralized structure, and this flexibility is as much a result of good leadership as it is of good program policy design.

In the past the OST was not noted for its flexibility or its ability to respond to changing situations, either in the political atmosphere of the White House or to the larger necessity of including political, economic and social factors in the construction of the reports and recommendations. In part this lack of flexibility was due to a "bureaucratization" of the OST. As staffs grew larger, the utility of their output diminished. Despite the disadvantages mentioned earlier, the strict proscription of large staffs in the current White House will reduce the likelihood of bureaucratic ossification taking place on the staff level.

Curbing Task Forces

The constituency nature of government, which includes not only agencies, Congress, and the concerned public, but various factions within the scientific community as well, means that science policy will have a tendency to become pluralistic, decentralized, and fractionated. It also means that there will be constant pressure for the creation of specialized committees in FCCSET and on lower interagency levels to represent those interests. In recognition of this danger and despite the increased use of outside resources, the OSTP has attempted to curb the use of interagency task forces. The task forces were especially characteristic of the Johnson Administration. These task forces and the OST committee system tended to proliferate in number and missions, creating new goals as they accomplished their initial objectives. These committees were like hydras; an attempt to kill a committee would induce it to resurge and diversify.

Under the present system, the number of committees working with OSTP has been severely restricted. Precise aims and specific termination dates are assigned. At any given time there are usually five or six committees in operation with an average lifespan of 12 to 18 months.

Trimming committees also have been used with some success in FCCSET, the chief formal government-wide coordinating mechanism for science and technology issues. FCCSET operates as a sub-cabinet group under Press's chairmanship and is composed of chief officials for R&D in the various government bureaucracies. After an OSTP review, a number of committees were pared down or eliminated—in the latter case, the committee's responsibilities were reassigned to a lead agency. The lead agency coordinates activity on the issue with occasional reports back to OSTP. This decentralization concept conforms with Carter's emphasis on program management by agencies where possible. Special problems are dealt with by an ad hoc FCCSET committee when necessary.

Although hampered by limited resources, Press has sought the opinions of outside experts in reviewing the working of these committees. An example of this took place with the Interagency Review Group on Nuclear Waste Management (IRG). The task force report has been reviewed by the National Research Council's committee on radioactive waste management, by other experts such as Harvey Brooks and David Deese and public comments have been invited. The IRG also exemplifies greater attempts to attain public participation and win broad support for the resultant policy.[12]

Quality assurance is problematical in any organization and no less so in the OSTP. Outside semi-public review is a current attempt to accomplish this objective. Before the White House science office's disbandment by Nixon, this function was served by the President's Science Advisory Committee (PSAC) which was a small body of some of the most talented ''policy-oriented'' scientists in the nation. PSAC was able to insure the report quality of its own panels and of the OST and its panels, and was also able to inject something of a general public policy orientation to the more parochial special interest reports. At times the PSAC impact on the reports was substantial, especially in areas of technical quality or delineation of options. Today Press is operating almost exclusively through panels without utilizing any central review groups, relying instead on a piecemeal approach to review and evaluation.

Under contemporary conditions it would be almost impossible to have a group similar to PSAC that could operate effectively. In large part this is due to the Freedom of Information Act, which allows citizen access to governmental meetings and documents under certain conditions. PSAC was only willing and able to fulfill its role because it was an anonymous one. The critical procedure of review and criticism cannot function in the public eye because of the reprisals that would be levied against the reviewers. Clearly people are willing to say things in private that they would not say for the public record.

Problems in Critical Reviews

The problems inherent in critical reviews of politically sensitive projects or reports are illustrated by an incident that occurred during the Kennedy administration. PSAC studied Project Rover, a nuclear powered rocket, strongly supported by the Congress's Joint Committee on Atomic Energy. Several industry people were on the panel reviewing Rover and these individuals concluded that the project was nonsensical and said as much in their report. This report enraged a powerful committee member, Clinton Anderson, who attempted to discover who had been on the panel. He wanted this information in order to carry out reprisals against the firms employing these members. The reprisals were to take the form of a "bill of attainder" in the defense and space appropriations measures saying that these particular companies would receive no contracts. Whether or not Anderson could have carried out his threat remains moot, but the threat alone is enough to demonstrate that political considerations can easily erode objectivity unless there are appropriate safeguards.

The loss of a group to perform a PSAC-like role is a major weakness which transcends the quality assurance and review function. It was generally believed that PSAC served as a lobby for basic research. But, in areas not involving its own interests, PSAC earnestly sought to represent the President's interests in line with the technological realities of the situation. PSAC did give advice that was contrary to the prevailing conventional wisdom within an administration, and took positions contrary to the expressed desires of a President. This was true in the case of the Skybolt missile, the antiballistic missile, and the supersonic transport plane. (In each of these cases the administration in office would probably have done better to heed instead of ignore PSAC's advice.) Yet in these situations it was clear that PSAC was representing the President's interests as opposed to any parochial interests. This continued to be true even though the membership of PSAC became more and more diverse as the years went by.[13]

Under the current arrangement of relying on specialized groups, this unity of Presidential focus is much more difficult to insure. And even if Presidential interests remain the paramount orientation, it will be more difficult for people to believe that this is in fact the case. This in turn means that the impact and credibility of OSTP reports will be diminished since the primary audience for the reports is the President and his staff. In that rarified atmosphere there is no tolerance for special interests that try to assume the mantle of Presidential legitimacy. The value of the Science Advisers had to be proven again and again with each new administration or operative. Initially, in each administration they were perceived as useless, or worse as representing special interests.

The political staff of the President has always been competitive with the science advisory staff. The only products any of these advisers have to sell are information, opinion, and advice (these are usually indistinguishable), and the only market is "the President's ear." Being highly political both in nature and job description, the Presidential advisers have an inside track over the Science Advisers. In the exercise of sheer power over policy, the Science Adviser is at a great disadvantage, although he may carry the day on the strength of the technical merits to his case.

In this situation, an appropriate analogy is that the scientific advice office is like a transplanted organ: the body's defense mechanisms are always present, waiting until they are no longer suppressed to reject the foreign body (science advisory apparatus). A President, of course, can either encourage or discourage those rejection mechanisms. President Nixon quite obviously encouraged those mechanisms which ultimately led to the office's reorganization out of existence.[14] The chronic resistance of White House staff to Science Advisers and staffs is motivated by more than considerations involving power politics. The scientists are seen as being overspecialized and their advice as too esoteric and narrow for the Presidential level. They are also perceived as indulging in special interest pleading and ax grinding, and as unwilling to respect the sensitive and privileged nature of White House matters.[15]

Overcoming the Doubts

Press has sought to overcome these doubts about scientists in the White House, and has met with significant success. OSTP staff has been able to work closely with the OMB on science budgets and has been able to share in some national security discussions. An approach to enhancing cross-agency communication has been tried: an OSTP assistant director, Ben Huberman, has a joint appointment to the National Security Council staff. This move symbolizes the efforts to integrate the OSTP more closely with the mainstream of presidential concerns.[16]

A particular dilemma for the science advisers is that unlike the other Presidential advisers they always appear to have a constituency, whether or not they are actually representing that constituency. This problem became most obvioius during the Vietnam war. The scientific and academic communities were centers of protest over the war, and the nearest face of this community was the White House scientists. To many advisers, dealing with these scientists became the "moral equivalent of trading with the enemy."

There are certain advantages to the Science Adviser's position vis-a-vis other Presidential advisers. To some staff members, the Science Advisers can be seen as "neutral," since they are not competing for influence in their areas. Of the coterie of advisers, it is almost invariably only the Science Advisers that know anything about technical matters, and are able to give sound advice on these subjects. Because of the seeming neutrality in this situation, the adviser's position might actually be enhanced. This was certainly the case when Press was chosen to select an advisory panel to overhaul the White House information system. Given the aphorism in Washington that information is power, it is notable that Press was selected to carry out this sensitive assignment. This assignment fortuitously coincided with Press's attempt to gain more office space for his staff.

The contradiction between thorough study and analysis and the policymaker's need for immediate information and recommendations has accompanied every science policy office. It is neither new nor resolvable in the foreseeable future. But the contemporary arrangement of the OSTP seems to intensify this contradiction. Many of the reports must be produced quickly in order to be of a value to the policymaker. This is difficult first of all because the resources for the study exist outside the staff, and must be drawn together and organized in order to be utilized. Secondly, the short time frame for the study is incompatible with the proper functioning of review mechanisms. This is particularly true of the NRC which has an institutional review mechanism in place to review any reports and recommendations generated by that organization. Thus, by the time policy requests are passed down to staff, time compression is even more severe.

To a limited extent, this issue has been circumvented by Press. He anticipates what the likely issues are going to be, and then thoroughly prepares himself and his staff on a small number of issues. On selected issues the OSTP has become highly expert, but this narrow focus consequently means that Press's and OSTP's overall impact is reduced. Press simply does not, and cannot, become engaged in the broad array of issues confronting the President. The narrow focus is nonetheless a source of great strength on the issues which he does become involved in. His superior substantive knowledge on a particular issue helps him prevail even against Cabinet level officials. This has occurred in a number of cases, for instance in defense policy (weapons systems) and space policy (satellite programs). Obviously, when Press has been able to get the support of the rest of the governmental officials on an issue, his job of selling a program to the President is eased considerably.

The Institutional Factors

There are institutional factors that work to reduce the Science Adviser's effectiveness. Specialized White House advisory agencies, both scientific and non-scientific, tend to wane quickly; in fact, it is often counterproductive for narrowly-focused groups to have functionaries closely linked to the President. Long-term observation demonstrates that the Presidents grow hostile to special-topic advisory staffs, despite the usually warm initial reception. This characterizes the fate not only of the previous OST, space, and marine councils, but also the Office of Drug Abuse Policy, the Office of Telecommunication Policy, the Council on Environmental Quality, and the consumer-affairs advisers. There are both institutional and psychological reasons why this takes place.

The special-topic adviser represents an imbalance that automatically discounts his advice; the scientist, for example, is concerned with particular issues and even more so with particular solutions to those issues (i.e., techni-scientific answers). While this might be acceptable in critical periods such as World War II and Sputnik, it does not seem to serve the President well in ordinary times. Balanced and objective advice is important to the President, his staff, and the OMB. The belief that the special topic group's advice will have to be counterbalanced adds an additional burden to the advice they receive, hence the advice (and its source) is downgraded.

A Tension-Charged Relationship

The more focused and specific the specialist's domain is, the more tension-charged will be his relationship with the President. Consequently, the "half-life" of his impact on Presidential decisions will be foreshortened or lengthened depending on the specificity or diffuseness of the perspective he is representing. This has been the key to the viability and power of the OMB and also to the long-term impact of the Council of Economic Advisers. The OMB is useful because its views cut across the whole spectrum of governmental activities; it represents no single perspective or interest. Likewise, the economy is the linchpin of the nation. Hence the Science Adviser and his staff must be certain to assure that no taint of special interest appears in their recommendation and that all conflict of interest is eschewed. Conventional wisdom dictates that the Science Adviser's usefulness is predicated entirely on his personal rapport with the President, the "sine qua non." Although this perception does contain an element of truth, it mistakes effects for causes. The personal relationship grows out of the President's appraisal of the Science Adviser; more specifically, the relationship hinges on what the President feels the Science Adviser can do for him politically, or to keep him out of trouble politically. While compatibility of style is important, the President still has to feel that he needs the Science Adviser. If he does, the personal relationship will flourish. Obviously some advisers are able to demonstrate their utility more easily than others and, while this is partially a result of personal characteristics, it also depends on the particular period and events which structure the political needs of the President and the ability of the Science Adviser to respond to those needs. By extending this point it becomes discernible that when R&D budgets are growing, the Science Adviser will be more useful to the President than when they are shrinking (at which time he is likely to be a liability to the President).

All the former Presidential Science Advisers have echoed George Kistiakowsky's sentiment that "the Science Adviser to the President first and foremost is a servant of the President."[17] In order to securely maintain his rapport with the President, the Science Adviser must repeatedly demonstrate that he is indeed "first and foremost" a servant. But in order to be effective there are a number of fine lines the adviser must read; servitude itself is only a necessary, not a sufficient, cause of effectiveness.

The Science Adviser must deal with conflicting roles of representing agency programs and being a neutral and disinterested analyst of those programs. On the one hand, if the Science Adviser too rigidly serves as a policy analyst, the agencies will stop using him and sharing their programs with him; he will be perceived as just one more budgeteer, an opponent—and probably a highly knowledgeable and influential one. On the other hand, if he encourages and supports agency programs, his credibility with the OMB and White House staff will be diminished. Either choice results in diminished effectiveness. Similarly, the Science Adviser must be able to foster the feeling among the scientific community that he is representing their interest in order to be able to communicate freely with that community. Without this, he will be hampered in service to the President. Should the Science Adviser be perceived as the exponent of the scientific community, however, his usefulness would be terminated: "It would be a complete political disaster" to be "the spokesman of the scientific community in the White House," said Kistiakowsky.[18] The need for balance and diplomacy means that there is no simple recipe for a Science Adviser's effectiveness. Each Science Adviser must carve

out his own niche within the flow of the dynamic and powerful forces surrounding
the central position in the US political system, or be swept away by them.

References

1. In this paper, science policy is used generally to include "high technology" and scientific research and development. Science advice includes both science for policy and policy for science. See H. Brooks *The Government of Science* (Cambridge, Mass.: MIT Press, 1968).
2. US Congress, House Committee on Appropriations, "Department of Housing and Urban Development—Independent Agencies Appropriations for 1980," Part 6, Office of Science and Technology Policy (1979), p. 60.
3. Critical assessments are contained in the US Congress, Senate Committee on Commerce, Science and Transportation, "Oversight on OSTP" (1979), in *Chemical and Engineering News*, July 16, 1979, p. 16 ff., in a Congressional Research Service report on the Office of Science and Technology prepared by Dorothy Bates, and an American Society for Public Administrators report prepared by Edward Wenk, Jr.
4. *Ibid.*, p. 54.
5. US Congress, House Committee on Appropriations, "Department of Housing and Urban Development—Independent Agencies Appropriations for 1979," Part 5, Office of Science and Technology Policy (1978), p. 22.
6. US Bureau of the Budget, "Report to the President on Government Contracting for Research and Development," 1962.
7. US Congress, House Committee on Science and Technology, "Interagency Coordination of Federal Scientific Research and Development: Special Oversight Hearings," 1976. p. 31.
8. *Ibid.*, p. 44-49. Harold Seidman, *Politics, Position and Power.* London: Oxford University Press, 1970.
9. J. E. Katz, *Presidential Politics and Science Policy*, New York: Praeger Press, 1978, pp. 205-208. An almost identical series of events occurred when a Presidential Review Memorandum was prepared by the Department of Commerce entitled "Domestic Policy Review of Industrial or Technological Innovation" (The White House, 1979).
10. US Government Accounting Office, "Federal Materials Research and Development: Modernizing Institutions and Management" (OSP-76-9, December 2, 1975), p. 7.
11. *Op. cit.,* House Committee on Science and Technology, p. 107.
12. US Congress, House Committee on Science and Technology, "National Science and Technology Policy Issues, 1979." Part I, 1979, p. 12.
13. *Op. cit.*, Katz.
14. D. Beckler, "The Precarious Life of Science in the White House." *Daedalus* 103, Summer 1974, pp. 115-34.
15. J. R. Killian, Jr., *Sputnik, Scientists and Eisenhower.* Cambridge, Mass.: MIT Press, 1977), p. 65.
16. Huberman is not the first person with a joint appointment between OST(P) and NSC. Spurgeon Keeny held a similar position during the Kennedy and Johnson Administrations.
17. Franklin Institute, *Science Policies for the Decade Ahead* (Philadelphia: Franklin Institute Press, 1976), p. 61.
18. *Ibid.*, p. 62.

Science Advice in the White House

The Genesis of the President's Science Advisers and the National Science Foundation

Detlev W. Bronk

There are reasons for hope that the recent precarious role of science and scientists in the national government may be redefined and strengthened by President Ford and Congressional acts. There may again be an agency within the White House through which scientists will be able to participate in formulating laws and national policies that involve scientific considerations. Accordingly, it is timely and of especial interest to recall the origins of two closely related institutions that have had profound influence on national policies and on the development of science in this country during the past 25 years: the President's Science Adviser and his Science Advisory Committee, and the National Science Foundation.

1950-1951 was a remarkable period. It was the beginning of 12 years during which there was strong support of science by Presidents Truman, Eisenhower, and Kennedy. Science flourished in all branches of the government. Vannevar Bush was still the vigorous, wise creator and catalyst of scientific institutions.

The historic achievements of the Office of Scientific Research and Development (OSRD) during World War II had left a heritage of scientists and respect for science in the Department of Defense. Each of the services was developing an agency for science and technology with civilian advisory committees, such as the Office of Naval Research and the Naval Research Advisory Committee. Over all was the Research and Development Board (RDB) that had recently been created by Bush. The Korean War was a reminder of the military importance of science.

Reprinted from *Science*, Vol. 186, pp. 116-121, 11 October 1974. Copyright © 1974 by the American Association for the Advancement of Science.

Detlev W. Bronk (1897-1975), biophysicist and humanist, was President of the Rockefeller University, President of Johns Hopkins University, professor at the University of Pennsylvania, President of the National Academy of Sciences (1952-60), and Chairman of the National Research Council, Chairman of the Executive Committee of the National Science Board, President of the American Association for the Advancement of Science, and achieved many other distinctions. He was a member of the President's Science Advisory Committee (1956-63). He was an ensign in the United States Navy in World War I.

The Atomic Energy Commission was firmly established and was beginning to support much research in universities as well as within the commission.

Supplementing the postdoctoral fellowships which they had provided for 30 years, the National Academy of Sciences and the National Research Council conducted the first country-wide program of fellowships for graduate study in the sciences with financial assistance from The Rockefeller Foundation at first and then the Atomic Energy Commission. The academy and the research council were at the start of a new era of unprecedented initiative in the development of science within universities and executive agencies of the federal government.

The National Science Foundation (NSF) was created in the spring of 1950 by Act of Congress after five years of discussion regarding its role and structure.

Truman Seeks Advice

Early in the autumn of 1950 I received this letter from David Stowe, Administrative Assistant to President Truman:

> At the request of the President a review of scientific research of military significance and of the organization of the government for promotion of scientific activities generally is being undertaken by William T. Golden who is serving for the purpose as a Special Consultant to the Director of the Bureau of the Budget.
>
> Mr. Golden wishes the benefit of an informal discussion with you and will communicate with you . . . . This study is important to the Government and your assistance will be greatly appreciated.

William Golden was an investment banker with a lively interest in science; now he is widely known and appreciated among scientists as treasurer of the American Association for the Advancement of Science and as trustee of many scientific institutions. During 1946-1949 he had been assistant to Atomic Energy Commissioner Lewis Strauss. It was that experience and the wide contacts with scientists and government officials he thus formed that fitted him for the study conducted for Truman.

In my first meeting with Golden, he said that his study would deal with the organization of scientific research and development within the government and the interrelationship of such agencies as the Research and Development Board of the Department of Defense, the National Science Foundation which was about to be activated, and whatever agency was to be responsible for those functions which had been performed by the Office of Scientific Research and Development during World War II. He described those organizations as "the three segments of my study."

We discussed the general dissatisfaction with the Research and Development Board which was then being investigated by a review committee of which James Killian was chairman. Golden and I agreed that the National Science Foundation, whose board was soon to have its first meeting, "should confine its activities entirely to non-military matters" except perhaps in time of war. Golden asked what I thought about the plan he was formulating for the establishment of a Scientific Adviser to the President whose functions would be "to keep fully informed on all major scientific research and development activities of a military character in all Government agencies so engaged.'' Should the adviser be supplemented by an advisory committee? Who did I think would be a suitable Science Adviser to the President?

Although those were Golden's primary concerns, he asked discerning questions about the National Academy of Sciences and its National Research Council; he was enthusiastic about our plans for vitalizing them and their relations to all agencies of government.

During the 6 months before and after that conversation, Golden had discussions with 165 scientists and engineers from universities, industries, and government and with nonscientific government officials. He thus heard a broad spectrum of opinions regarding the National Academy of Sciences, the Research and Development Board, the National Science Foundation, the proposed Scientific Adviser to the President, the interrelations among these institutions, and their several roles in the government.

As 1950 drew to a close, Golden was assured that he had widespread support for his proposal that there be appointed a President's Science Adviser. He had the unanimous approval of the Killian committee that was reviewing the Research and Development Board.

Accordingly, Golden sent to President Truman a memorandum: ''Mobilizing science for a war; a Scientific Adviser to the President.'' The memorandum recommended prompt appointment of an outstanding scientific leader as Science Adviser to the President. His functions would be:

a) To inform himself and keep informed on all scientific research and development programs of military significance within the several independent Government departments so engaged.
b) To plan for and stand ready promptly to initiate a civilian Scientific Research Agency, roughly comparable to the Office of Scientific Research and Development (OSRD) of World War II.
c) To be available to give the President independent and comprehensive advice on scientific matters inside and outside the Government, particularly those of military significance.

Before 2 weeks had gone by, unexpected opposition to Golden's proposals developed within the newly created National Science Board of the National Science Foundation.

National Science Board

The National Science Foundation was established by Act of Congress in May 1950. The Act had the unusual provision that the ultimate power to disburse funds made available each year by Congress was lodged in a National Science Board composed of citizens appointed by the President and confirmed by the Senate. Because of this unprecedented power, the White House under the leadership of John Steelman, Assistant to the President, solicited advice widely as to the composition of the board with regard to scientific competence; university, industrial, or political affiliation; religious, racial, and geographical representation.

The first meeting of the National Science Board was finally held during December in the Cabinet Room of the White House; the presidents of Harvard, Wisconsin, and Johns Hopkins—James B. Conant, Edwin Broun Fred, and Bronk— were elected chairman, vice-chairman, and chairman of the executive committee, respectively. In the course of the meeting, the President greeted each member of the board, then asked, "What have you fellows and Sophie Aberle been talking about?" Conant replied that we had been discussing possible directors of the foundation whom we would then recommend to him. With a smile, Truman said, "That should be easy, someone who can get along with me." He then went on for 10 or 15 minutes discussing his hopes for the foundation, what it could, should, and should not do. He ended, "You may have trouble getting money out of those fellows over in Congress. I will help."

A month later the board met again. Its purpose was to consider the post of director, but the discussion soon turned to Golden's recent recommendation to the President. Lee A. DuBridge and I had been members of the Killian committee that had unanimously supported the proposal to create a Science Adviser to the President. We were dismayed to hear a majority of our fellow members on the board strongly oppose the proposal.

In reporting this to the Bureau of the Budget, Conant told of the board's concern that the appointment of a Science Adviser to the President with an advisory committee would lower the status of the foundation and obstruct its congressional appropriations.

This conflict between the National Science Board and the proponents of a Science Adviser to the President with a Science Advisory Committee caused much concern in scientific and government circles. William Webster, chairman of the Research and Development Board, was especially vehement in his criticism of the National Science Board's objection, which he described as typical of "scientists' vacillation and naivete." This was unjustified because most of the members of the National Science Board had not been involved in discussions with Golden preceding his recommendations. Their desire that the foundation take an active part in furthering military science was indeed supported by the NSF Act of 1950 which states that the foundation's duty is "to secure the national defense" as well as "to promote the progress of science; to advance the national health, prosperity, and welfare; and for other purposes."

The conflict was soon resolved, but the role of the National Science Foundation in military science and its relation to the Department of Defense was again an issue a year later, as it has been recently. Congressman Wolverton, who had been a member of the committee that sponsored the NSF Act, wrote to the acting director of the Office of Defense Mobilization (ODM): "It is my recollection that the principal function of the foundation relates to adequate coordination and stimulation of scientific matters relating to the national defense." And so he asked, "What is this new Science Advisory Committee, how does it differ from the NSF and why cannot NSF act in the advisory capacity of this new committee?" The ODM replied that "there is no overlapping of functions . . . coordination with the activities of the NSF is assured by the appointment of four members of the NSB and the Director of the NSF to this new Science Advisory Committee."

Conant, DuBridge, and I continued to urge our colleagues on the National Science Board to concentrate the foundation's initial activities on basic research and on a fellowship program. We were aided by a lengthy "Memorandum on Program for the National Science Foundation" that was prepared by Golden and sent through the director of the Bureau of the Budget to all members of the National Science Board. It began:

It is well to reiterate the preeminent need from a long term viewpoint, for advancing basic scientific knowledge. To promote such activities is the primary purpose of the National Science Foundation. To this end provision is being made for a representative of the NSF to be a member of the newly created Advisory Committee on Defense Scientific Research. This latter committee located within the ODM and reporting to the Defense Mobilizer and to the President will serve as a focus in the mobilization program for the representation of the scientific community and further will serve as the central point for knowledge of the government as a whole in scientific research and development of military significance. Membership of the committee will consist of ex officio representatives of the appropriate governmental agencies plus a representative selection of distinguished scientists at large.

It may be worth repeating that in accordance with the spirit of the Act, as well as the judgment of substantially all scientists with whom I have discussed the question, the National Science Foundation should confine its activities to furthering basic scientific studies and that it should not dilute its effectiveness by supporting studies of directly military or other applied character. To do so would seriously impair the long-term mission of the National Science Foundation without materially contributing to the war effort, since such work can better be done by other agencies. In the long run, of course, additions to basic scientific knowledge will contribute, as previously indicated, to both the wartime and peacetime strength of the country; but short-term results are not to be looked for.

The question of appropriations to the National Science Foundation is important but will not become a matter for immediate consideration until the Board itself analyzes its undertakings and prepares a recommended program for the near-term and long-term future. As a matter of interest, the Act as passed authorized direct appropriations not to exceed $500,000 for the FY [fiscal year] ending June 30, 1951, and not to exceed $15 million for each FY thereafter."

At its February meeting, the board agreed that the foundation should not become involved in military research; opposition to the appointment of a Science Adviser to the President was dropped.

After a few months of lengthy deliberations by the board, Conant reported to Congress that

[O]ne of the purposes of the National Science Foundation is to provide in every section of the country educational and research facilities which will assist the development of scientific pioneers . . . . There must be all over the United States intense efforts to discover latent scientific talent and provide for its adequate development. This means strengthening many institutions which have not developed their full potentialities as scientific centers, it means assisting promising young men and women who have completed their college education but require postgraduate training in order to become leaders in science and engineering. To this end a fellowship program has been placed high on the list of priorities of the National Science Board . . . . Measured solely in terms of a contribution to national defense in a period of lengthy partial mobilization, I, for one, have no question but that the money will be well spent.

On 6 April the President appointed Alan Waterman, formerly deputy chief and chief scientist of the Office of Naval Research, first director of the foundation.

Search for a Science Adviser

Activation of the Golden report was delayed by the process of choosing and gaining the acceptance of a Science Adviser to the President. It was a difficult, critical position to fill. Bush wisely commented that the value of the post turned on who the man was, who was the President, and how they got along together. Conant thought that an advisory committee would be better because it was unlikely that a man with a sufficient range of competence could be found.

During six months of conversations with more than 150 persons, Golden had been asking who should be the Adviser. The range of suggestions he discussed with me was remarkable for its diversity, and for the widely conflicting judgments on personal qualifications. Only DuBridge and Mervin Kelly were generally approved. But DuBridge would not leave the presidency of Cal Tech and Kelly was committed to Bell Telephone Laboratories, of which he was soon to become president. Conant's preference for an advisory committee was slowly gaining favor.

During Golden's search he conferred with General Lucius Clay, assistant director of Defense Mobilization. Clay stated flatly that he did not like the title Scientific Adviser to the President and that the adviser and his committee should be located in the Office of Defense Mobilization and that the adviser should be called "Assistant to the Director of ODM for Scientific Matters." This was a step down from Golden's concept of a Presidential Adviser which he had been urging for six months and which had received wide support. General Clay had sufficient influence to prevail.

The concept of Science Adviser to the President was retained in part. In the draft of a letter that was to be sent by the President to the still-to-be-chosen chairman of an "Advisory Committee on Defense Scientific Research," the President said that he would welcome the recommendations of the committee and would call upon it for advice from time to time. The committee was to include the president of the National Academy of Sciences, the chairman of the Research and Development Board, the chairman of the Inter-Departmental Committee on Scientific Research and Development, the director of the National Science Foundation, and also a number of eminent scientists and engineers.

This new proposal for an Advisory Committee on Defense Scientific Research with a full-time chairman was approved by Bush who thought it "far better than a single Scientific Adviser." Alfred Loomis, who had been one of the leaders of National Defense Research Committee, also thought that it would be desirable to have the functions that were to have been those of a Science Adviser to the President placed in the Office of Defense Mobilization although "university scientists might prefer the prestige of a presidential appointment."

During the weeks following General Clay's suggestion that a Committee on Defense Scientific Research and its chairman be placed in the Office of Defense Mobilization, the search for the Adviser continued. After Kelly declined the President's appointment, Oliver Buckley, who was Kelly's superior, and was soon to retire as president of Bell Laboratories, was widely discussed. Buckley was an able administrator, a highly respected scientist who had been active in the Office of Scientific Research and Development, was a member of the General Advisory Committee of the Atomic Energy Commission, and had been offered the post of chairman of the Research and Development Board following Karl Compton's resignation. After much deliberation he finally agreed to serve although he insisted that he be designated chairman of the Science Advisory Committee of ODM, a title he preferred to that proposed by Golden.

Late in April Truman appointed Buckley and a committee comprising Waterman, Webster and Bronk as representatives of the National Science Foundation, the Research and Development Board, and the National Academy of Sciences, as well as Conant, Hugh L. Dryden, DuBridge, Killian, Robert Loeb, Robert Oppenheimer, and Charles Thomas. In his letter of appointment, the President stressed the role of the chairman and members of the committee as *advisers to himself* as well as to the director of the Office of Defense Mobilization (ODM) "in the achievement of continued progress in scientific research and development. The successful performance of the committee's functions can be of great value to this country, both during this period of emergency and in future years."

The ODM Science Advisory Committee

The Science Advisory Committee convened in May 1951; throughout the year Buckley was chairman it met each month, usually with full attendance. From its beginning the chairman proposed that the committee be "advisory, not operating; have no budgetary responsibilities; work with and through existing agencies; avoid fanfare and minimize public appearances." Having thus defined the committee's principles, it was not surprising that Buckley should have written to the members: "By its structure and location, the contribution of the committee is limited largely to policy and other general matters. It cannot be relied on as the principal source of imaginative, technical leadership in the government."

With characteristic but perhaps too great modesty, the chairman looked to the members of the committee for initiative in bringing matters to the committee's attention. In consideration of problems which came to him, he leaned heavily on the advice of the members and especially on the four Washington members of the committee: the president of the National Academy of Sciences, the director of the National Science Foundation, and the chairmen of the Research and Development Board and the Interdepartmental Committee on Scientific Research and Development. This was consistent with his wise policy of strengthening existing agencies by bringing them into effective relations with the Executive Office of the White House through the Science Advisory Committee.

At the end of the first year Buckley reported to the President:

Since the committee is composed of members with a great diversity of ties to other government agencies as well as to scientific and educational institutions outside the Government, its meetings have proved to be an excellent focus for interchange of views and development of opinion. It is in this way that the Committee has principally been effective [in providing relations between scientists and the federal government].

It has been an added privilege of the Chairman to serve as a member of the staff of the Director of the ODM, and in this capacity a variety of tasks have been performed.

With informal operations of this type, the Committee has exercised a helpful influence in scientific affairs without interfering with other agencies in the conduct of their normal functions which we have been endeavoring to facilitate.

Because of these principles of working through existing agencies and avoiding publicity, it was widely thought and still said that the committee was "useful, but of little effectiveness, a status that was not overcome by Buckley's successors." With that I cannot agree. Coming soon after the notable success of the Office of Scientific Development, the Science Advisory Committee suffered by contrast, but throughout five years it had an important role in nurturing science within the government.

At the end of the first year in June 1952, Buckley resigned because of a growing illness which was ultimately fatal. He was succeeded by DuBridge as part-time chairman.

Soon after the change in leadership the committee spent three long days at the Institute for Advanced Study as guests of Robert Oppenheimer in a searching, informal discussion of the urgent problems that confronted the scientific community. After critically debating whether there was need for the committee and after appraising its value to science and the government, it was decided that the committee should be continued because it provided a useful deliberative group that could be briefed on projects that were wider in scope than any existing service or agency. However, because the Office of Defense Mobilization was primarily concerned with production and controls, it was not considered a suitable home for the committee. It was suggested that the committee would be more useful to the President if it were attached more closely to him. It could thus provide for the

Science Advisory Committee, Office of Defense Mobilization. Seated (left to right): Arthur S. Flemming (director, Office of Defense Mobilization), President Dwight D. Eisenhower, Lee A. DuBridge (chairman), and Isidor I. Rabi. Standing (left to right): Emanuel R. Piore, Oliver E. Buckley, Alan T. Waterman, James B. Fisk, Detlev W. Bronk, Bruce S. Old, James R. Killian, Jr., David Z. Beckler, Robert F. Bacher, Jerrold R. Zacharias, and Charles C. Lauritsen.

National Security Council scientific assessments of situations involved in decisions the President was required to make. The suggestion was not accepted.

The committee continued in the Office of Defense Mobilization for another five years. Nevertheless, it had increasingly direct and personal relations with President Eisenhower, who had a lively interest in furthering science as a major element in the conduct of government. Once when I told him of scientists' appreciation for his understanding and support, he reminded me that he was a graduate of the first school of engineering in the country; he liked to think of himself as one of us.

DuBridge was chairman of the Science Advisory Committee throughout four years; he was followed by I. I. Rabi for somewhat more than a year. Because of their wide range of interests and competence, the scope of the committee's activities widened. The agendas and minutes of its meetings during those years recount how scientists were learning to play their proper role in government.

I am reminded, for instance, that our chairman told the President through the director of the Office of Defense Mobilization:

We are seriously concerned that the operations of the Office of Science Adviser in the State Department are in some danger of being reduced to the point of ineffectiveness. . . . We feel that scientific liaison with friendly countries is essential to well rounded cultural relations. We hope that a way can be found for the Secretary of State to recognize this, support the office and the scientific attaches in our embassies abroad.

During a time when there were alarmist attacks on the loyalty of scientists, DuBridge protested to Vice President Nixon that the committee

was seriously concerned that the extensive security and loyalty review programs carried on by many agencies of the Federal Government were being carried to extremes that involve costs and dangers to national security far greater than are warranted . . . those attacks are reducing the availability of key scientists for important posts in the Government.

DuBridge added that not one single American scientist of the thousands engaged in security programs during the past 15 years had been convicted or seriously accused of espionage or treason.

In a letter to Killian, I find evidence of Eisenhower's deep interest in the committee's activities:

I understand that you have been asked by the Science Advisory Committee of the ODM to direct a study of the country's technological capabilities to meet some of its current problems. This project grew out of suggestions which I made to the SAC and I am naturally very keenly interested in it. The results will be of great value to the Government. Accordingly, I hope very much that you will find it possible to free yourself of your many other heavy responsibilities for a period long enough to undertake this important assignment, and that others whom you choose to be members of your staff will also be able to devote time to the work.

This is not the place for a further account of the many important and timely activities of the Science Advisory Committee nor of the wise guidance by David Beckler, its executive secretary.

But it should be emphasized that scientists for the first time had effective contact with the Executive Office of the President. And had it not been for the continued existence of the Science Advisory Committee, it would not have been available for transformation into the President's Science Advisory Committee with a Science Adviser to the President.

International Geophysical Year

One of the greatest cooperative endeavors ever undertaken by scientists from many nations was a program of planned research during the International Geophysical Year (IGY) of July 1957—December 1958. Under the general direction of the Special Committee for the IGY (CSAGI) of the International Council of Scientific Unions, a wide diversity of research was initiated in many fields and places including the Antarctic, the ionosphere, and space. The United States was represented by the National Academy of Sciences; the National Science Foundation provided much of the financial support.

At a meeting of CSAGI in Rome during October 1954, a resolution was adopted recommending that "in view of the advanced state of rocket techniques . . . thought should be given to the launching of small satellite vehicles" during the IGY. The Soviet Union and the United States supported the proposal.

The US National Committee for the IGY that had been organized by the academy promptly recommended that the US initiate a scientific satellite program. The Eisenhower administration enthusiastically agreed to support the project and directed the National Science Foundation to provide the necessary funding. The Defense Department was made responsible for providing the rocketry needed to place a satellite in orbit, but without interfering with the top priority ballistic missile program.

By the following summer, plans for our satellite program were sufficiently advanced to justify a public announcement. At the White House in July 1955, President Eisenhower, together with representatives of NAS and NSF, reported that "plans are going forward for the launching of small, unmanned, earth-circling satellites as part of the United States participation in the International Geophysical Year. Data that is collected will be made available to all scientists throughout the world."

Because the development and launching of the satellite vehicle (Vanguard) had been assigned to the Naval Research Laboratory, there was controversy and friction with the Army and Air Force. And Secretary of Defense Charles Wilson did not give the program enthusiastic support. On the occasion of a meeting of the National Security Council, he saw Waterman and me outside the Cabinet Room and told us that he was going to recommend that the program, which he considered a "scientific boondoggle," be abolished. When Wilson made his proposal, Eisenhower asked me what I thought would be the international reaction to our cancellation of the satellite program. I replied, "It was your decision, Mr. President, to announce our program at a press conference in the White House. Cancellation of the program within a year will bring much criticism from scientists throughout the world." After a moment of thought, the President announced that the program would be doubled: 12 Vanguards instead of six.

Early in October 1957 a week-long, international conference on rockets and satellites, sponsored by CSAGI, was held at the National Academy of Sciences. Lloyd Berkner presided. Among the social events was a cocktail party at the Soviet Embassy on the last evening. During the party a correspondent of the *New York Times* drew Lloyd Berkner and Hugh Odishaw of our committee aside and whispered, "I have just had a call from our New York office saying that a cable from Moscow reports that the Russians put a satellite in orbit about an hour ago." Berkner quickly stood on a chair, tapped on a glass for silence, then announced to our Russian hosts and their hundred guests that Russia had put a satellite in orbit for the first time in history. There was awed surprise, "then all hell broke loose."

The final session at the academy next morning was quite different from the series of routine resolutions that had been planned. There was much excited discussion of how the Russian feat would affect the future of the IGY satellite program and why had the Russians kept their plan secret. Finally it came time for me to bid our guests farewell "at a time when I can congratulate our Russian colleagues and our sister Academy of Sciences of the USSR on their great achievement of yesterday." I went on:

Friendly competition as well as cooperation is a stimulus to achievement in scientific endeavor as in other forms of human effort. Because scientists are human, they naturally wish to be the first to achieve success in a scientific undertaking to which they are committed. But because scientists are humane explorers on the frontiers of knowledge, they rejoice in new discoveries made by their colleagues.

There was little rejoicing in the country at large. Some belittled the Russian achievement as did an admiral who was connected with the Vanguard project: "Why all the excitement? They have only fired a hunk of iron into the sky. Anyone can do that." Which we were still to do. And the press aroused widespread fear of Sputnik as a military threat and symbol of their scientific superiority. Only President Eisenhower and most scientists approved my cablegram to the president of the Russian Academy: "This is a brilliant contribution to the furtherance of science for which scientists everywhere will be grateful."

On the following Tuesday afternoon Sherman Adams, Assistant to the President, called me at the academy: Would I come over to the White House to discuss with the President, Press Secretary Hagerty, and himself what Eisenhower should say at his press conference next day about satellites and security.

The President began by saying that he was not surprised that the Russians had failed, in characteristic fashion, to reveal their plans despite the IGY agreement to do so. But he was surprised by the ungenerous attitude of so many Americans and our press. He recalled how grateful we were to English scientists for the discovery of penicillin and how much it meant to his troops in World War II. We recalled his remark in connection with the opening of the IGY: "The most important result of the IGY is the demonstration of the ability of peoples of all nations to work together harmoniously for the common good." And so we decided that his remarks to the press should begin: "We congratulate Russian scientists upon having put their satellite into orbit." He concluded: "Our satellite program has never been considered as a race with other nations . . . . We are carrying the program forward in keeping with our arrangements with the international scientific community.''

After his statement to the press was agreed on, Eisenhower turned to a discussion of the sudden, irrational furor over the status of American versus Russian science. He again recalled the discovery of penicillin. ''I heard no one complain that the English achievement belittled the quality of science in America.'' What could be done by other than words, he asked, to assure the country that American science was indeed vigorous and was respected and supported at the highest levels of government. That led to the role of the Science Advisory Committee of the Office of Defense Mobilization. I told him of Golden's original proposal that there be a full-time Science Adviser to the President supported by an advisory committee of eminent scientists, both located within the White House. I urged him to consult at once with Rabi, the chairman of the Science Advisory Committee, develop closer relations with the committee, and give to the public assurance that it was indeed a committee advisory to himself. All this he did, and much more with the wise advice of Rabi and the committee.

A month later in a broadcast entitled ''Science and national security,'' Eisenhower said:

I have made sure that the very best thought and advice that the scientific community can supply, heretofore provided to me on an informal basis, will now be fully organized and formalized so that no gap can occur. The purpose is to make it possible for me, personally, whenever there appears to be any unnecessary delay in our development system, to act promptly and decisively.

To that end, I have created a new office called the office of Special Assistant to the President for Science and Technology. This man, who will be aided by a staff of scientists and a strong Advisory Group of outstanding experts reporting to him and to me, will have the active responsibility of helping me follow through on the program of scientific improvement of our defenses.

I am glad to be able to tell you that this position has been accepted by Dr. James R. Killian, President of the Massachusetts Institute of Technology. He is a man who holds my confidence, and enjoys the confidence of his colleagues in the scientific and engineering world, and in the government . . . .

In conclusion, although I am now stressing the influence of science on defense, I am not forgetting that there is much more to science than its function in strengthening our defense, and much more to our defense than the part played by science. The peaceful contributions of science—to healing, to enriching life, to freeing the spirit—these are the most important products of the conquest of nature's secrets. And as to our security, the spiritual powers of a nation—its underlying religious faith, its self-reliance, its capacity for intelligent sacrifice—these are the most important stones in any defense structure.

Seven years had passed since William Golden recommended to Truman that he appoint a Science Adviser to himself and a Presidential Science Advisory Committee.

Index of Names

Note: Page 10 lists names of all Science Advisers; pages 11-16 list names of members of Presidential Science Advisory Committees.